AF818448

Section 12

Psychrometry, Evaporative Cooling, and Solids Drying*

Larry R. Genskow *Technical Director, Corporate Engineering Technologies, The Procter & Gamble Company; Advisory Associate Editor,* Drying Technology—An International Journal; *Member, International Advisory Committee, International Drying Symposia (Section Editor)*

Wayne E. Beimesch, Ph.D. *Technical Associate Director, Corporate Engineering, The Procter & Gamble Company; Member, The Controlled Release Society; Member, Institute for Liquid Atomization and Spray Systems*

John P. Hecht, Ph.D. *Senior Engineer, The Procter & Gamble Company*

Ian C. Kemp, M.A. (Cantab), C.Eng. *Senior Technical Manager, GlaxoSmithKline; Fellow, Institution of Chemical Engineers; Associate Member, Institution of Mechanical Engineers*

Tim Langrish, D.Phil. *School of Chemical and Biomolecular Engineering, The University of Sydney (Australia)*

Christian Schwartzbach, M.Sc. *Manager, Technology Development (retired), Niro A/S*

(Francis) Lee Smith, Ph.D., M.Eng. *Principal, Wilcrest Consulting Associates, Houston, Texas; Member, American Institute of Chemical Engineers, Society of American Value Engineers, Water Environment Federation, Air and Waste Management Association (Biofiltration)*

PSYCHROMETRY

EVAPORATIVE COOLING

*The contributions of Paul Y. McCormick, George A. Schurr, and Eno Bagnoli of E. I. du Pont de Nemours & Co., and Charles G. Moyers and Glenn W. Baldwin of Union Carbide Corporation to material that was used from the fifth to seventh editions are acknowledged.
The assistance of Kwok-Lun Ho, Ph.D., Principal Engineering Consultant, in the preparation of the present section is acknowledged.

SOLIDS-DRYING FUNDAMENTALS

Nomenclature and Units

Symbol	Definition	SI units	U.S. Customary System units
A	Area	m^2	ft^2
a_w	Water activity	—	—
a_w^{vapor}	Activity of water in the vapor phase	—	—
a_w^{solid}	Activity of water in the solid	—	—
c	Concentration	kg/m^3	lb/ft^3
C_P	Specific heat capacity at constant pressure	J/(kg·K)	Btu/(lb·°F)
C_w	Concentration of water in the solid	kg/m^3	lbm/ft^3
$\mathcal{D}(w)$	Diffusion coefficient of water in a solid or liquid as a function of moisture content	m^2/s	ft^2/s
$\mathcal{D}$	Diffusion coefficient	m^2/s	ft^2/s
d	Diameter (particle)	m	in
E	Power	W	Btu/h
F	Solids or liquid mass flow rate	kg/s	lb/h
F	Mass flux of water at surface	$kg/(m^2 \cdot s)$	$lbm/(ft^2 \cdot s)$
f	Relative drying rate	—	—
G	Gas mass flow rate	kg/s	lb/h
g	Acceleration due to gravity, 9.81 m/s^2	m/s^2	ft/s^2
H	Enthalpy of a pure substance	J/kg	Btu/lb
ΔH_{vap}	Heat of vaporization	J/kg	Btu/lb
h	Heat-transfer coefficient	$W/(m^2 \cdot K)$	$Btu/(ft^2 \cdot h \cdot °F)$
I	Humid enthalpy (dry substance and associated moisture or vapor)	J/kg	Btu/lb
J	Mass flux (of evaporating liquid)	$kg/(m^2 \cdot s)$	$lb/(ft^2 \cdot h)$
k	Mass-transfer coefficient	m/s	$lb/(ft^2 \cdot h \cdot atm)$
k_{air}	Thermal conductivity of air	W/(m·k)	Btu/(ft·h·°F)
k_c	Mass-transfer coefficient for a concentration driving force	m/s	ft^2/s
k_p	Mass transfer coefficient for a partial pressure driving force	$kg/(m^2 \cdot s)$	$lbm/(ft^3 \cdot s)$

Nomenclature and Units *(Concluded)*

Symbol	Definition	SI units	U.S. Customary System units
L	Length; length of drying layer	m	ft
M	Molecular weight	kg/mol	lb/mol
m	Mass	kg	lb
m_{solids}	Mass of dry solids	kg	lbm
N	Specific drying rate ($-dX/dt$)	1/s	1/s
N	Rotational speed (drum, impeller, etc.)	1/s	rpm
P	Total pressure	kg/(m·s^2)	lbf/in^2
P_w^{bulk}	Partial pressure of water vapor in the air far from the drying material	kg/m·s^2	lbf/in^2
$P_w^{surface}$	Partial pressure of water vapor in the air at the solid interface	kg/m·s^2	lbf/in^2
p	Partial pressure/vapor pressure of component	kg/(m·s^2)	lbf/in^2
p_{pure}^{sat}	Pure component vapor pressure	kg/(m·s^2)	lbf/in^2
p_w, air	Partial pressure of water vapor in air	kg/(m·s^2)	lbf/in^2
Q	Heat-transfer rate	W	Btu/h
q	Heat flux	W/m^2	Btu/(ft^2·h)
R	Universal gas constant, 8314 J/(kmol·K)	J/(mol·K)	Btu/(mol·°F)
R	Droplet radius	m	ft
r	Radius; radial coordinate	m	ft
RH	Relative humidity	—	—
S	Percentage saturation	—	—
s	Solid-fixed coordinate	Depends on geometry	
T	Absolute temperature	K	°R
T, t	Temperature	°C	°F
t	Time	s	h
U	Velocity	m/s	ft/s
u	Mass of water/mass of dry solid	—	—
V	Volume	m^3	ft^3
V	Air velocity	m/s	ft/s
v	Specific volume	m^3/kg	ft^3/lb
$v_{droplet}$	Droplet volume	m^3	ft^3
w	Wet-basis moisture content	—	—
$w_{avg\ dry\text{-}basis}$	Average wet-basis moisture content	—	—
X	Solids moisture content (dry basis)	—	—
Y	Mass ratio	—	—
z	Distance coordinate	m	ft
	Dimensionless groups		
Ar	Archimedes number, $(gd_P^3\rho_G/\mu^2)(\rho_P - \rho_G)$	—	—
Bi	Biot number, $h\cdot L/\kappa$	—	—
Gr	Grashof number, $L^3\cdot\rho^2\cdot\beta g\Delta T/\mu^2$	—	—
Nu	Nusselt number, hd_P/κ	—	—
Pr	Prandtl number, $\mu C_P/\kappa$	—	—
Re	Reynolds number, $\rho d_P U/\mu$	—	—
Sc	Schmidt number, $\mu/\rho\mathcal{D}$	—	—
Sh	Sherwood number, $k_Y d_P/\mathcal{D}$	—	—
Le	Lewis = Sc/Pr	—	—
	Greek letters		
α	Slope	—	—
β	Psychrometric ratio	—	—
ε	Voidage (void fraction)	—	—
ζ	Dimensionless distance	—	—
η	Efficiency	—	—
θ	Dimensionless time	—	—
κ	Thermal conductivity	W/(m·K)	Btu/(ft·h·°F)
λ	Latent heat of evaporation	J/kg	Btu/lb
μ	Absolute viscosity	kg/(m·s)	lb/(ft·s)
μ_{air}	Viscosity of air	kg/(m·s)	lbm/(ft·s)
ρ	Density	kg/m^3	lb/ft^3
ρ_{air}	Air density	kg/m^3	lbm/ft^3
ρ_s	Mass concentration of solids	kg/m^3	lbm/ft^3
ρ_s^o	Density of dry solid	kg/m^3	lbm/ft^3
ρ_w^o	Density of pure water	kg/m^3	lbm/ft^3
τ	Residence time of solids	s	h
Φ	Characteristic (dimensionless) moisture content	—	—
ψ	Relative humidity	%	%

PSYCHROMETRY

General References *ASHRAE 2002 Handbook: Fundamentals,* SI Edition, American Society of Heating, Refrigeration and Air-Conditioning Engineers, Atlanta, Ga., 2002, Chap. 6, "Psychrometrics," Chap. 19.2, "Sorbents and Desiccants." *Aspen Process Manual* (Internet knowledge base), Aspen Technology, 2000 onward. *Humidity and Dewpoint.* British Standard BS 1339 (rev.). Humidity and dewpoint, Pt. 1 (2002); Terms, definitions and formulae, Pt. 2 (2005); Psychrometric calculations and tables (including spreadsheet), Pt. 3 (2004); Guide to humidity measurement. British Standards Institution, Gunnersbury, United Kingdom. Cook and DuMont, *Process Drying Practice,* McGraw-Hill, New York, 1991, Chap. 6. Keey, *Drying of Loose and Particulate Materials,* Hemisphere, New York, 1992. Poling, Prausnitz, and O'Connell, *The Properties of Gases and Liquids,* 5th ed., McGraw-Hill, New York, 2000. Earlier editions: 1st/2d editions, Reid and Sherwood (1958/1966); 3d ed., Reid, Prausnitz, and Sherwood (1977); 4th ed., Reid, Prausnitz, and Poling (1986). Soininen, "A Perspectively Transformed Psychrometric Chart and Its Application to Drying Calculations," *Drying Technol.* **4**(2): 295–305 (1986). Sonntag, "Important New Values of the Physical Constants of 1986, Vapor Pressure Formulations Based on the ITS-90, and Psychrometer Formulae," *Zeitschrift für Meteorologie,* **40**(5):340–344 (1990). Treybal, *Mass-Transfer Operations,* 3d ed., McGraw-Hill, New York, 1980. Wexler, *Humidity and Moisture,* vol. 1, Reinhold, New York, 1965.

Psychrometry is concerned with the determination of the properties of gas-vapor mixtures. These are important in calculations for humidification and dehumidification, particularly in cooling towers, air-conditioning systems, and dryers. The first two cases involve the air-water vapor system at near-ambient conditions, but dryers normally operate at elevated temperatures and may also use elevated or subatmospheric pressures and other gas-solvent systems.

Principles involved in determining the properties of other systems are the same as with air-water vapor, with one major exception. Whereas the psychrometric ratio (ratio of heat-transfer coefficient to product of mass-transfer coefficient and humid heat, terms defined in the following subsection) for the air-water system can be taken as 1, the ratio for other systems in general does not equal 1. This has the effect of making the adiabatic saturation temperature different from the wet-bulb temperature. Thus, for systems other than air-water vapor, accurate calculation of psychrometric and drying problems is complicated by the necessity for point-to-point calculation of the temperature of the evaporating surface. For example, for the air-water system, the temperature of the evaporating surface will be constant during the constant-rate drying period even though the temperature and humidity of the gas stream change. For other systems, the temperature of the evaporating surface would change.

TABLE 12-1 Interconversion Formulas for Air-Water System, to 3 Significant Figures

T = temperature in **kelvins** (K); P = total pressure in **pascals** (Pa or N/m²)

Convert from:	Y (or ppm$_w$)*	y	p	Y_v
Convert to: Absolute humidity (mixing ratio) Y (kg·kg⁻¹)	1	$Y = \frac{0.622Y}{1-Y}$	$Y = \frac{0.622p}{P-p}$	$Y = \frac{0.622}{0.002167P/(Y_vT)-1}$
Mole fraction y (mol·mol⁻¹)	$y = \frac{Y}{0.622+Y}$	1	$y = \frac{p}{P}$	$y = \frac{461.5Y_vT}{P}$
Vapor pressure p (Pa)	$p = \frac{PY}{0.622+Y}$	$p = yP$	1	$p = 461.5Y_vT$
Volumetric humidity Y_v (kg·m⁻³)	$Y_v = \frac{0.002167PY}{(0.622+Y)T}$	$Y_v = \frac{0.002167yP}{T}$	$Y_v = \frac{0.002167p}{T}$	1

TERMINOLOGY

Terminology and nomenclature pertinent to psychrometry are given below. There is often considerable confusion between dry and wet basis, and between mass, molar, and volumetric quantities, in both definitions and calculations. Dry- and wet-basis humidity are similar at ambient conditions but can differ significantly at elevated humidities, e.g., in dryer exhaust streams. Complete interconversion formulas between four key humidity parameters are given in Table 12-1 for the air-water system and in Table 12-2 for a general gas-vapor system.

Definitions related to humidity, vapor pressure, saturation, and volume are as follows; the most useful are absolute humidity, vapor pressure, and relative humidity.

Absolute humidity Y Mass of water (or solvent) vapor carried by unit mass of dry air (or other carrier gas). It is also known as the **mixing ratio**, *mass ratio,* or *dry-basis humidity*. Preferred units are lb/lb or kg/kg, but g/kg and gr/lb are often used, as are ppm$_w$ and ppb$_w$ (parts per million/billion by weight); ppm$_w$ = 10^6Y, ppb$_w$ = 10^9Y.

Specific humidity Y_W Mass of vapor per unit mass of gas-vapor mixture. Also known as mass fraction or wet-basis humidity, and much more rarely used than dry-basis absolute humidity. $Y_W = Y/(1+Y)$; $Y = Y_W/(1-Y_W)$.

Mole ratio z Number of moles of vapor per mole of gas (dry basis), mol/mol; $z = (M_g/M_v)Y$, where M_v = molecular weight of vapor and M_g = molecular weight of gas. It may also be expressed as ppm$_v$ and ppb$_v$ (parts per million/billion by volume); ppm$_v$ = 10^6z, ppb$_v$ = 10^9z.

Mole fraction y Number of moles of vapor per mole of gas-vapor mixture (wet basis); $y = z/(1+z)$; $z = y/(1-y)$. If a mixture contains m_v kg and n_v mol of vapor (e.g., water) and m_g kg and n_g mol of noncondensible gas (e.g., air), with $m_v = n_vM_v$ and $m_g = n_gM_g$, then the four quantities above are defined by

$$Y = \frac{m_v}{m_g} \qquad Y_w = \frac{m_v}{m_g + m_v} \qquad z = \frac{n_v}{n_g} \qquad y = \frac{n_v}{n_g + n_v}$$

Volumetric humidity Y_v Mass of vapor per unit volume of gas-vapor mixture. It is sometimes, confusingly, called the absolute humidity, but it is really a vapor concentration; preferred units are kg/m³ or lb/ft³, but g/m³ and gr/ft³ are also used. It is inconvenient for calculations because it depends on temperature and pressure and on the units system; absolute humidity Y is always preferable for heat and mass balances. It is proportional to the specific humidity (wet basis); $Y_V = Y_W\rho_g$, where ρ_g is the **humid gas density** (mass of gas-vapor mixture per unit volume, wet basis). Also

$$Y_v = \frac{M_vPn_v}{RT(n_g + n_v)}$$

Vapor pressure p Partial pressure of vapor in gas-vapor mixture, and is proportional to the mole fraction of vapor; $p = yP$, where P = total pressure, in the same units as p (Pa, N/m², bar, atm, or psi). Hence

$$p = \frac{n_v}{n_g + n_v}P$$

Saturation vapor pressure p_s Pressure exerted by pure vapor at a given temperature. When the vapor partial pressure p in the gas-vapor mixture at a given temperature equals the saturation vapor pressure p_s at the same temperature, the air is **saturated** and the absolute humidity is designated the **saturation humidity** Y_s.

Relative humidity RH or Ψ The partial pressure of vapor divided by the saturation vapor pressure at the given temperature, usually expressed as a percentage. Thus RH = $100p/p_s$.

Percentage absolute humidity (percentage saturation) S Ratio of absolute humidity to saturation humidity, given by $S = 100Y/Y_s = 100p(P-p_s)/[p_s(P-p)]$. It is much less commonly used than relative humidity.

Dew point T_{dew}, or saturation temperature Temperature at which a given mixture of water vapor and air becomes saturated on cooling; i.e., the temperature at which water exerts a vapor pressure equal to the partial pressure of water vapor in the given mixture.

TABLE 12-2 Interconversion Formulas for a General Gas-Vapor System

M_g, M_v = molal mass of gas and vapor, respectively; R = 8314 J/(kmol·K); T = temperature in **kelvins** (K); P = total pressure in **pascals** (Pa or N/m²)

Convert from:	Y (or ppm$_w$)	y	p	Y_v
Convert to:				
Absolute humidity (mixing ratio) Y (kg·kg⁻¹)	1	$Y = \frac{M_vy}{M_g(1-Y)}$	$Y = \frac{pM_v}{(P-p)M_g}$	$Y = \frac{M_v}{M_g(PM_v/Y_vRT-1)}$
Mole fraction y (mol·mol⁻¹)	$y = \frac{Y}{M_v/M_g+Y}$	1	$y = \frac{p}{P}$	$y = \frac{Y_vRT}{PM_v}$
Vapor pressure p (Pa)	$p = \frac{PY}{M_v/M_g+Y}$	$p = yP$	1	$p = \frac{Y_vRT}{M_v}$
Volumetric humidity Y_v (kg·m⁻³)	$Y_v = \frac{M_v}{RT}\frac{PY}{M_v/M_g+Y}$	$Y_v = \frac{M_vyP}{RT}$	$Y_v = \frac{M_vp}{RT}$	1

Humid volume v Volume in cubic meters (cubic feet) of 1 kg (1 lb) of dry air and the water vapor it contains.

Saturated volume v_s Humid volume when the air is saturated.

Terms related to heat balances are as follows:

Humid heat C_s Heat capacity of unit mass of dry air and the moisture it contains. $C_s = C_{Pg} + C_{Pv}Y$, where C_{Pg} and C_{Pv} are the heat capacities of dry air and water vapor, respectively, and both are assumed constant. For approximate engineering calculations at near-ambient temperatures, in SI units, $C_s = 1 + 1.9Y$ kJ/(kg·K) and in U.S. units, $C_s = 0.24 + 0.45Y$ (Btu/(lb·°F).

Humid enthalpy H Heat content at a given temperature T of unit mass of dry air and the moisture it contains, relative to a datum temperature T_0, usually 0°C. As water is liquid at 0°C, the humid enthalpy also contains a term for the latent heat of water. If heat capacity is invariant with temperature, $H = (C_{Pg} + C_{Pv}Y)(T - T_0) + \lambda_0 Y$, where λ_0 is the latent heat of water at 0°C, 2501 kJ/kg (1075 Btu/lb). In practice, for accurate calculations, it is often easier to obtain the vapor enthalpy H_v from steam tables, when $H = H_g + H_v = C_{Pg}T + H_v$.

Adiabatic saturation temperature T_{as} Final temperature reached by a small quantity of vapor-gas mixture into which water is evaporating. It is sometimes called the thermodynamic wet-bulb temperature.

Wet-bulb temperature T_{wb} Dynamic equilibrium temperature attained by a liquid surface from which water is evaporating into a flowing airstream when the rate of heat transfer to the surface by convection equals the rate of mass transfer away from the surface. It is very close to the adiabatic saturation temperature for the air-water system, but not for most other vapor-gas systems; see later.

CALCULATION FORMULAS

Table 12-1 gives formulas for conversion between absolute humidity, mole fraction, vapor pressure, and volumetric humidity for the air-water system, and Table 12-2 does likewise for a general gas-vapor system. Where relationships are not included in the definitions, they are given below.

In U.S. units, the formulas are the same except for the volumetric humidity Y_v. Because of the danger of confusion with pressure units, it is recommended that in both Tables 12-1 and 12-2, Y_v be calculated in SI units and then converted.

Volumetric humidity is also related to absolute humidity and humid gas density by

$$Y_v = Y_W \rho_g = \frac{Y}{1+Y}\rho_g \qquad (12\text{-}1)$$

Two further useful formulas are as follows:

Parameter	General vapor-gas system	Air-water system, SI units, to 3 significant figures	Eq. no.
Density of humid gas (moist air) ρ_g (kg/m^3)	$\rho_g = \frac{M_g}{RT}\left(P - p + \frac{M_v}{M_g}p\right)$	$\rho_g = \frac{P - 0.378p}{287.1T}$	(12-2)
Humid volume v per unit mass of dry air (m^3/kg)	$v = \frac{RT}{M_g(P-p)} = \frac{RT}{P} \times \left(\frac{1}{M_g} + \frac{Y}{M_v}\right)$	$v = \frac{461.5T}{P}(0.622 + Y)$	(12-3)

From Eq. (12-2), the density of dry air at 0°C (273.15 K) and 1 atm (101,325 Pa) is 1.292 kg/m^3 (0.08065 lb/ft^3). Note that the density of moist air is always lower than that of dry air.

Equation (12-3) gives the humid volume of dry air at 0°C (273.15 K) and 1 atm as 0.774 m^3/kg (12.4 ft^3/lb). For moist air, humid volume is not the reciprocal of humid gas density; $v = (1 + Y)/\rho_g$.

The **saturation vapor pressure** of water is given by Sonntag (1990) in pascals (N/m^2) at absolute temperature T (K).

Over water:

$$\ln p_s = -6096.9385T^{-1} + 21.2409642 - 2.711193 \times 10^{-2}T + 1.673952 \times 10^{-5}T^2 + 2.433502 \ln T \qquad (12\text{-}4a)$$

Over ice:

$$\ln p_s = -6024.5282T^{-1} + 29.32707 + 1.0613868 \times 10^{-2}T - 1.3198825 \times 10^{-5}T^2 - 0.49382577 \ln T \qquad (12\text{-}4b)$$

Simpler equations for saturation vapor pressure are the Antoine equation and Magnus formula. These are slightly less accurate, but easier to calculate and also easily reversible to give T in terms of p. For the Antoine equation, given below, coefficients for numerous other solvent-gas systems are given in Poling, Prausnitz, and O'Connell, *The Properties of Gases and Liquids*, 5th ed., McGraw-Hill, 2000.

$$\ln p_S = C_0 - \frac{C_1}{T - C_2} \qquad T = \frac{C_1}{C_0 - \ln p_S} + C_2 \qquad (12\text{-}5)$$

Values for Antoine coefficients for the air-water system are given in Table 12-3. The standard values give vapor pressure within 0.1 percent of steam tables over the range 50 to 100°C, but an error of nearly 3 percent at 0°C. The alternative coefficients give a close fit at 0 and 100°C and an error of less than 1.2 percent over the intervening range.

The Sonntag equation strictly only applies to water vapor with no other gases present (i.e., in a partial vacuum). The vapor pressure of a gas mixture, e.g., water vapor in air, is given by multiplying the pure liquid vapor pressure by an enhancement factor f, for which various equations are available (see British Standard BS 1339 Part 1, 2002). However, the correction is typically less than 0.5 percent, except at elevated pressures, and it is therefore usually neglected for engineering calculations.

RELATIONSHIP BETWEEN WET-BULB AND ADIABATIC SATURATION TEMPERATURES

If a stream of air is intimately mixed with a quantity of water in an adiabatic system, the temperature of the air will drop and its humidity will increase. If the equilibration time or the number of transfer units approaches infinity, the air-water mixture will reach saturation. The **adiabatic saturation temperature** T_{as} is given by a heat balance between the initial unsaturated vapor-gas mixture and the final saturated mixture at thermal equilibrium:

$$C_s(T - T_{as}) = \lambda_{as}(Y_{as} - Y) \qquad (12\text{-}6)$$

This equation has to be reversed and solved iteratively to obtain Y_{as} (absolute humidity at adiabatic saturation) and hence T_{as} (the calculation is divergent in the opposite direction). Approximate direct formulas are available from various sources, e.g., British Standard BS 1339 (2002) and Liley (*Int. J. Mech. Engg. Educ.* 21(2), 1993). The latent heat of evaporation evaluated at the adiabatic saturation temperature is λ_{as},

TABLE 12-3 Alternative Sets of Values for Antoine Coefficients for the Air-Water System

		C_0	C_1	C_2		C_0	C_1	C_2
Standard values	p in Pa	23.1963	3816.44	46.13 K	p in mmHg	18.3036	3816.44	46.13 K
Alternative values	p in Pa	23.19	3830	44.83 K	p in mmHg	18.3	3830	44.87 K

which may be obtained from steam tables; humid heat C_s is evaluated at initial humidity Y. On a psychrometric chart, the adiabatic saturation process almost exactly follows a **constant-enthalpy line**, as the sensible heat given up by the gas-vapor mixture exactly balances the latent heat of the liquid that evaporates back into the mixture. The only difference is due to the sensible heat added to the water to take it from the datum temperature to T_{as}. The adiabatic saturation line differs from the constant-enthalpy line as follows, where C_{PL} is the specific heat capacity of the liquid:

$$H_{as} - H = C_{PL}T_{as}(Y_{as} - Y) \qquad (12\text{-}7)$$

Equation (12-7) is useful for calculating the adiabatic saturation line for a given T_{as} and gives an alternative iterative method for finding T_{as}, given T and Y; compared with Eq. (12-6), it is slightly more accurate and converges faster, but the calculation is more cumbersome.

The **wet-bulb temperature** is the temperature attained by a fully wetted surface, such as the wick of a wet-bulb thermometer or a droplet or wet particle undergoing drying, in contact with a flowing unsaturated gas stream. It is regulated by the rates of vapor-phase heat and mass transfer to and from the wet bulb. Assuming mass transfer is controlled by diffusion effects and heat transfer is purely convective:

$$h(T - T_{wb}) = k_y \lambda_{wb} (Y_{wb} - Y) \qquad (12\text{-}8)$$

where k_y is the corrected mass-transfer coefficient [kg/(m^2·s)], h is the heat-transfer coefficient [kW/(m^2·K)], Y_{wb} is the saturation mixing ratio at t_{wb}, and λ_{wb} is the latent heat (kJ/kg) evaluated at T_{wb}. Again, this equation must be solved iteratively to obtain T_{wb} and Y_{wb}.

In practice, for any practical psychrometer or wetted droplet or particle, there is significant extra heat transfer from radiation. For an Assmann psychrometer at near-ambient conditions, this is approximately 10 percent. This means that any measured real value of T_{wb} is slightly higher than the "pure convective" value in the definition. It is often more convenient to obtain wet-bulb conditions from adiabatic saturation conditions (which are much easier to calculate) by the following formula:

$$\frac{T - T_{wb}}{Y_{wb} - Y} = \frac{T - T_{as}}{Y_{as} - Y}\beta \qquad (12\text{-}9)$$

where the psychrometric ratio $\beta = \overline{C_s}k_y/h$ and $\overline{C_s}$ is the mean value of the humid heat over the range from T_{as} to T.

The advantage of using β is that it is approximately constant over normal ranges of temperature and pressure for any given pair of vapor and gas values. This avoids having to estimate values of heat- and mass-transfer coefficients α and k_y from uncertain correlations. For the air-water system, considering convective heat transfer alone, $\beta \sim 1.1$. In practice, there is an additional contribution from radiation, and β is very close to 1. As a result, the wet-bulb and adiabatic saturation temperatures differ by less than 1°C for the air-water system at near-ambient conditions (0 to 100°C, $Y < 0.1$ kg/kg) and can be taken as equal for normal calculation purposes. Indeed, typically the T_{wb} measured by a practical psychrometer or at a wetted solid surface is closer to T_{as} than to the "pure convective" value of T_{wb}.

However, for nearly all other vapor-gas systems, particularly for organic solvents, $\beta < 1$, and hence $T_{wb} > T_{as}$. This is illustrated in Fig. 12-5. For these systems the psychrometric ratio may be obtained by determining h/k_y from heat- and mass-transfer analogies such as the Chilton-Colburn analogy. The basic form of the equation is

$$\beta = \left(\frac{\text{Sc}}{\text{Pr}}\right)^n = \text{Le}^{-n} \qquad (12\text{-}10)$$

Sc is the Schmidt number for mass-transfer properties, Pr is the Prandtl number for heat-transfer properties, and Le is the Lewis number $\kappa/(C_s\rho_g \mathcal{D})$, where κ is the gas thermal conductivity and $\mathcal{D}$ is the diffusion coefficient for the vapor through the gas. Experimental and theoretical values of the exponent n range from 0.56 [Bedingfield and Drew, *Ind. Eng. Chem*, **42**:1164 (1950)] to $\frac{2}{3} = 0.667$ [Chilton and Colburn, *Ind. Eng. Chem.*, **26**:1183 (1934)]. A detailed discussion is given by Keey (1992). Values of β for any system can be estimated from the specific heats, diffusion coefficients, and other data given in Sec. 2. See the example below.

For calculation of wet-bulb (and adiabatic saturation) conditions, the most commonly used formula in industry is the **psychrometer equation**. This is a simple, linear formula that gives vapor pressure directly if the wet-bulb temperature is known, and is therefore ideal for calculating humidity from a wet-bulb measurement using a psychrometer, although the calculation of wet-bulb temperature from humidity still requires an iteration.

$$p = p_{wb} - AP(T - T_{wb}) \qquad (12\text{-}11)$$

where A is the psychrometer coefficient. For the air-water system, the following formulas based on equations given by Sonntag [*Zeitschrift für Meteorologie*, **40**(5): 340–344 (1990)] may be used to give A for T_{wb} up to 30°C; they are based on extensive experimental data for Assmann psychrometers.

Over water (wet-bulb temperature):

$$A = 6.5 \times 10^{-4}(1 + 0.000944T_{wb}) \qquad (12\text{-}12a)$$

Over ice (ice-bulb temperature):

$$A_i = 5.72 \times 10^{-4} \qquad (12\text{-}12b)$$

For other vapor-gas systems, A is given by

$$A = \frac{M_g C_s}{M_V \beta \lambda_{wb}} \qquad (12\text{-}13)$$

Here β is the psychrometric coefficient for the system. As a cross-check, for the air-water system at 20°C wet-bulb temperature, 50°C dry-bulb temperature, and absolute humidity 0.002 kg/kg, $C_s = (1.006 + 1.9 \times 0.002) = 1.01$ kJ/(kg·K) and $\lambda_{wb} = 2454$ kJ/kg. Since $M_g = 28.97$ kg/kmol and $M_v = 18.02$ kg/kmol, Eq. (12-12) gives A as $6.617 \times 10^{-4}/\beta$, compared with Sonntag's value of 6.653×10^{-4} at this temperature, giving a value for the psychrometric coefficient β of 0.995; that is, $\beta \approx 1$, as expected for the air-water system.

PSYCHROMETRIC CHARTS

Psychrometric charts are plots of humidity, temperature, enthalpy, and other useful parameters of a gas-vapor mixture. They are helpful for rapid estimates of conditions and for visualization of process operations such as humidification and drying. They apply to a given system at a given pressure, the most common of course being air-water at atmospheric pressure. There are four types, of which the Grosvenor and Mollier types are most widely used:

The **Grosvenor chart** plots temperature (abscissa) against humidity (ordinate). Standard charts produced by ASHRAE and other groups, or by computer programs, are usually of this type. The saturation line is a curve from bottom left to top right, and curves for constant relative humidity are approximately parallel to this. Lines from top left to bottom right may be of either constant wet-bulb temperature or constant enthalpy, depending on the chart. The two are not quite identical, so if only one is shown, correction factors are required for the other parameter. Examples are shown in Figs. 12-1 (SI units), 12-2*a* (U.S. Customary System units, medium temperature), and 12-2*b* (U.S. Customary System units, high temperature).

The **Bowen chart** is a plot of enthalpy (abscissa) against humidity (ordinate). It is convenient to be able to read enthalpy directly, especially for near-adiabatic convective drying where the operating line approximately follows a line of constant enthalpy. However, it is very difficult to read accurately because the key information is compressed in a narrow band near the saturation line. See Cook and DuMont, *Process Drying Practice*, McGraw-Hill, New York, 1991, chap. 6.

The **Mollier chart** plots humidity (abscissa) against enthalpy (lines sloping diagonally from top left to bottom right). Lines of constant temperature are shallow curves at a small slope to the horizontal. The chart is nonorthogonal (no horizontal lines) and hence a little difficult to plot and interpret initially. However, the area of greatest interest is expanded, and they are therefore easy to read accurately. They tend to cover a wider

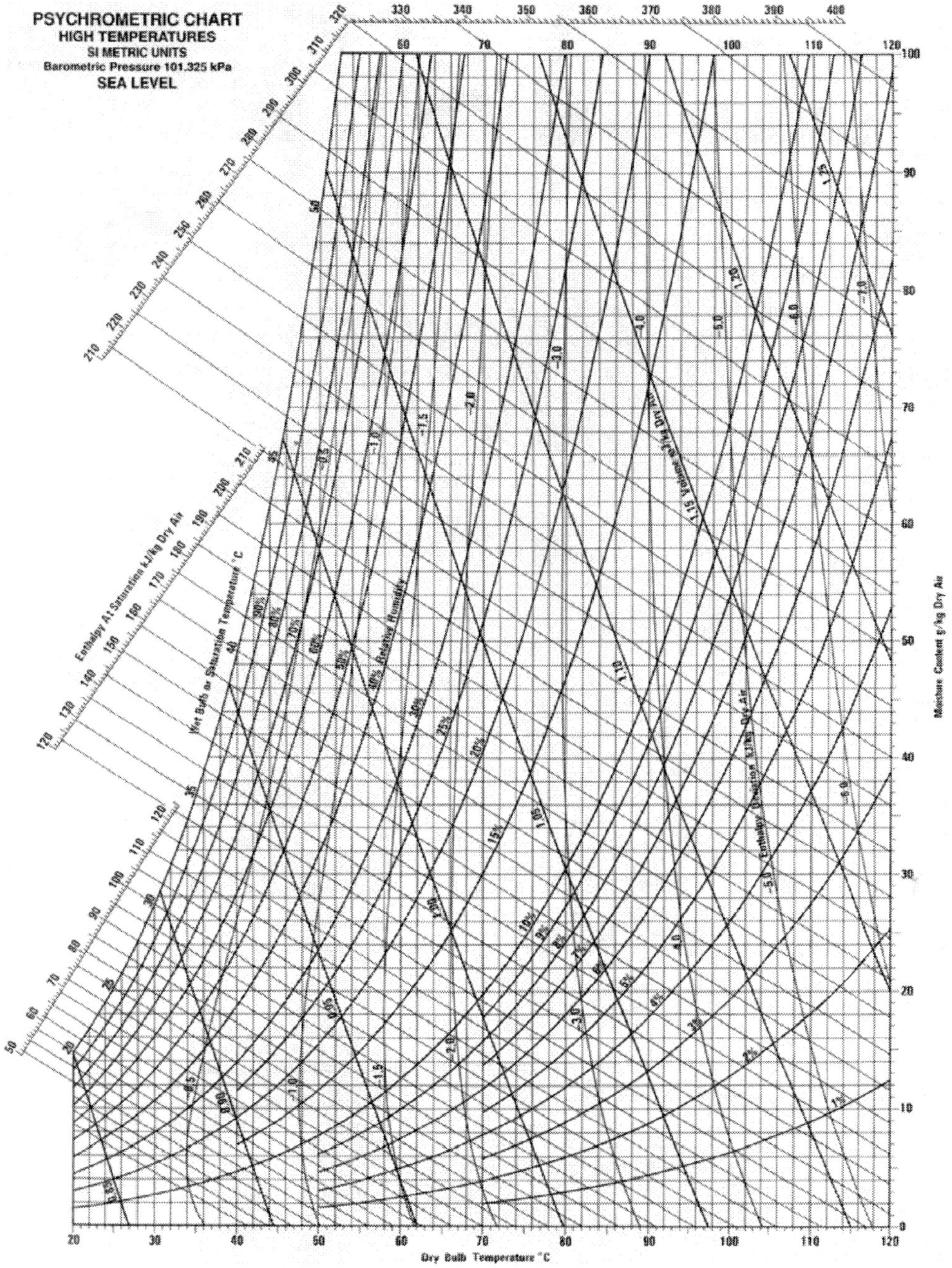

FIG. 12-1 Grosvenor psychrometric chart for the air-water system at standard atmospheric pressure, 101,325 Pa, SI units. (*Courtesy Carrier Corporation.*)

temperature range than Grosvenor charts, so are useful for dryer calculations. The slope of the enthalpy lines is normally $-1/\lambda$, where λ is the latent heat of evaporation. Adiabatic saturation lines are not quite parallel to constant-enthalpy lines and are slightly curved; the deviation increases as humidity increases. Figure 12-3 shows an example.

The **Salen-Soininen** perspectively transformed **chart** is a triangular plot. It is tricky to plot and read, but covers a much wider range of humidity than do the other types of chart (up to 2 kg/kg) and is thus very effective for high-humidity mixtures and calculations near the boiling point, e.g., in pulp and paper drying. See Soininen, *Drying Technol.* **4**(2): 295–305 (1986).

Figure 12-4 shows a psychrometric chart for combustion products in air. The thermodynamic properties of moist air are given in Table 12-1. Figure 12-4 shows a number of useful additional relationships, e.g., specific volume and latent heat variation with temperature. Accurate figures should always be obtained from physical properties tables or by calculation using the formulas given earlier, and these charts should only be used as a quick check for verification.

In the past, psychrometric charts have been used to perform quite precise calculations. To do this, additive corrections are often required for enthalpy of added water or ice, and for variations in barometric pressure from the standard level (101,325 Pa, 14.696 lbf/in^2, 760 mmHg, 29.921 inHg). It is preferable to use formulas, which give an accurate figure at any set of conditions. Psychrometric charts and tables can be used as a rough cross-check that the result has been calculated correctly. Table 12-4 gives values of saturation humidity, specific volume, enthalpy, and entropy of saturated moist air at selected conditions. Below the freezing point, these become virtually identical to the values for dry air, as saturation humidity is very low. For pressure corrections, an altitude increase of approximately 900 ft gives a pressure decrease of 1 inHg (0.034 bar). For a recorded wet-bulb temperature of 50°F (10°C), this gives an increase in humidity of 1.9 gr/lb (0.00027 kg/kg) and the enthalpy increases by 0.29 Btu/lb (0.68 kJ/kg). This correction increases roughly proportionately for further changes in pressure, but climbs sharply as wet-bulb temperature is increased; when T_{wb} reaches 100°F (38°C), ΔY = 11.2 gr/lb (0.0016 kg/kg) and ΔH = 1.77 Btu/lb (4.12 kJ/kg). Equivalent, more detailed tables in SI units can be found in the ASHRAE Handbook.

Examples Illustrating Use of Psychrometric Charts In these examples the following nomenclature is used:

t = dry-bulb temperatures, °F
t_w = wet-bulb temperature, °F
t_d = dewpoint temperature, °F
H = moisture content, lb water/lb dry air
ΔH = moisture added to or rejected from the airstream, lb water/lb dry air
h' = enthalpy at saturation, Btu/lb dry air
D = enthalpy deviation, Btu/lb dry air
$h = h' + D$ = true enthalpy, Btu/lb dry air
h_w = enthalpy of water added to or rejected from system, Btu/lb dry air
q_a = heat added to system, Btu/lb dry air
q_r = heat removed from system, Btu/lb dry air

Subscripts 1, 2, 3, etc., indicate entering and subsequent states.

Example 1: Determination of Moist Air Properties Find the properties of moist air when the dry-bulb temperature is 80°F and the wet-bulb temperature is 67°F.

Solution: Read directly from Fig. 12-2*a* (Fig. 12-6*a* shows the solution diagrammatically).

Moisture content H = 78 gr/lb dry air
= 0.011 lb water/lb dry air
Enthalpy at saturation h' = 31.6 Btu/lb dry air
Enthalpy deviation D = −0.1 Btu/lb dry air
True enthalpy h = 31.5 Btu/lb dry air
Specific volume v = 13.8 ft^3/lb dry air
Relative humidity = 51 percent
Dew point t_d = 60.3°F

Example 2: Air Heating Air is heated by a steam coil from 30°F dry-bulb temperature and 80 percent relative humidity to 75°F dry-bulb temperature. Find the relative humidity, wet-bulb temperature, and dew point of the heated air. Determine the quantity of heat added per pound of dry air.

Solution: Reading directly from the psychrometric chart (Fig. 12-2*a*),

Relative humidity = 15 percent
Wet-bulb temperature = 51.5°F
Dew point = 25.2°F

The enthalpy of the inlet air is obtained from Fig. 12-2*a* as $h_1 = h_1' + D_1$ = 10.1 + 0.06 = 10.16 Btu/lb dry air; at the exit, $h_2 = h_2' + D_2$ = 21.1 − 0.1 = 21 Btu/lb dry air. The heat added equals the enthalpy difference, or

$$q_a = \Delta h = h_2 - h_1 = 21 - 10.16 = 10.84 \text{ Btu/lb dry air}$$

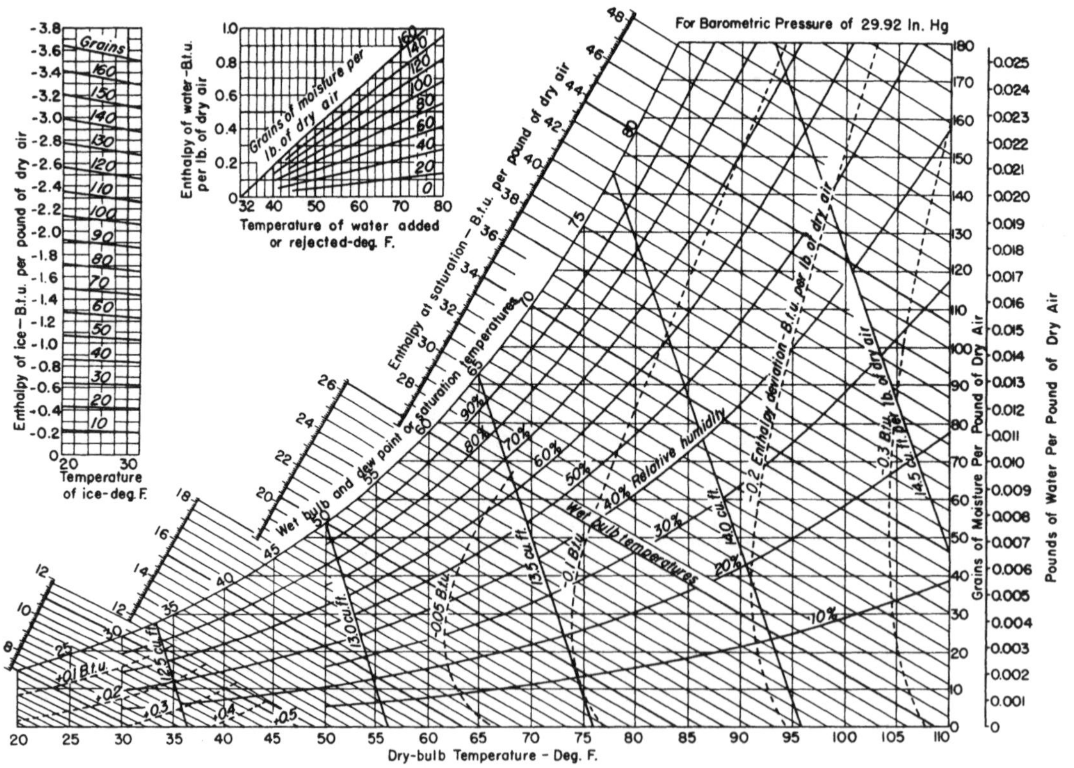

FIG. 12-2*a* Grosvenor psychrometric chart (medium temperature) for the air-water system at standard atmospheric pressure, 29.92 inHg, U.S. Customary units. (*Courtesy Carrier Corporation.*)

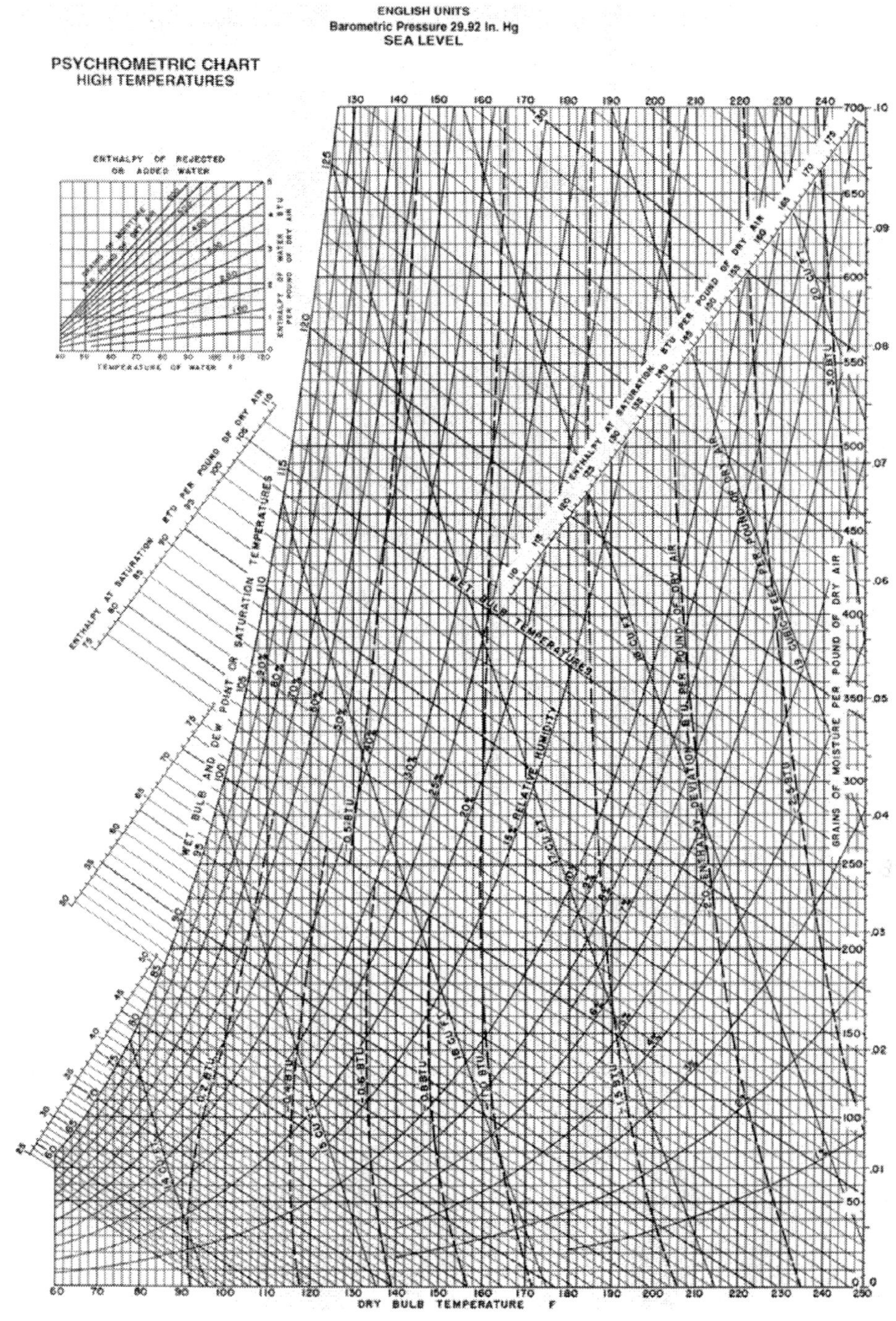

FIG. 12-2*b* Grosvenor psychrometric chart (high-temperature) for the air-water system at standard atmospheric pressure, 29.92 inHg, U.S. Customary units. (*Source: Carrier Corporation.*)

If the enthalpy deviation is ignored, the heat added q_a is $\Delta h = 21.1 - 10.1 = 11$ Btu/lb dry air, or the result is 1.5 percent high. Figure 12-6*b* shows the heating path on the psychrometric chart.

Example 3: Evaporative Cooling Air at 95°F dry-bulb temperature and 70°F wet-bulb temperature contacts a water spray, where its relative humidity is increased to 90 percent. The spray water is recirculated; makeup water enters at 70°F. Determine exit dry-bulb temperature, wet-bulb temperature, change in enthalpy of the air, and quantity of moisture added per pound of dry air.

Solution: Figure 12-6*c* shows the path on a psychrometric chart. The leaving dry-bulb temperature is obtained directly from Fig. 12-2*a* as 72.2°F. Since the spray water enters at the wet-bulb temperature of 70°F and there is no heat added to or removed from it, this is by definition an adiabatic process and there

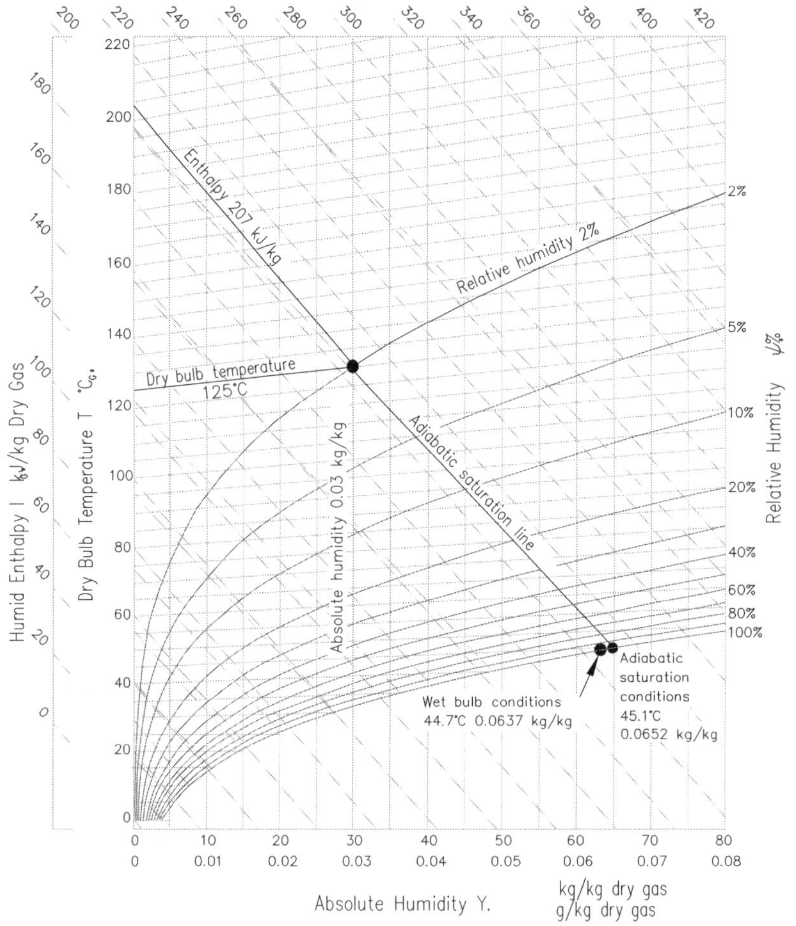

FIG. 12-3 Mollier psychrometric chart for the air-water system at standard atmospheric pressure, 101,325 Pa SI units, plots humidity (abscissa) against enthalpy (lines sloping diagonally from top left to bottom right). (*Source: Aspen Technology.*)

will be no change in wet-bulb temperature. The only change in enthalpy is that from the heat content of the makeup water. This can be demonstrated as follows:

$$\text{Inlet moisture } H_1 = 70 \text{ gr/lb dry air}$$
$$\text{Exit moisture } H_2 = 107 \text{ gr/lb dry air}$$
$$\Delta H = 37 \text{ gr/lb dry air}$$
$$\text{Inlet enthalpy } h_1 = h_1' + D_1 = 34.1 - 0.22 = 33.88 \text{ Btu/lb dry air}$$
$$\text{Exit enthalpy } h_2 = h_2' + D_2 = 34.1 - 0.02 = 34.08 \text{ Btu/lb dry air}$$
$$\text{Enthalpy of added water } h_w = 0.2 \text{ Btu/lb dry air (from small diagram, 37 gr at 70°F)}$$

Then

$$q_a = h_2 - h_1 + h_w = 34.08 - 33.88 + 0.2 = 0$$

Example 4: Cooling and Dehumidification Find the cooling load per pound of dry air resulting from infiltration of room air at 80°F dry-bulb temperature and 67°F wet-bulb temperature into a cooler maintained at 30°F dry-bulb and 28°F wet-bulb temperature, where moisture freezes on the coil, which is maintained at 20°F.

Solution: The path followed on a psychrometric chart is shown in Fig. 12-6*d*.

$$\text{Inlet enthalpy } h_1 = h_1' + D_1 = 31.62 - 0.1 = 31.52 \text{ Btu/lb dry air}$$
$$\text{Exit enthalpy } h_2 = h_2' + D_2 = 10.1 + 0.06 = 10.16 \text{ Btu/lb dry air}$$
$$\text{Inlet moisture } H_1 = 78 \text{ gr/lb dry air}$$
$$\text{Exit moisture } H_2 = 19 \text{ gr/lb dry air}$$
$$\text{Moisture rejected } \Delta H = 59 \text{ gr/lb dry air}$$
$$\text{Enthalpy of rejected moisture} = -1.26 \text{ Btu/lb dry air (from small diagram of Fig. 12-2a)}$$
$$\text{Cooling load } q_r = 31.52 - 10.16 + 1.26 = 22.62 \text{ Btu/lb dry air}$$

Note that if the enthalpy deviations were ignored, the calculated cooling load would be about 5 percent low.

Example 5: Cooling Tower Determine water consumption and amount of heat dissipated per 1000 ft³/min of entering air at 90°F dry-bulb temperature and 70°F wet-bulb temperature when the air leaves saturated at 110°F and the makeup water is at 75°F.

Solution: The path followed is shown in Fig. 12-6*e*.

$$\text{Exit moisture } H_2 = 416 \text{ gr/lb dry air}$$
$$\text{Inlet moisture } H_1 = 78 \text{ gr/lb dry air}$$
$$\text{Moisture added } \Delta H = 338 \text{ gr/lb dry air}$$
$$\text{Enthalpy of added moisture } h_w = 2.1 \text{ Btu/lb dry air (from small diagram of Fig. 12-2b)}$$

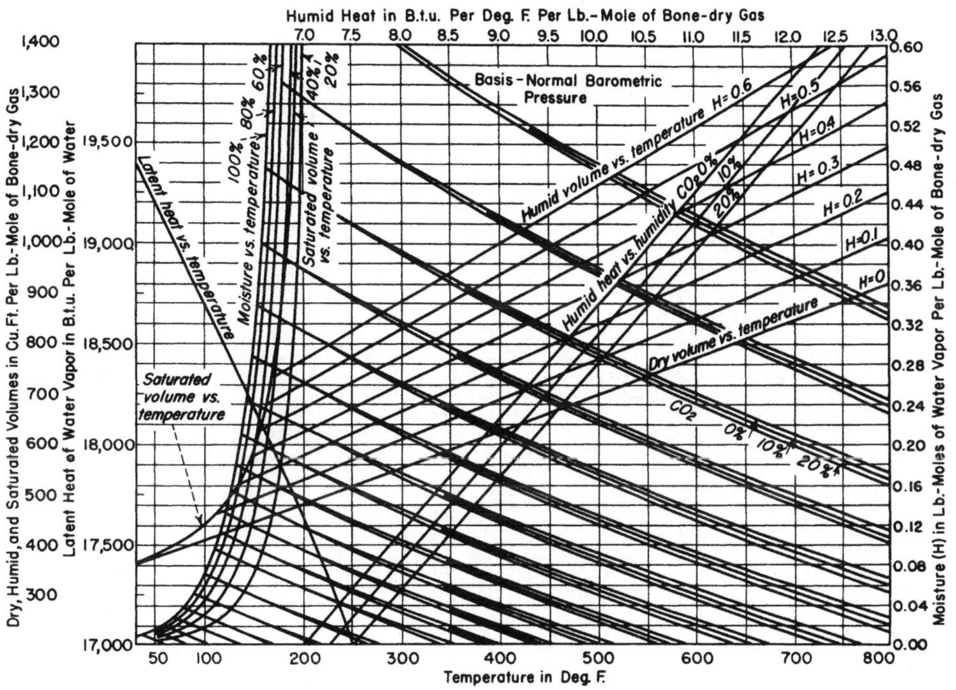

FIG. 12-4 Grosvenor psychrometric chart for air and flue gases at high temperatures, molar units [*Hatta,* Chem. Metall. Eng., *37:64 (1930)*].

TABLE 12-4 Thermodynamic Properties of Saturated Air (U.S. Customary Units, at Standard Atmospheric Pressure, 29.921 inHg)

Temp. T, °F	Saturation humidity H_s	Volume, ft³/lb dry air v_a	v_{as}	v_s	Enthalpy, Btu/lb dry air h_a	h_{as}	h_s	Entropy, Btu/(°F·lb dry air) s_a	s_{as}	s_s	Condensed water: Enthalpy, Btu/lb h_w	Condensed water: Entropy, Btu/(lb·°F) s_w	Condensed water: Vapor pressure, inHg p_s	Temp. T, °F
−150	6.932×10^{-9}	7.775	.000	7.775	36.088	.000	36.088	0.09508	.00000	0.09508	218.77	0.4800	3.301×10^{-6}	−150
−100	9.772×10^{-7}	9.046	.000	9.046	24.037	.001	24.036	0.05897	.00000	0.05897	201.23	0.4277	4.666×10^{-5}	−100
−50	4.163×10^{-5}	10.313	.001	10.314	12.012	.043	11.969	0.02766	.00012	0.02754	181.29	0.3758	1.991×10^{-3}	−50
0	7.872×10^{-4}	11.578	.015	11.593	0.000	.835	0.835	0.00000	.00192	0.00192	158.93	0.3244	0.037645×10^{-2}	0
10	1.315×10^{-3}	11.831	.025	11.856	2.402	1.401	3.803	.00518	.00314	.00832	154.17	0.3141	0.062858	10
20	2.152×10^{-3}	12.084	.042	12.126	4.804	2.302	7.106	.01023	.00504	.01527	149.31	0.3039	0.10272	20
30	3.454×10^{-3}	12.338	.068	12.406	7.206	3.709	10.915	.01519	.00796	.02315	144.36	0.2936	0.16452	30
32	3.788×10^{-3}	12.388	.075	12.463	7.686	4.072	11.758	.01617	.00870	.02487	143.36	0.2916	0.18035	32
32*	3.788×10^{-3}	12.388	.075	12.463	7.686	4.072	11.758	.01617	.00870	.02487	0.04	0.0000	0.18037	32*
40	5.213×10^{-3}	12.590	.105	12.695	9.608	5.622	15.230	.02005	.01183	.03188	8.09	.0162	.24767	40
50	7.658×10^{-3}	12.843	.158	13.001	12.010	8.291	20.301	.02481	.01711	.04192	18.11	.0361	.36240	50
60	1.108×10^{-2}	13.096	.233	13.329	14.413	12.05	26.46	.02948	.02441	.05389	28.12	.0555	.52159	60
70	1.582×10^{-2}	13.348	.339	13.687	16.816	17.27	34.09	.03405	.03437	.06842	38.11	.0746	.73915	70
80	2.233×10^{-2}	13.601	0.486	14.087	19.221	24.47	43.69	0.03854	0.04784	0.08638	48.10	0.0933	1.0323	80
90	3.118×10^{-2}	13.853	.692	14.545	21.625	34.31	55.93	.04295	.06596	.10890	58.08	.1116	1.4219	90
100	4.319×10^{-2}	14.106	.975	15.081	24.029	47.70	71.73	.04729	.09016	.13745	68.06	.1296	1.9333	100
110	5.944×10^{-2}	14.359	1.365	15.724	26.434	65.91	92.34	.05155	.1226	.1742	78.03	.1472	2.5966	110
120	8.149×10^{-2}	14.611	1.905	16.516	28.841	90.70	119.54	.05573	.1659	.2216	88.01	.1646	3.4474	120
130	0.1116	14.864	2.652	17.516	31.248	124.7	155.9	.05985	.2245	.2844	98.00	.1817	4.5272	130
140	0.1534	15.117	3.702	18.819	33.655	172.0	205.7	.06390	.3047	.3686	107.99	.1985	5.8838	140
150	0.2125	15.369	5.211	20.580	36.063	239.2	275.3	.06787	.4169	.4848	117.99	.2150	7.5722	150
160	0.2990	15.622	7.446	23.068	38.472	337.8	376.3	.07179	.5793	.6511	128.00	.2313	9.6556	160
170	0.4327	15.874	10.938	26.812	40.882	490.6	531.5	.07565	.8273	.9030	138.01	.2473	12.203	170
180	0.6578	16.127	16.870	32.997	43.292	748.5	791.8	.07946	1.240	1.319	148.03	.2631	15.294	180
190	1.099	16.379	28.580	44.959	45.704	1255	1301	.08320	2.039	2.122	158.07	.2786	19.017	190
200	2.295	16.632	60.510	77.142	48.119	2629	2677	.08689	4.179	4.266	168.11	.2940	23.468	200

NOTE: Compiled by John A. Goff and S. Gratch. See also Keenan and Kaye. *Thermodynamic Properties of Air*, Wiley, New York, 1945. Enthalpy of dry air taken as zero at 0°F. Enthalpy of liquid water taken as zero at 32°F.

To convert British thermal units per pound to joules per kilogram, multiply by 2326; to convert British thermal units per pound dry air-degree Fahrenheit to joules per kilogram-kelvin, multiply by 4186.8; and to convert cubic feet per pound to cubic meters per kilogram, multiply by 0.0624.

*Entrapolated to represent metastable equilibrium with undercooled liquid.

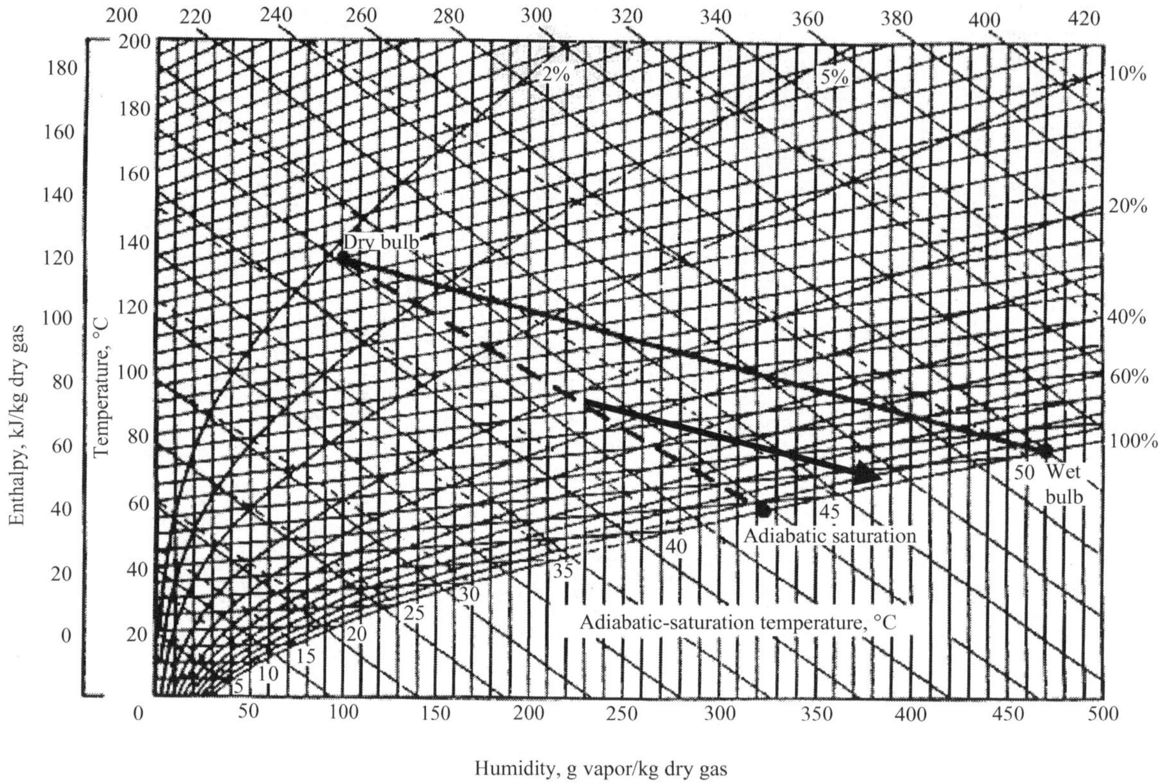

FIG. 12-5 Mollier chart showing changes in T_{wb} during an adiabatic saturation process for an organic system (nitrogen-toluene).

If greater precision is desired, h_w can be calculated as

$$h_w = (338/7000)(1)(75 - 32)$$
$$= 2.08 \text{ Btu/lb dry air}$$

Enthalpy of inlet air $h_1 = h_1' + D_1 = 34.1 - 0.18$
$= 33.92$ Btu/lb dry air

Enthalpy of exit air $h_2 = h_2' + D_2 = 92.34 + 0$
$= 92.34$ Btu/lb dry air

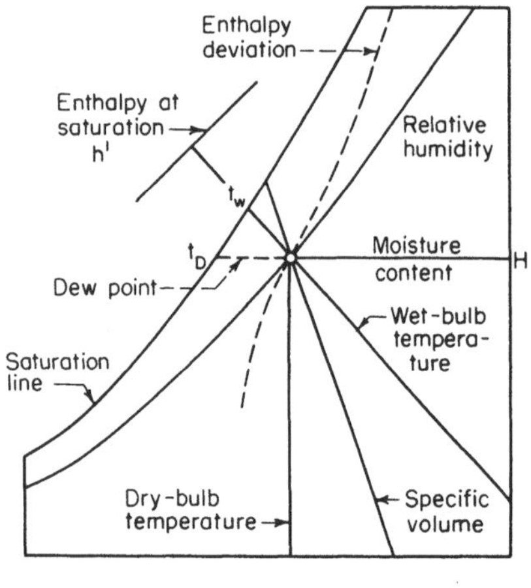

FIG. 12-6a Diagram of psychrometric chart showing the properties of moist air.

Heat dissipated $= h_2 - h_1 - h_w$
$= 92.34 - 33.92 - 2.08$
$= 56.34$ Btu/lb dry air

Specific volume of inlet air $= 14.1$ ft^3/lb dry air

$$\text{Total heat dissipated} = \frac{(1000)(56.34)}{14.1} = 3990 \text{ Btu/min}$$

Example 6: Recirculating Dryer A dryer is removing 100 lb water/h from the material being dried. The air entering the dryer has a dry-bulb temperature of 180°F and a wet-bulb temperature of 110°F. The air leaves the dryer at 140°F. A portion of the air is recirculated after mixing with room air having a dry-bulb temperature of 75°F and a relative humidity of 60 percent. Determine the quantity of air required, recirculation rate, and load on the preheater if it is assumed that the system is adiabatic. Neglect heatup of the feed and of the conveying equipment.

Solution: The path followed is shown in Fig. 12-6*f*.

Humidity of room air $H_1 = 0.0113$ lb/lb dry air

Humidity of air entering dryer $H_3 = 0.0418$ lb/lb dry air

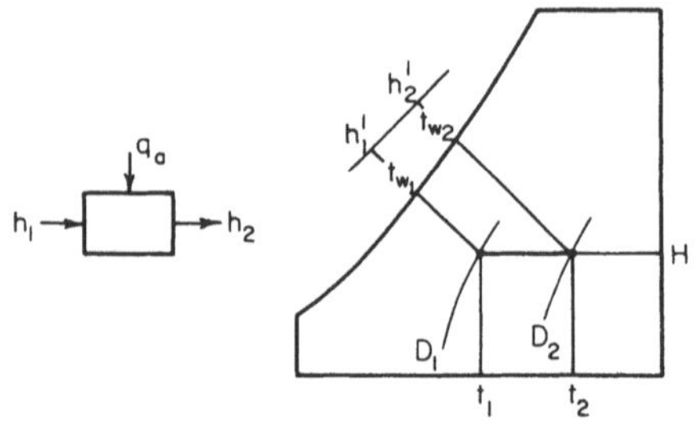

FIG. 12-6b Heating process

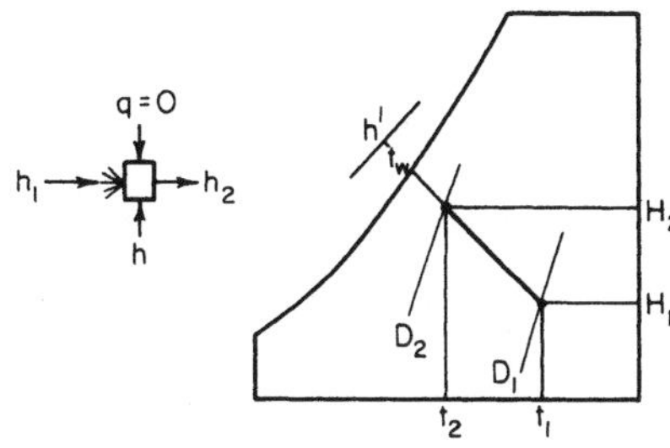

FIG. 12-6c Spray or evaporative cooling.

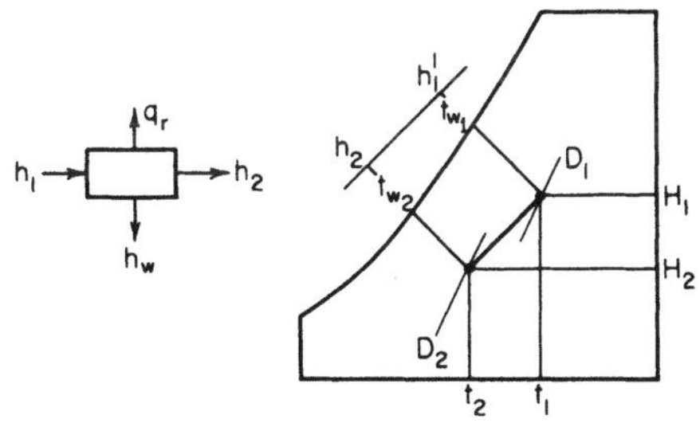

FIG. 12-6d Cooling and dehumidifying process.

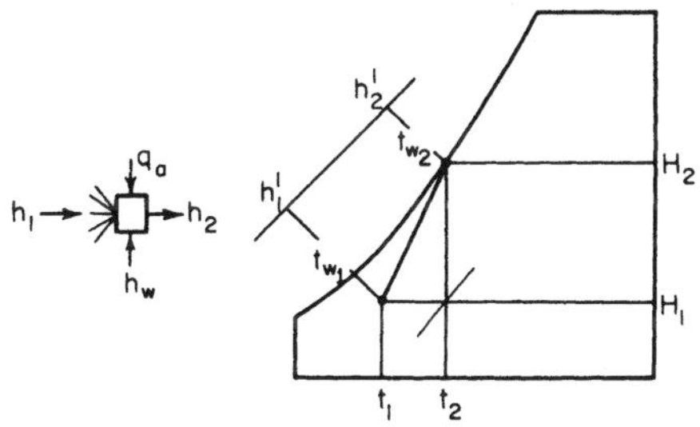

FIG. 12-6e Cooling tower.

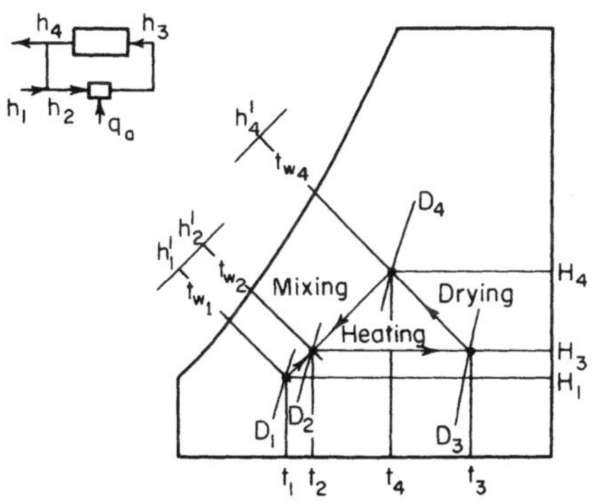

FIG. 12-6f Drying process with recirculation.

$$\text{Humidity of air leaving dryer } H_4 = 0.0518 \text{ lb/lb dry air}$$
$$\text{Enthalpy of room air } h_1 = 30.2 - 0.3 = 29.9 \text{ Btu/lb dry air}$$
$$\text{Enthalpy of entering air } h_3 = 92.5 - 1.3 = 91.2 \text{ Btu/lb dry air}$$
$$\text{Enthalpy of leaving air } h_4 = 92.5 - 0.55 = 91.95 \text{ Btu/lb dry air}$$

Quantity of air required is 100/(0.0518 − 0.0418) = 10,000 lb dry air/h. At the dryer inlet the specific volume is 17.1 ft^3/lb dry air. Air volume is (10,000)(17.1)/60 = 2850 ft^3/min. Fraction exhausted is

$$\frac{X}{W_a} = \frac{0.0518 - 0.0418}{0.0518 - 0.0113} = 0.247$$

where X = quantity of fresh air and W_a = total airflow. Thus 75.3 percent of the air is recirculated. Load on the preheater is obtained from an enthalpy balance

$$q_a = 10{,}000(91.2) - 2470(29.9) - 7530(91.95) = 146{,}000 \text{ Btu/h}$$

PSYCHROMETRIC CALCULATIONS

Table 12-5 gives the steps required to perform the most common humidity calculations, using the formulas given earlier.

Methods (i) to (iii) are used to find the humidity and dew point from temperature readings in a wet- and dry-bulb psychrometer.

Method (iv) is used to find the humidity and dew point from a relative humidity measurement at a given temperature.

Methods (v) and (vi) give the adiabatic saturation and wet-bulb temperatures from absolute humidity (or relative humidity) at a given temperature.

Method (vii) gives the absolute and relative humidity from a dew point measurement.

Method (viii) allows the calculation of all the main parameters if the absolute humidity is known, e.g., from a mass balance on a process plant.

Method (ix) converts the volumetric form of absolute humidity to the mass form (mixing ratio).

Method (x) allows the dew point to be corrected for pressure. The basis is that the mole fraction $y = p/P$ is the same for a given mixture composition at all values of total pressure P. In particular, the dew point measured in a compressed air duct can be converted to the dew point at atmospheric pressure, from which the humidity can be calculated. It is necessary to check that the temperature change associated with compression or expansion does not bring the dry-bulb temperature to a point where condensation can occur. Also, at these elevated pressures, it is strongly advisable to apply the enhancement factor (see BS 1339).

Psychrometric Software and Tables As an alternative to using charts or individual calculations, lookup tables have been published for many years for common psychrometric conversions, e.g., to find relative humidity given the dry-bulb and wet-bulb temperatures. These were often very extensive. To give precise coverage of T_{wb} in 1°C or 0.1°C steps, a complete table would be needed for each individual dry-bulb temperature.

Software is available that will perform calculations of humidity parameters for any point value, and for plotting psychrometric charts. Moreover, British Standard BS 1339 Part 2 (2006) provides functions as macros which can be embedded into any Excel-compatible spreadsheet. Users can therefore generate their own tables for any desired combination of parameters as well as perform point calculations. Hence, the need for published lookup tables has been eliminated. However, this software, like the previous lookup tables, is only valid for the air-water system. For other vapor-gas systems, the equations given in previous sections must be used.

Software may be effectively used to draw psychrometric charts or perform calculations. A wide variety of other psychrometric software may be found on the Internet, but quality varies considerably; the

TABLE 12-5 Calculation Methods for Various Humidity Parameters

	Known	Required	Method
i.	T, T_{wb}	Y	Find saturation vapor pressure p_{wb} at wet-bulb temperature T_{wb} from Eq. (12-4). Find actual vapor pressure p at dry-bulb temperature T from psychrometer equation (12-11). Find mixing ratio Y by conversion from p (Table 12-1).
ii.	T, T_{wb}	T_{dp}, d_v	Find p if necessary by method (i) above. Find dew point T_{dp} from Eq. (12-4) by calculating the T corresponding to p [iteration required; Antoine equation (12-5) gives a first estimate]. Calculate volumetric humidity Y_v, using Eq. (12-1).
iii.	T, T_{wb}	%RH (ψ)	Use method (i) to find p. Find saturation vapor pressure p_s at T from Eq. (12-4). Now relative humidity %RH = $100p/p_s$.
iv.	T, %RH	Y, d_v	Find saturation vapor pressure p_s at T from Eq. (12-4). Actual vapor pressure $p = p_s$(%RH/100). Convert to Y (Table 12-1). Find Y_v from Eq. (12-1).
v.	T, %RH (or T, Y)	T_{as}	Use method (iv) to find p and Y. Make an initial estimate of T_{as}, say, using a psychrometric chart. Calculate Y_{as} from Eq. (12-6). Find p from Table 12-1 and T_{as} from Antoine equation (12-5). Repeat until iteration converges (e.g., using spreadsheet). Alternative method: Evaluate enthalpy H_{est} at these conditions and H at initial conditions. Find H_{as} from Eq. (12-7) and compare with H_{est}. Make new estimate of Y_{as} which would give H_{est} equal to H_{as}. Find p from Table 12-1 and T_{as} from Antoine equation (12-5). Reevaluate H_{as} from Eq. (12-7) and iterate to refine value of Y_{as}.
vi.	T, %RH (or T, Y)	T_{wb}	Use method (iv) to find p and Y. Make an initial estimate of T_{wb}, e.g., using a psychrometric chart, or (for air-water system) by estimating adiabatic saturation temperature T_{as}. Find p_{wb} from psychrometer equation (12-11). Calculate new value of T_{wb} corresponding to p_{wb} by reversing Eq. (12-4) or using the Antoine equation (12-5). Repeat last two steps to solve iteratively for T_{wb} (computer program is preferable method).
vii.	T, T_{dp}	Y, %RH	Find saturation vapor pressure at dew point T_{dp} from Eq. (12-4); this is the actual vapor pressure p. Find Y from Table 12-1. Find saturation vapor pressure p_s at dry-bulb temperature T from Eq. (12-4). Now %RH = $100p/p_s$.
viii.	T, Y	T_{dp}, d_v, %RH, T_{wb}	Find p by conversion from Y (Table 12-1). Then use method (ii), (iii), or (v) as appropriate.
ix.	T, Y_v	Y	Find specific humidity Y_W from Eqs. (12-2) and (12-1). Convert to absolute humidity Y using $Y = Y_W/(1 - Y_W)$.
x.	T_{dp} at P_1 (elevated)	T_{dp} at P_2 (ambient)	Find vapor pressure p_1 at T_{dp} and P_1 from Eq. (12-4), Convert to vapor pressure p_2 at new pressure P_2 by the formula $p_2 = p_1P_2/P_1$. Find new dew point T_{dp} from Eq. (12-4) by calculating the T corresponding to p_2 [iteration required as in (ii)].

source and basis of the calculation methods should be carefully checked before using the results. In particular, most methods only apply for the air-water system at moderate temperatures (below 100°C). For high-temperature dryer calculations, only software stated as suitable for this range should be used.

Reliable sources include the following:

1. The American Society of Agricultural Engineers (ASAE): http://www.asae.org. Psychrometric data in chart and equation form in both SI and English units. Charts for temperature ranges of −35 to 600°F in USCS units and −10 to 120°C in SI units. Equations and calculation procedures. Air-water system and Grosvenor (temperature-humidity) charts only.

2. The American Society of Heating, Refrigerating and Air-Conditioning Engineers (ASHRAE): http://www.ashrae.org. Psychrometric Analysis CD with energy calculations and creation of custom charts at virtually any altitude or pressure. Detailed scientific basis given in ASHRAE Handbook. Air-water system and Grosvenor charts only.

3. Carrier Corporation, a United Technologies Company: http://www.training.carrier.com. PSYCH+, computerized psychrometric chart and instructional guide, including design of air conditioning processes and/or cycles. Printed psychrometric charts also supplied. Air-water system and Grosvenor charts only.

4. Linric Company: http://www.linric.com. PsycPro generates custom psychrometric charts in English (USCS) or metric (SI) units, based on ASHRAE formulas. Air-water system and Grosvenor charts only.

5. Aspen Technology: http://www.aspentech.com. PSYCHIC, one of the Process Tools, generates customized psychrometric charts. Mollier and Bowen enthalpy-humidity charts are produced in addition to Grosvenor. Any gas-vapor system can be handled as well as air-water; data supplied for common organic solvents. Can draw operating lines and spot points, as shown in Fig. 12-7.

6. British Standards Institution: http://www.bsonline.bsi-global.com. British Standard BS 1339 Part 2 is a spreadsheet-based software program providing functions based on latest internationally agreed upon standards. It calculates all key psychrometric parameters and can produce a wide range of psychrometric tables. Users can embed the functions in their own spreadsheets to do psychrometric calculations. Air-water system only (although BS 1339 Part 1 text gives full calculation methods for other gas-vapor systems). SI (metric) units. It does not plot psychrometric charts.

7. Akton Associates provides digital versions of psychrometry charts.

Psychrometric Calculations—Worked Examples

Example 7: Determination of Moist Air Properties An air-water mixture is found from the heat and mass balance to be at 60°C (333 K) and 0.025 kg/kg (25 g/kg) absolute humidity. Calculate the other main parameters for the mixture. Take atmospheric pressure as 101,325 Pa.

Method: Consult item (vi) in Table 12-5 for the calculation methodology.

From the initial terminology section, specific humidity Y_W = 0.02439 kg/kg, mole ratio z = 0.0402 kmol/kmol, mole fraction y = 0.03864 kmol/kmol.

From Table 12-1, vapor pressure p = 3915 Pa (0.03915 bar) and volumetric humidity Y_v = 0.02547 kg/m³. Dew point is given by the temperature corresponding to p at saturation. From the reversed Antoine equation (12-5), T_{dp} = 3830/(23.19 − ln 3915) + 44.83 = 301.58 K = 28.43°C.

Relative humidity is the ratio of actual vapor pressure to saturation vapor pressure at dry-bulb temperature. From the Antoine equation (12-5), p_s = exp [23.19 − 3830/(333.15 − 44.83)] = 20,053 Pa (new coefficients), or p_s = exp [23.1963 − 3816.44/(333.15 − 46.13)] = 19,921 Pa (old coefficients).

From Sonntag equation (12-4), p_s = 19,948 Pa; difference from Antoine is less than 0.5 percent. Relative humidity = 100 × 3915/19,948 = 19.6 percent. From a psychrometric chart, e.g., Fig. 12-1, a humidity of 0.025 kg/kg at T = 60°C lies very close to the adiabatic saturation line for 35°C. Hence a good first estimate for T_{as} and T_{wb} will be 35°C. Refining the estimate of T_{wb} by using the psychrometer equation and iterating gives

$$p_{wb} = 3915 + 6.46 \times 10^{-4}\,(1.033)(101{,}325)\,(60 - 35) = 5605$$

From the Antoine equation,

$$T_{wb} = 3830/(23.19 - \ln 5605) + 44.83 = 307.9\text{ K} = 34.75°\text{C}$$

Second iteration:

$$p_{wb} = 3915 + 6.46 \times 10^{-4}(1.033)(101{,}325)(60 - 34.75) = 5622$$
$$T_{wb} = 307.96\text{ K} = 34.81°\text{C}.$$

To a sensible level of precision, T_{wb} = 34.8°C.

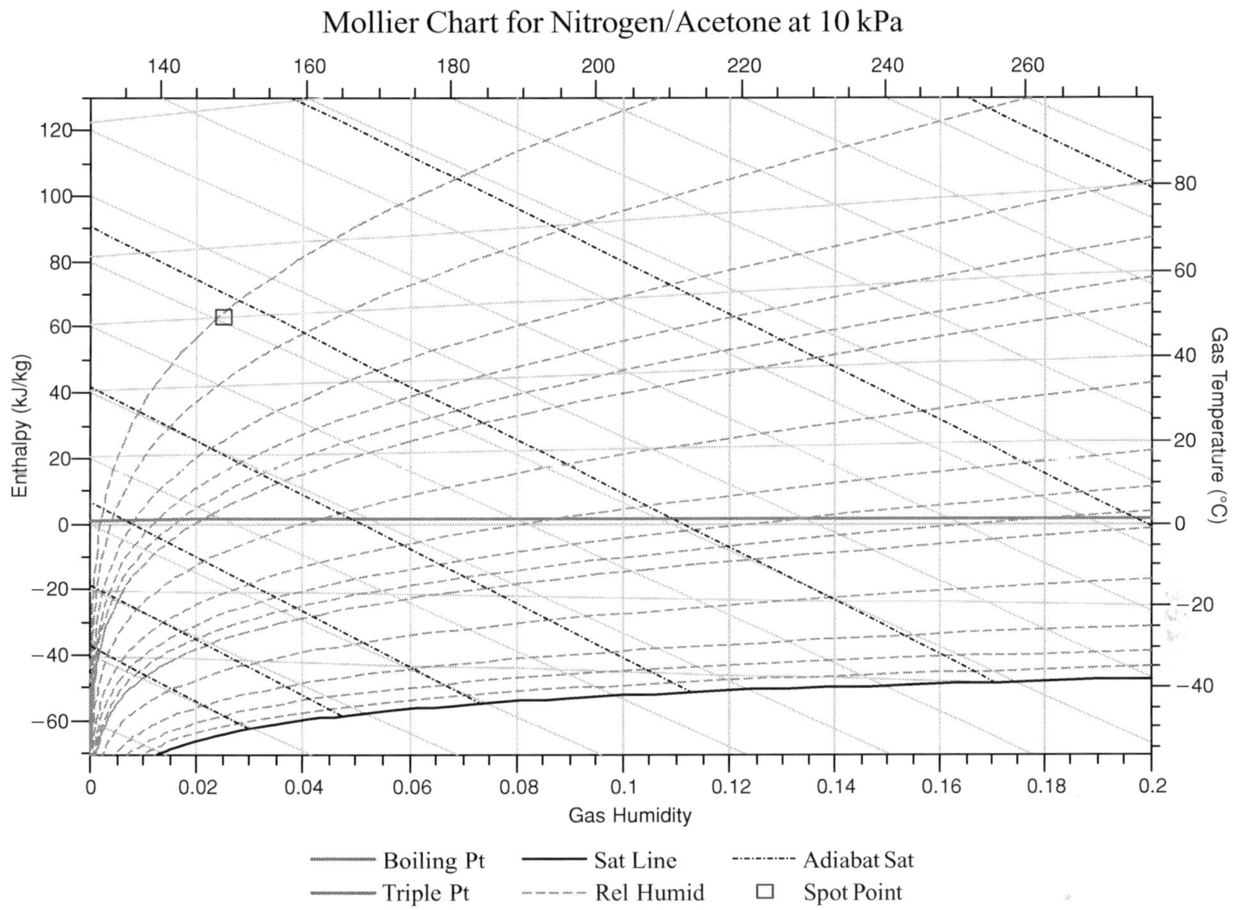

FIG. 12-7 Mollier psychrometric chart (from PSYCHIC software program) showing determination of adiabatic saturation temperature plots humidity (abscissa) against enthalpy (lines sloping diagonally from top left to bottom right). (*Courtesy AspenTech.*)

From Table 12-1 $Y_{wb} = 5622 \times 0.622/(101{,}325 - 5622) = 0.0365(4)$ kg/kg.

Enthalpy of original hot air is approximately given by $H = (C_{Pg} + C_{Pv}Y)(T - T_0) + \lambda_0 Y = (1 + 1.9 \times 0.025) \times 60 + 2501 \times 0.025 = 62.85 + 62.5 = 125.35$ kJ/kg. A more accurate calculation can be obtained from steam tables; $C_{Pg} = 1.005$ kJ/(kg·K) over this range, H_v at 60°C = 2608.8 kJ/kg, $H = 60.3 + 65.22 = 125.52$ kJ/kg.

Calculation (v), method 1: if $T_{as} = 34.8$, from Eq. (12-6), with $C_s = 1 + 1.9 \times 0.025 = 1.048$ kJ/(kg·K), $\lambda_{as} = 2419$ kJ/kg (steam tables), $Y_{as} = 0.025 + 1.048/2419\ (60 - 34.8) = 0.0359(2)$ kg/kg. From Table 12-1, $p = 5530$ Pa. From the Antoine equation (12-5), $T_{as} = 3830/(23.19 - \ln 5530) + 44.83 = 307.65$ K = 34.52°C. Repeat until iteration converges (e.g., using spreadsheet). Final value $T_{as} = 34.57$°C, $Y_{as} = 0.0360$ kg/kg.

Enthalpy check: From Eq. (12-7), $H_{as} - H = 4.1868 \times 34.57 \times (0.036 - 0.025) = 1.59$ kJ/kg. So $H_{as} = 127.11$ kJ/kg. Compare H_{as} calculated from enthalpies; H_g at 34.57°C = 2564 kJ/kg, $H_{est} = 34.90 + 92.29 = 127.19$ kJ/kg. The iteration has converged successfully.

Note that T_{as} is 0.2°C lower than T_{wb} and Y_{as} is 0.0005 kg/kg lower than Y_{wb}, both negligible differences.

Example 8: Calculation of Humidity and Wet-Bulb Condition A dryer exhaust which can be taken as an air-water mixture at 70°C (343.15 K) is measured to have a relative humidity of 25 percent. Calculate the humidity parameters and wet-bulb conditions for the mixture. Pressure is 1 bar (100,000 Pa).

Method: Consult item (v) in Table 12-5 for the calculation methodology.

From the Antoine equation (12-5), using standard coefficients (which give a better fit in this temperature range), $p_s = \exp[23.1963 - 3816.44/(343.15 - 46.13)] = 31{,}170$ Pa. Actual vapor pressure p = 25 percent of 31,170 = 7792 Pa (0.078 bar).

From Table 12-1, absolute humidity $Y = 0.05256$ kg/kg and volumetric humidity $Y_v = 0.0492$ kg/m³. From the terminology section, mole fraction $y = 0.0779$ kmol/kmol, mole ratio $z = 0.0845$ kmol/kmol, specific humidity $Y_w = 0.04994$ kg/kg.

$$\text{Dew point } T_{dp} = 3816.44/(23.1963 - \ln 7792) + 46.13 = 314.22 \text{ K} = 41.07°\text{C}.$$

From the psychrometric chart, a humidity of 0.0526 kg/kg at $T = 70$°C falls just below the adiabatic saturation line for 45°C. Estimate T_{as} and T_{wb} as 45°C. Refining the estimate of T_{wb} by using the psychrometer equation and iterating gives

$$p_{wb} = 7792 + 6.46 \times 10^{-4}\,(1.0425)(10^5)(70 - 45) = 9476$$

From the Antoine equation,

$$T_{wb} = 3816.44/(23.1963 - \ln 9476) + 46.13 = 317.96 \text{ K} = 44.81°\text{C}$$

Second iteration (taking $T_{wb} = 44.8$):

$$p_{wb} = 9489 \qquad T_{wb} = 317.99 \text{ K} = 44.84°\text{C}$$

The iteration has converged.

Example 9: Calculation of Psychrometric Properties of Acetone/Nitrogen Mixture A mixture of nitrogen N_2 and acetone CH_3COCH_3 is found from the heat and mass balance to be at 60°C (333 K) and 0.025 kg/kg (25 g/kg) absolute humidity (same conditions as in Example 7). Calculate the other main parameters for the mixture. The system is under vacuum at 100 mbar (0.1 bar, 10,000 Pa).

Additional data for acetone and nitrogen are obtained from *The Properties of Gases and Liquids* (Prausnitz et al.). Molecular weight (molal mass) M_g for nitrogen = 28.01 kg/kmol; M_v for acetone = 58.08 kg/kmol. Antoine coefficients for acetone are 16.6513, 2940.46, and 35.93, with p_s in **mmHg** and T in **K**. Specific heat capacity of nitrogen is approximately 1.014 kJ/(kg·K). Latent heat of acetone is 501.1 kJ/kg at the boiling point. The psychrometric ratio for the nitrogen-acetone system is not given, but the diffusion coefficient $\mathcal{D}$ can be roughly evaluated as 1.34×10^{-5}, compared to 2.20×10^{-5} for water in air. As the psychrometric ratio is linked to $\mathcal{D}^{2/3}$, it can be estimated as 0.72, which is in line with tabulated values for similar organic solvents (e.g., propanol).

Method: Consult item (vi) in Table 12-5 for the calculation methodology.

From the terminology, specific humidity $Y_W = 0.02439$ kg/kg, the same as in Example 7. Mole ratio $z = 0.0121$ kmol/kmol, mole fraction $y = 0.01191$ kmol/kmol—lower than in Example 7 because molecular weights are different.

From the Antoine equation (12-5),

$$\ln p_s = C_0 - \frac{C_1}{T - C_2} = 16.6513 - \frac{2940.46}{T - 35.93}$$

Since $T = 60°C$, $\ln p_s = 6.758$, $p_s = 861.0$ mmHg. Hence $p_s = 1.148$ bar $= 1.148 \times 10^5$ Pa. The saturation vapor pressure is higher than atmospheric pressure; this means that acetone at 60°C must be above its normal boiling point. Check; T_{bp} for acetone = 56.5°C.

Vapor pressure $p = yP = 0.01191 \times 10{,}000 = 119.1$ Pa (0.001191 bar)—much lower than before because of the reduced total pressure. This is 0.89 mmHg. Volumetric humidity $Y_v = 0.0025$ kg/m³—again substantially lower than at 1 atm.

Dew point is the temperature where p_s equals p'. From the reversed Antoine equation (12-5),

$$T = \frac{C_1}{C_0 - \ln p_s} + C_2$$

so

$$T_{dp} = \frac{2940}{16.6513 - \ln 0.89} + 35.93 = 211.27 \text{ K} = -61.88°C$$

This very low dew point is due to the low boiling point of acetone and the low concentration.

Relative humidity is the ratio of actual vapor pressure to saturation vapor pressure at dry-bulb temperature. So $p = 119.1$ Pa, $p_s = 1.148 \times 10^5$ Pa, RH = 0.104 percent—again very low.

A special psychrometric chart would need to be constructed for the acetone-nitrogen system to get first estimates (this can be done using PSYCHIC, as shown in Fig. 12-7). A humidity of 0.025 kg/kg at $T = 60°C$ lies just below the adiabatic saturation line for − 40°C. The wet-bulb temperature will not be the same as T_{as} for this system; as the psychrometric ratio β is less than 1, T_{wb} should be significantly above T_{as}. However, let us assume no good first estimate is available and simply take T_{wb} to be 0°C initially.

When using the psychrometer equation, we will need to use Eq. (12-13) to obtain the value of the psychrometer coefficient. Using the tabulated values above, we obtain $A = 0.00135$, about double the value for air-water. We must remember that the estimate will be very rough because of the uncertainty in the value of β. Refining the estimate of T_{wb} by using the psychrometer equation and iterating gives

$$p_{wb} = 119.1 + 1.35 \times 10^{-3} (10^4)(60 - 0) = 932.3 \text{ Pa} = 7.0 \text{ mmHg}$$

From the Antoine equation,

$$T_{wb} = 2940/(16.6513 - \ln 7) + 35.93 = 235.84 \text{ K} = -37.3°C$$

Second iteration:

$$p_{wb} = 119.1 + 1.35 \times 10^{-3} (10^4)(60 + 37.3) = 1433 \text{ Pa} = 10.7 \text{ mmHg}$$

$$T_{wb} = 241.85 \text{ K} = -31.3°C$$

Third iteration:

$$p_{wb} = 119.1 + 1.35 \times 10^{-3} (10^4)(60 + 31.3) = 1352 \text{ Pa} = 10.1 \text{ mmHg}$$

$$T_{wb} = 241.0 \text{ K} = -32.1°C$$

The iteration has converged successfully, despite the poor initial guess. The wet-bulb temperature is −32°C; given the levels of error in the calculation, it will be meaningless to express this to any greater level of precision.

In a similar way, adiabatic saturation temperature can be calculated from Eq. (12-6) by taking the first guess as −40°C and assuming the humid heat to be 1.05 kJ/(kg·K) including the vapor:

$$Y_{as} = Y + \frac{C_s}{\lambda_{as}}(T - T_{as})$$

$$= 0.025 + \left(\frac{1.05}{501.1}\right)(60 + 40) = 0.235 \text{ kg/kg}$$

From Table 12-2,

$$p_{as} = 1018 \text{ Pa} = 7.63 \text{ mmHg}$$

From Antoine,

$$T_{as} = 237.05 \text{ K} = -36.1°C$$

Second iteration:

$Y_{as} = 0.025 + (1.05/501.1)(60 + 36.1) = 0.226$ kg/kg $\quad p_{as} = 984$ Pa $= 7.38$ mmHg

From Antoine,

$$T_{as} = 236.6 \text{ K} = -36.6°C$$

This has converged. A more accurate figure could be obtained with more refined estimates for C_s and λ_{wb}.

MEASUREMENT OF HUMIDITY

Dew Point Method The dew point of wet air is measured directly by observing the temperature at which moisture begins to form on an artificially cooled, polished surface.

Optical dew point hygrometers employing this method are the most commonly used fundamental technique for determining humidity. Uncertainties in temperature measurement of the polished surface, gradients across the surface, and the appearance or disappearance of fog have been much reduced in modern instruments. Automatic mirror cooling, e.g., thermoelectric, is more accurate and reliable than older methods using evaporation of a low-boiling solvent such as ether, or external coolants (e.g., vaporization of solid carbon dioxide or liquid air, or water cooling). Contamination effects have also been reduced or compensated for, but regular recalibration is still required, at least once a year.

Wet-Bulb Method In the past, probably the most commonly used method for determining the humidity of a gas stream was the measurement of wet- and dry-bulb temperatures. The wet-bulb temperature is measured by contacting the air with a thermometer whose bulb is covered by a wick saturated with water. If the process is adiabatic, the thermometer bulb attains the wet-bulb temperature. When the wet- and dry-bulb temperatures are known, the humidity is readily obtained from charts such as Figs. 12-1 through 12-4. To obtain reliable information, care must be exercised to ensure that the wet-bulb thermometer remains wet and that radiation to the bulb is minimized. The latter is accomplished by making the relative velocity between wick and gas stream high [a velocity of 4.6 m/s (15 ft/s) is usually adequate for commonly used thermometers] or by the use of radiation shielding. In the **Assmann psychrometer** the air is drawn past the bulbs by a motor-driven fan. Making sure that the wick remains wet is a mechanical problem, and the method used depends to a large extent on the particular arrangement. Again, as with the dew point method, errors associated with the measurement of temperature can cause difficulty.

For measurement of atmospheric humidities the **sling** or **whirling psychrometer** is widely used to give a quick and cheap, but inaccurate, estimate. A wet- and dry-bulb thermometer is mounted in a sling which is whirled manually to give the desired gas velocity across the bulb.

In addition to the mercury-in-glass thermometer, other temperature-sensing elements may be used for psychrometers. These include resistance thermometers, thermocouples, bimetal thermometers, and thermistors.

Electric hygrometers have been the fastest-growing form of humidity measurement in recent years. They measure the electrical resistance, capacitance, or impedance of a film of moisture-absorbing materials exposed to the gas. A wide variety of sensing elements have been used. Often, it is relative humidity which is measured.

Mechanical hygrometers utilizing materials such as human hair, wood fiber, and plastics have been used to measure humidity. These methods rely on a change in dimension with humidity. They are not suitable for process use.

Other hygrometric techniques in process and laboratory use include electrolytic and piezoelectric hygrometers, infrared and mass spectroscopy, and vapor pressure measurement, e.g., by a Pirani gauge.

The **gravimetric method** is accepted as the most accurate humidity-measuring technique. In this method a known quantity of gas is passed over a moisture-absorbing chemical such as phosphorus pentoxide, and the increase in weight is determined. It is mainly used for calibrating standards and measurements of gases with SO_x present.

EVAPORATIVE COOLING

GENERAL REFERENCES: *2005 ASHRAE Handbook of Fundamentals,* "Climatic Design Information," Chap. 28, ASHRAE, Atlanta, Ga.; *ASHRAE Handbook and Product Directory: Equipment,* ASHRAE, Atlanta, 2001.

INTRODUCTION

Evaporative cooling, using recirculated cooling water systems, is the method most widely used throughout the process industries for employing water to remove process waste heat, rejecting that waste heat into the environment. Maintenance considerations (water-side fouling control), through control of makeup water quality and control of cooling water chemistry, form one reason for this preference. Environmental considerations, by minimizing consumption of potable water, minimizing the generation and release of contaminated cooling water, and controlling the release into the environment of chemicals from leaking heat exchangers (HX), form the second major reason.

Local ambient climatic conditions, particularly the maximum summer wet-bulb temperature, determine the design of the evaporative equipment. Typically, the wet-bulb temperature used for design is the 0.4 percent value, as listed in the *ASHRAE Handbook of Fundamentals,* equivalent to 35-h exceedance per year on average.

The first subsection below presents the classic cooling tower (CT), the evaporative cooling technology most widely used today. The second subsection presents the wet surface air cooler (WSAC), a more recently perfected technology, combining within one piece of equipment the functions of cooling tower, circulated cooling water system, and HX tube bundle. The most common application for WSACs is in the direct cooling of process streams. However, the closed-circuit cooling tower, employing WSACs for cooling the circulated cooling water (replacing the CT), is an important alternative WSAC application, presented at the end of this section.

To minimize the total annualized costs for evaporative cooling is a complex engineering task in itself, separate from classic process design (Sec. 24, "Minimizing the Annualized Costs for Process Energy"). The evaluation and the selection of the best option for process cooling impact many aspects of how the overall project will be optimally designed (utilities supply, reaction and separations design, pinch analyses, 3D process layout, plot plan, etc.). Therefore, evaluation and selection of the evaporative cooling technology system should be performed at the start of the project design cycle, during conceptual engineering (Sec. 9, "Process Economics," "Value Improving Practices"), when the potential to influence project costs is at a maximum value (Sec. 9, *VIP Figure 9-33*). The relative savings achievable for selection of the optimum heat rejection technology option can frequently exceed 25 percent, for the installed cost for the technology alone.

PRINCIPLES

The processes of cooling water are among the oldest known. Usually water is cooled by exposing its surface to air. Some of the processes are slow, such as the cooling of water on the surface of a pond; others are comparatively fast, such as the spraying of water into air. These processes all involve the exposure of water surface to air in varying degrees.

The heat-transfer process involves (1) latent heat transfer owing to vaporization of a small portion of the water and (2) sensible heat transfer owing to the difference in temperatures of water and air. Approximately 80 percent of this heat transfer is due to latent heat and 20 percent to sensible heat.

Theoretical possible heat removal per pound of air circulated in a cooling tower depends on the temperature and moisture content of air. An indication of the moisture content of the air is its wet-bulb temperature. Ideally, then, the wet-bulb temperature is the lowest theoretical temperature to which the water can be cooled. Practically, the cold water temperature approaches but does not equal the air wet-bulb temperature in a cooling tower; this is so because it is impossible to contact all the water with fresh air as the water drops through the wetted fill surface to the basin. The magnitude of approach to the wet-bulb temperature is dependent on the tower design. Important factors are air-to-water contact time, amount of fill surface, and breakup of water into droplets. In actual practice, cooling towers are seldom designed for approaches closer than 2.8°C (5°F).

COOLING TOWERS*

GENERAL REFERENCES: *Counterflow Cooling Tower Performance,* Pritchard Corporation, Kansas City, Mo., 1957; Hensley, "Cooling Tower Energy," *Heat Piping Air Cond.* (October 1981); Kelley and Swenson, *Chem. Eng. Prog.* **52:** 263 (1956); McAdams, *Heat Transmission,* 3d ed., McGraw-Hill, New York, 1954, pp. 356–365; Merkel, *Z. Ver. Dtsch. Ing. Forsch.,* no. 275 (1925); *The Parallel Path Wet-Dry Cooling Tower,* Marley Co., Mission Woods, Kan., 1972; *Performance Curves,* Cooling Tower Institute, Houston, Tex., 1967; *Plume Abatement and Water Conservation with Wet-Dry Cooling Tower,* Marley Co., Mission Woods, Kan., 1973; Tech. Bull. R-54-P-5, R-58-P-5, Marley Co., Mission Woods, Kan., 1957; Wood and Betts, *Engineer,* **189**(4912), **377**(4913), **349** (1950); Zivi and Brand, *Refrig. Eng.,* **64**(8): 31–34, 90 (1956); Hensley, *Cooling Tower Fundamentals,* 2d ed., Marley Cooling Technologies, 1998; Mortensen and Gagliardo, *Impact of Recycled Water Use in Cooling Towers,* TP-04-12, Cooling Technology Institute, 2004; www.cti.org; www.ashrae.org; www.marleyct.com.

Cooling Tower Theory The most generally accepted theory of the cooling tower heat-transfer process is that developed by Merkel (op. cit.). This analysis is based upon **enthalpy potential difference** as the driving force.

Each particle of water is assumed to be surrounded by a film of air, and the enthalpy difference between the film and surrounding air provides the driving force for the cooling process. In the integrated form the Merkel equation is

$$\frac{KaV}{L} = \int_{T_2}^{T_1} \frac{C_L dT}{h' - h} \qquad (12\text{-}14a)$$

where K = mass-transfer coefficient, lb water/(h·ft²); a = contact area, ft²/ft³ tower volume; V = active cooling volume, ft³/ft² of plan area; L = water rate, lb/(h·ft²); C_L = heat capacity of water, Btu/(lb·°F); h'= enthalpy of saturated air at water temperature, Btu/lb; h = enthalpy of

*The contributions of Ken Mortensen, and coworkers, of Marley Cooling Technologies, Overland Park, Kansas, toward the review and update of this subsection are acknowledged.

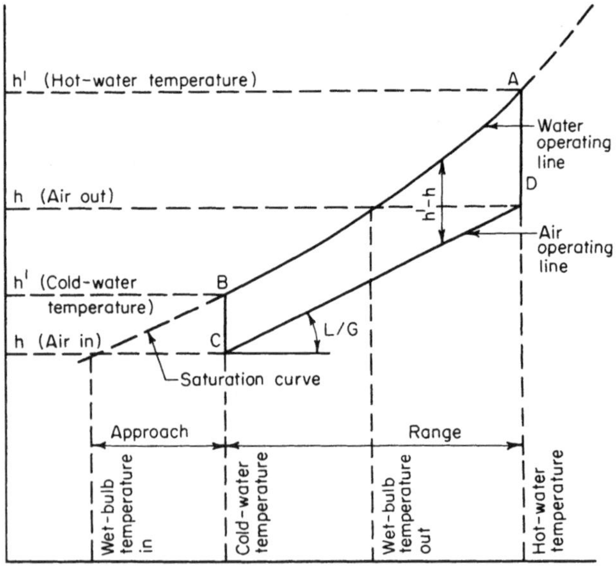

FIG. 12-8a Cooling-tower process heat balance. (*Marley Co.*)

airstream, Btu/lb; and T_1 and T_2 = entering and leaving water temperatures, °F. The right-hand side of Eq. (12-14*a*) is entirely in terms of air and water properties and is independent of tower dimensions.

Figure 12-8*a* illustrates water and air relationships and the driving potential which exist in a counterflow tower, where air flows parallel but opposite in direction to water flow. An understanding of this diagram is important in visualizing the cooling tower process.

The water operating line is shown by line *AB* and is fixed by the inlet and outlet tower water temperatures. The air operating line begins at *C*, vertically below *B* and at a point having an enthalpy corresponding to that of the entering wet-bulb temperature. Line *BC* represents the initial driving force $h' - h$. In cooling water at 1°F, the enthalpy per pound of air is increased 1 Btu multiplied by the ratio of pounds of water to pound of air. The liquid-gas ratio *L/G* is the slope of the operating line. The air leaving the tower is represented by point *D*. The cooling range is the projected length of line *CD* on the temperature scale. The cooling tower approach is shown on the diagram as the difference between the cold water temperature leaving the tower and the ambient wet-bulb temperature.

The coordinates refer directly to the temperature and enthalpy of any point on the water operating line but refer directly only to the enthalpy of a point on the air operating line. The corresponding wet-bulb temperature of any point on *CD* is found by projecting the point horizontally to the saturation curve, then vertically to the temperature coordinate. The integral [Eq. (12-14*a*)] is represented by the area *ABCD* in the diagram. This value is known as the **tower characteristic,** varying with the *L/G* ratio.

For example, an increase in entering wet-bulb temperature moves the origin *C* upward, and the line *CD* shifts to the right to maintain a constant *KaV/L*. If the cooling range increases, line *CD* lengthens. At a constant wet-bulb temperature, equilibrium is established by moving the line to the right to maintain a constant *KaV/L*. On the other hand, a change in *L/G* ratio changes the slope of *CD*, and the tower comes to equilibrium with a new *KaV/L*.

To predict tower performance, it is necessary to know the required tower characteristics for fixed ambient and water conditions. The tower characteristic *KaV/L* can be determined by integration. The Chebyshev method is normally used for numerically evaluating the integral, whereby

$$\frac{KaV}{L} = \int_{T_2}^{T_1} \frac{dT}{h_w - h_a} \cong \frac{T_1 - T_2}{4}\left(\frac{1}{\Delta h_1} + \frac{1}{\Delta h_2} + \frac{1}{\Delta h_3} + \frac{1}{\Delta h_4}\right)$$

where h_w = enthalpy of air-water vapor mixture at bulk water temperature, Btu/lb dry air

h_a = enthalpy of air-water vapor mixture at wet-bulb temperature, Btu/lb dry air

Δh_1 = value of $h_w - h_a$ at $T_2 + 0.1(T_1 - T_2)$

Δh_2 = value of $h_w - h_a$ at $T_2 + 0.4(T_1 - T_2)$

Δh_3 = value of $h_w - h_a$ at $T_1 - 0.4(T_1 - T_2)$

Δh_4 = value of $h_w - h_a$ at $T_1 - 0.1(T_1 - T_2)$

Example 10: Calculation of Mass-Transfer Coefficient Group

Determine the theoretically required *KaV/L* value for a cooling duty from 105°F inlet water, 85°F outlet water, 78°F ambient wet-bulb temperature, and an *L/G* ratio of 0.97.

From the air-water vapor-mixture tables, the enthalpy h_1 of the ambient air at 78°F wet-bulb temperature is 41.58 Btu/lb.

$$h_2 \text{ (leaving air)} = 41.58 + 0.97(105 - 85) = 60.98 \text{ Btu/lb}$$

T, °F	h_{water}	h_{air}	$h_w - h_a$	$1/\Delta h$
$T_2 = 85$	49.43	$h_1 = 41.58$		
$T_2 + 0.1(20) = 87$	51.93	$h_1 + 0.1L/G(20) = 43.52$	$\Delta h_1 = 8.41$	0.119
$T_2 + 0.4(20) = 93$	60.25	$h_1 + 0.4L/G(20) = 49.34$	$\Delta h_2 = 10.91$	0.092
$T_1 - 0.4(20) = 97$	66.55	$h_2 - 0.4L/G(20) = 53.22$	$\Delta h_3 = 13.33$	0.075
$T_1 - 0.1(20) = 103$	77.34	$h_2 - 0.1L/G(20) = 59.04$	$\Delta h_4 = 18.30$	0.055
$T_1 = 105$	81.34	$h_2 = 60.98$		0.341

$$\frac{KaV}{L} = \frac{105 - 85}{4}(0.341) = 1.71$$

A quicker but less accurate method is by the use of a nomograph (Fig. 12-8*b*) prepared by Wood and Betts (op. cit.).

Mechanical draft cooling towers normally are designed for *L/G* ratios ranging from 0.75 to 1.50; accordingly, the values of *KaV/L* vary from 0.50 to 2.50. With these ranges in mind, an example of the use of the nomograph will readily explain the effect of changing variables.

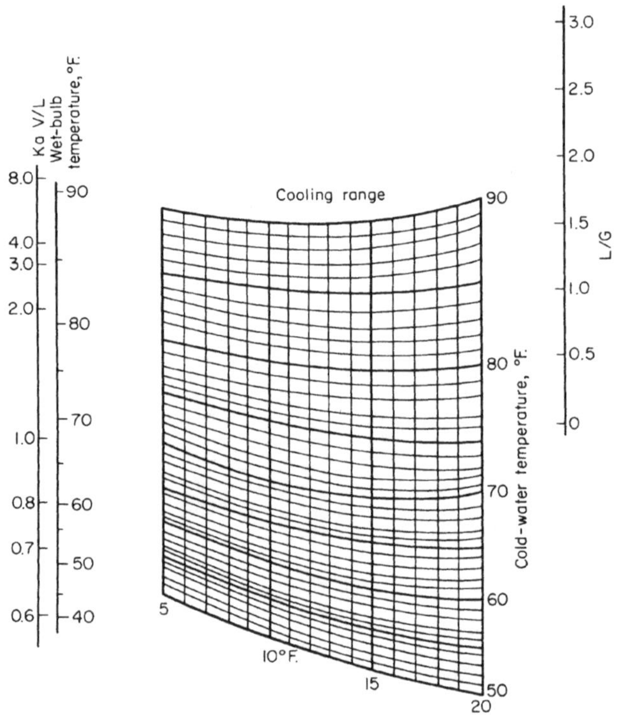

FIG. 12-8b Nomograph of cooling tower characteristics. [*Wood and Betts, Engineer,* ***189****(4912), 337 (1950).*]

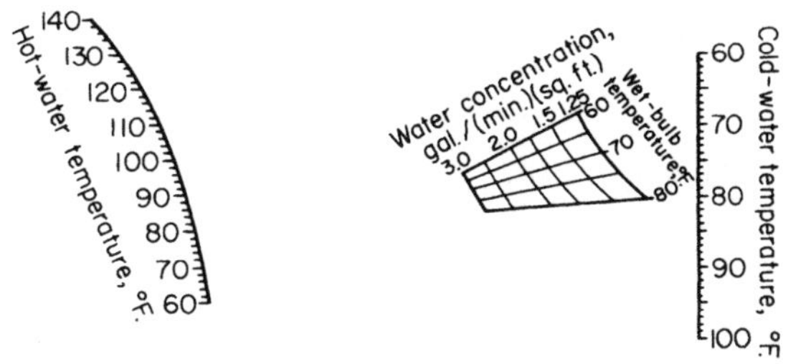

FIG. 12-8c Sizing chart for a counterflow induced-draft cooling tower. For induced-draft towers with (1) an upspray distributing system with 24 ft of fill or (2) a flume-type distributing system and 32 ft of fill. The chart will give approximations for towers of any height. *(Ecodyne Corp.)*

Example 11: Application of Nomograph for Cooling Tower Characteristics If a given tower is operating with 20°F range, a cold water temperature of 80°F, and a wet-bulb temperature of 70°F, a straight line may be drawn on the nomograph. If the *L/G* ratio is calculated to be 1.0, then *KaV/L* may be established by a line drawn through *L/G* 1.0 and parallel to the original line. The tower characteristic *KaV/L* is thus established at 1.42. If the wet-bulb temperature were to drop to 50°F, then *KaV/L* and *L/G* ratios may be assumed to remain constant. A new line parallel to the original will then show that for the same range the cold-water temperature will be 70°F.

The nomograph provides an approximate solution; degree of accuracy will vary with changes in cooling as well as from tower to tower. Once the theoretical cooling tower characteristic has been determined by numerical integration or from the nomograph for a given cooling duty, it is necessary to design the cooling tower fill and air distribution to meet the theoretical tower characteristic. The Pritchard Corporation (op. cit.) has developed performance data on various tower fill designs. These data are too extensive to include here, and those interested should consult this reference. See also Baker and Mart (Marley Co., Tech. Bull. R-52-P-10, Mission Woods, Kan.) and Zivi and Brand (loc. cit.).

Mechanical Draft Towers Two types of mechanical draft towers are in use today: the forced-draft and the induced-draft. In the **forced-draft tower** the fan is mounted at the base, and air is forced in at the bottom and discharged at low velocity through the top. This arrangement has the advantage of locating the fan and drive outside the tower, where it is convenient for inspection, maintenance, and repairs. Since the equipment is out of the hot, humid top area of the tower, the fan is not subjected to corrosive conditions. However, because of the low exit-air velocity, the forced-draft tower is subjected to excessive recirculation of the humid exhaust vapors back into the air intakes. Since the wet-bulb temperature of the exhaust air is considerably above the wet-bulb temperature of the ambient air, there is a decrease in performance evidenced by an increase in cold (leaving) water temperature.

The **induced-draft tower** is the most common type used in the United States. It is further classified into counterflow and cross-flow design, depending on the relative flow directions of water and air. Thermodynamically, the **counterflow arrangement** is more efficient, since the coldest water contacts the coldest air, thus obtaining maximum enthalpy potential. The greater the cooling ranges and the more difficult the approaches, the more distinct are the advantages of the counterflow type. For example, with an *L/G* ratio of 1, an ambient wet-bulb temperature of 25.5°C (78°F), and an inlet water temperature of 35°C (95°F), the counterflow tower requires a *KaV/L* characteristic of 1.75 for a 2.8°C (5°F) approach, while a cross-flow tower requires a characteristic of 2.25 for the same approach. However, if the approach is increased to 3.9°C (7°F), both types of tower have approximately the same required *KaV/L* (within 1 percent).

The **cross-flow tower** manufacturer may effectively reduce the tower characteristic at very low approaches by increasing the air quantity to give a lower *L/G* ratio. The increase in airflow is not necessarily achieved by increasing the air velocity but primarily by lengthening the tower to increase the airflow cross-sectional area. It appears then that the cross-flow fill can be made progressively longer in the direction perpendicular to the airflow and shorter in the direction of the airflow until it almost loses its inherent potential-difference disadvantage. However, as this is done, fan power consumption increases.

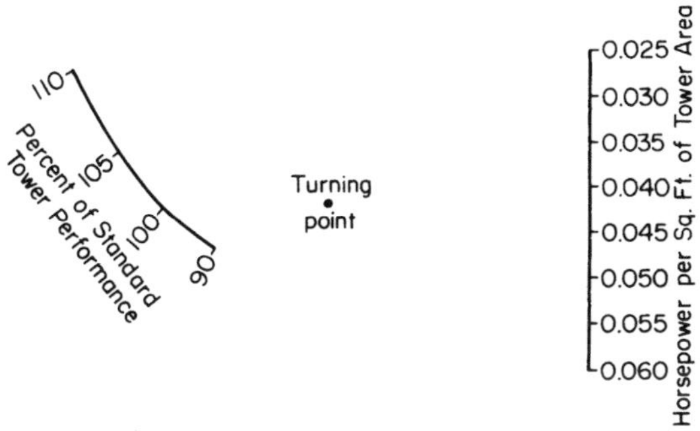

FIG. 12-8d Horsepower chart for a counterflow induced-draft cooling tower. [*Fluor Corp. (now Ecodyne Corp.)*]

Ultimately, the economic choice between counterflow and cross-flow is determined by the effectiveness of the fill, design conditions, water quality, and the costs of tower manufacture.

Performance of a given type of cooling tower is governed by the ratio of the weights of air to water and the time of contact between water and air. In commercial practice, the variation in the ratio of air to water is first obtained by keeping the air velocity constant at about 350 ft/(min·ft^2 of active tower area) and varying the water concentration, gal/(min·ft^2 of tower area). As a secondary operation, air velocity is varied to make the tower accommodate the cooling requirement.

Time of contact between water and air is governed largely by the time required for the water to discharge from the nozzles and fall through the tower to the basin. The time of contact is therefore obtained in a given type of unit by varying the height of the tower. Should the time of contact be insufficient, no amount of increase in the ratio of air to water will produce the desired cooling. It is therefore necessary to maintain a certain minimum height of cooling tower. When a wide approach of 8 to 11°C (15 to 20°F) to the wet-bulb temperature and a 13.9 to 19.4°C (25 to 35°F) cooling range are required, a relatively low cooling tower will suffice. A tower in which the water travels 4.6 to 6.1 m (15 to 20 ft) from the distributing system to the basin is sufficient. When a moderate approach and a cooling range of 13.9 to 19.4°C (25 to 35°F) are required, a tower in which the water travels 7.6 to 9.1 m (25 to 30 ft) is adequate. Where a close approach of 4.4°C (8°F) with a 13.9 to 19.4°C (25 to 35°F) cooling range is required, a tower in which the water travels from 10.7 to 12.2 m (35 to 40 ft) is required. It is usually not economical to design a cooling tower with an approach of less than 2.8°C (5°F).

Figure 12-8*c* shows the relationship of the hot water, cold water, and wet-bulb temperatures to the water concentration.* From this, the **minimum area** required for a given performance of a well-designed counterflow induced-draft cooling tower can be obtained. Figure 12-8*d* gives the horsepower per square foot of tower area required for a given performance. These curves do not apply to parallel or cross-flow cooling, since these processes are not so efficient as the counterflow process. Also, they do not apply when the approach to the cold water temperature is less than 2.8°C (5°F). These charts should be considered approximate and for preliminary estimates only. Since many factors not shown in the graphs must be included in the computation, the manufacturer should be consulted for final design recommendations.

The cooling performance of any tower containing a given depth of filling varies with the **water concentration.** It has been found that maximum contact and performance are obtained with a tower having a water concentration of 2 to 5 gal/(min·ft^2 of ground area). Thus the

*See also London, Mason, and Boelter, loc. cit.; Lichtenstein, loc. cit.; Simpson and Sherwood, *J. Am. Soc. Refrig. Eng.*, **52:**535, 574 (1946); Simons, *Chem. Metall. Eng.*, **49**(5):138; (6): 83 (1942);**46:** 208 (1939); and Hutchinson and Spivey, *Trans. Inst. Chem. Eng.*, **20:**14 (1942).

problem of calculating the size of a cooling tower becomes one of determining the proper concentration of water required to obtain the desired results. Once the necessary water concentration has been established, the tower area can be calculated by dividing the gallons per minute circulated by the water concentration in gallons per minute square foot. The required tower size then is a function of the following:

1. Cooling range (hot water temperature minus cold water temperature)
2. Approach to wet-bulb temperature (cold water temperature minus wet-bulb temperature)
3. Quantity of water to be cooled
4. Wet-bulb temperature
5. Air velocity through the cell
6. Tower height

Example 12: Application of Sizing and Horsepower Charts To illustrate the use of the charts, assume the following conditions:

Hot water temperature T_1,°F = 102
Cold water temperature T_2,°F = 78
Wet-bulb temperature t_w,°F = 70
Water rate, gal/min = 2000

A straight line in Fig. 12-8*c*, connecting the points representing the design water and wet-bulb temperature, shows that a water concentration of 2 gal/(ft²·min) is required. The area of the tower is calculated as 1000 ft² (quantity of water circulated divided by water concentration).

Fan horsepower is obtained from Fig. 12-8*d*. Connecting the point representing 100 percent of standard tower performance with the turning point and extending this straight line to the horsepower scale show that it will require 0.041 hp/ft² of actual effective tower area. For a tower area of 1000 ft², 41.0 fan hp is required to perform the necessary cooling.

Suppose that the actual commercial tower size has an area of only 910 ft². Within reasonable limits, the shortage of actual area can be compensated for by an increase in air velocity through the tower. However, this requires boosting fan horsepower to achieve 110 percent of standard tower performance. From Fig. 12-8*d*, the fan horsepower is found to be 0.057 hp/ft² of actual tower area, or 0.057 × 910 = 51.9 hp.

On the other hand, if the actual commercial tower area is 1110 ft², the cooling equivalent to 1000 ft² of standard tower area can be accomplished with less air and less fan horsepower. From Fig. 12-8*d*, the fan horsepower for a tower operating at 90 percent of standard performance is 0.031 hp/ft² of actual tower area, or 34.5 hp.

This example illustrates the sensitivity of fan horsepower to small changes in tower area. The importance of designing a tower that is slightly oversize in ground area and of providing plenty of fan capacity becomes immediately apparent.

Example 13: Application of Sizing Chart Assume the same cooling range and approach as used in Example 12 except that the wet-bulb temperature is lower. Design conditions would then be as follows:

Water rate, gal/min = 2000
Temperature range $T_1 - T_2$,°F = 24
Temperature approach $T_2 - t_w$,°F = 8
Hot water temperature T_1,°F = 92
Cold water temperature T_2,°F = 68
Wet-bulb temperature t_w,°F = 60

From Fig. 12-8*c*, the water concentration required to perform the cooling is 1.75 gal/(ft²·min), giving a tower area of 1145 ft² versus 1000 ft² for a 70°F wet-bulb temperature. This shows that the lower the wet-bulb temperature for the same cooling range and approach, the larger the area of the tower required and therefore the more difficult the cooling job.

Figure12-8*e* illustrates the type of performance curve furnished by the cooling tower manufacturer. This shows the variation in performance with changes in wet-bulb and hot water temperatures while the water quantity is maintained constant.

Cooling Tower Operation

Water Makeup Makeup requirements for a cooling tower consist of the summation of evaporation loss, drift loss, and blowdown. Therefore,

$$W_m = W_e + W_d + W_b \tag{12-14b}$$

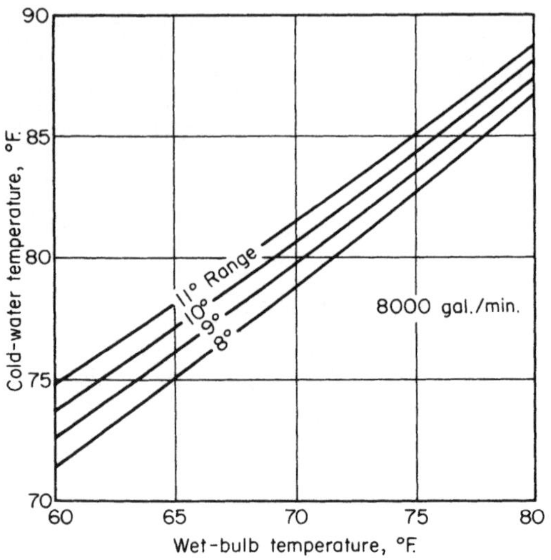

FIG. 12-8*e* Typical cooling-tower performance curve.

where W_m = makeup water, W_d = drift loss, and W_b = blowdown (consistent units: m³/h or gal/min).

Evaporation loss can be estimated by

$$W_e = 0.00085W_c(T_1 - T_2) \tag{12-14c}$$

where W_c = circulating water flow, m³/h or gal/min at tower inlet, and $T_1 - T_2$ = inlet water temperature minus outlet water temperature, °F. The 0.00085 evaporation constant is a good rule-of-thumb value. The actual evaporation rate will vary by season and climate.

Drift loss can be estimated by

$$W_d = 0.0002W_c$$

Drift is entrained water in the tower discharge vapors. Drift loss is a function of the drift eliminator design and is typically less than 0.02 percent of the water supplied to the tower with the new developments in eliminator design.

Blowdown discards a portion of the concentrated circulating water due to the evaporation process in order to lower the system solids concentration. The amount of blowdown can be calculated according to the number of cycles of concentration required to limit scale formation. "Cycles of concentration" is the ratio of dissolved solids in the recirculating water to dissolved solids in the makeup water. Since chlorides remain soluble on concentration, cycles of concentration are best expressed as the ratio of the chloride contents of the circulating and makeup waters. Thus, the blowdown quantities required are determined from

$$\text{Cycles of concentration} = \frac{W_e + W_b + W_d}{W_b + W_d} \tag{12-14d}$$

or

$$W_b = \frac{W_e - (\text{cycles} - 1)W_d}{\text{cycles} - 1} \tag{12-14e}$$

Cycles of concentration involved with cooling tower operation normally range from three to five cycles. For water qualities where operating water concentrations must be below 3 to control scaling, blowdown quantities will be large. The addition of acid or scale-inhibiting chemicals can limit scale formation at higher cycle levels with such a water, and will allow substantially reduced water usage for blowdown.

The blowdown equation (12-14*e*) translates to calculated percentages of the cooling system circulating water flow exiting to drain, as listed in Table 12-6. The blowdown percentage is based on the cycles targeted and the cooling range. The range is the difference between the system hot water and cold water temperatures.

TABLE 12.6 Blowdown (%)

Range,°F	2X	3X	4X	5X	6X
10	0.83	0.41	0.26	0.19	0.15
15	1.26	0.62	0.41	0.30	0.24
20	1.68	0.83	0.55	0.41	0.32
25	2.11	1.04	0.69	0.51	0.41
30	2.53	1.26	0.83	0.62	0.49

It is the open nature of evaporative cooling systems, bringing in external air and water continuously, that determines the unique water problems these systems exhibit. Cooling towers (1) concentrate solids by the mechanisms described above and (2) wash air. The result is a buildup of dissolved solids, suspended contaminants, organics, bacteria, and their food sources in the circulating cooling water. These unique evaporative water system problems must be specifically addressed to maintain cooling equipment in good working order.

Example 14: Calculation of Makeup Water Determine the amount of makeup required for a cooling tower with the following conditions:

Inlet water flow, m^3/h (gal/min)	2270 (10,000)
Inlet water temperature, °C (°F)	37.77 (100)
Outlet water temperature, °C (°F)	29.44 (85)
Drift loss, percent	0.02
Concentration cycles	5

Evaporation loss:

$$W_e,\ m^3/h = 0.00085 \times 2270 \times (37.77 - 29.44) \times (1.8°F/°C) = 28.9$$

$$W_e,\ gal/min = 127.5$$

Drift loss

$$W_d,\ m^3/h = 2270 \times 0.0002 = 0.45$$

$$W_d,\ gal/min = 2$$

Blowdown

$$W_b,\ m^3/h = 6.8$$

$$W_b,\ gal/min = 29.9$$

Makeup

$$W_m,\ m^3/h = 28.9 + 0.45 + 6.8 = 36.2$$

$$W_m,\ gal/min = 159.4$$

Fan Horsepower In evaluating cooling tower ownership and operating costs, fan horsepower requirements can be a significant factor. Large air quantities are circulated through cooling towers at exit velocities of about 10.2 m/s (2000 ft/min) maximum for induced-draft towers. Fan airflow quantities depend upon tower design factors, including such items as type of fill, tower configuration, and thermal performance conditions.

The effective output of the fan is the static air horsepower (SAHP), which is obtained by the following equation:

$$\text{SAHP} = -\frac{Q(h_s)(d)}{33{,}000(12)}$$

where Q = air volume, ft^3/min; h_s = static head, in of water; and d = density of water at ambient temperature, lb/ft^3.

Cooling tower fan horsepower can be reduced substantially as the ambient wet-bulb temperature decreases if two-speed fan motors are used. Theoretically, operating at half speed will reduce airflow by 50 percent while decreasing horsepower to one-eighth of that of full-speed operation. However, actual half-speed operation will require about 17 percent of the horsepower at full speed as a result of the inherent motor losses at lighter loads.

Figure 12-8*f* shows a typical plot of outlet water temperatures when a cooling tower is operated (1) in the fan-off position, (2) with the fan at half speed, and (3) with the fan at full speed. Note that at decreasing wet-bulb temperatures the water leaving the tower during half-speed operation could meet design water temperature requirements of, say, 85°F. For example, for a 60°F wet-bulb, 20°F range, a leaving-water temperature slightly below 85°F is obtained with design water flow over the tower. If the fan had a 100-hp motor, 83 hp would be saved when operating it at half speed. In calculating savings, one should not overlook the advantage of having colder tower water available for the overall water circulating system.

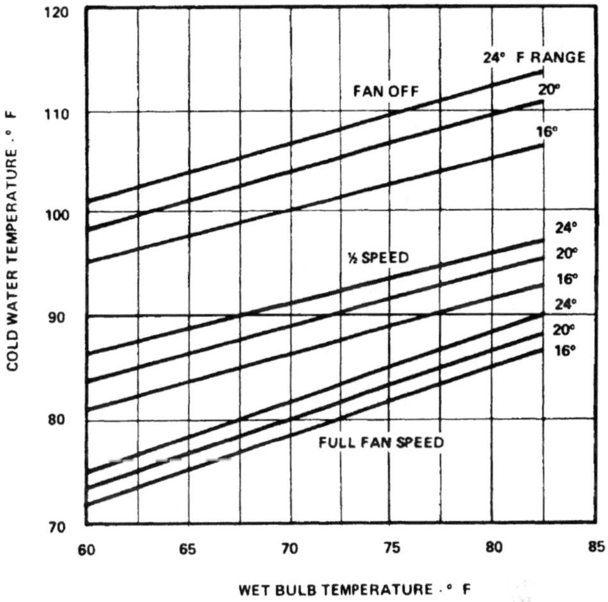

FIG. 12-8*f* Typical plot of cooling-tower performance at varying fan speeds.

Recent developments in cooling tower fan energy management also include automatic variable-pitch propeller-type fans and inverter-type devices to permit variable fan speeds. These schemes involve tracking the load at a constant outlet water temperature.

The variable-pitch arrangement at constant motor speed changes the pitch of the blades through a pneumatic signal from the leaving water temperature. As the thermal load and/or the ambient wet-bulb temperature decreases, the blade pitch reduces airflow and less fan energy is required.

Inverters make it possible to control a variable-speed fan by changing the frequency modulation. Standard alternating-current fan motors may be speed-regulated between 0 and 60 Hz. In using inverters for this application, it is important to avoid frequencies that would result in fan critical speeds.

Even though tower fan energy savings can result from these arrangements, they may not constitute the best system approach. Power plant steam condensers and refrigeration units, e.g., can take advantage of colder tower water to reduce power consumption. Invariably, these system savings are much larger than cooling tower fan savings with constant leaving water temperatures. A refrigeration unit condenser can utilize inlet water temperatures down to 12.8°C (55°F) to reduce compressor energy consumption by 25 to 30 percent.

Pumping Horsepower Another important factor in analyzing cooling tower selections, especially in medium to large sizes, is the portion of pump horsepower directly attributed to the cooling tower. A counterflow type of tower with spray nozzles will have a pumping head equal to static lift plus nozzle pressure loss. A cross-flow type of tower with gravity flow enables a pumping head to equal static lift. A reduction in tower height therefore reduces static lift, thus reducing pump horsepower:

$$\text{Pump bhp} = \frac{W_c h_t}{3960(\text{pump efficiency})} \qquad (12\text{-}14f)$$

where W_c = water recirculation rate, gal/min, and h_t = total head, ft.

Fogging and Plume Abatement A phenomenon that occurs in cooling tower operation is fogging, which produces a highly visible plume and possible icing hazards. Fogging results from mixing warm, highly saturated tower discharge air with cooler ambient air that lacks the capacity to absorb all the moisture as vapor. While in the past visible plumes have not been considered undesirable, properly locating towers to minimize possible sources of complaints has now received the necessary attention. In some instances, guyed high fan stacks have been used to reduce ground fog. Although tall stacks minimize the ground effects of plumes, they can do nothing about water vapor saturation or visibility. The persistence of plumes is much greater in periods of low ambient temperatures.

More recently, environmental aspects have caused public awareness and concern over any visible plume, although many laypersons misconstrue cooling tower discharge as harmful. This has resulted in a new development for plume abatement known as a wet-dry cooling tower configuration. Reducing the relative humidity or moisture content of the tower discharge stream will reduce the frequency of plume formation. Figure 12-8g shows a "parallel path" arrangement that has been demonstrated to be technically sound but at substantially increased tower investment. Ambient air travels in parallel streams through the top dry-surface section and the evaporative section. Both sections benefit thermally by receiving cooler ambient air with the wet and dry airstreams mixing after leaving their respective sections. Water flow is arranged in series, first flowing to the dry coil section and then to the evaporation fill section. A "series path" airflow arrangement, in which dry coil sections can be located before or after the air traverses the evaporative section, also can be used. However, series-path airflow has the disadvantage of water impingement, which could result in coil scaling and restricted airflow.

Wet-dry cooling towers incorporating these designs are being used for large-tower industrial applications. At present they are not available for commercial applications.

Thermal Performance The thermal performance of the evaporative cooling tower is critical to the overall efficiency of cooling systems. Modern electronic measurement instrumentation allows accurate verification of cooling tower capability. Testing and tracking of the cooling tower capability are a substantial consideration in measuring cooling system performance. Cooling tower testing is a complex activity that requires significant expertise in the art. Consult a competent testing company if such verification is desired.

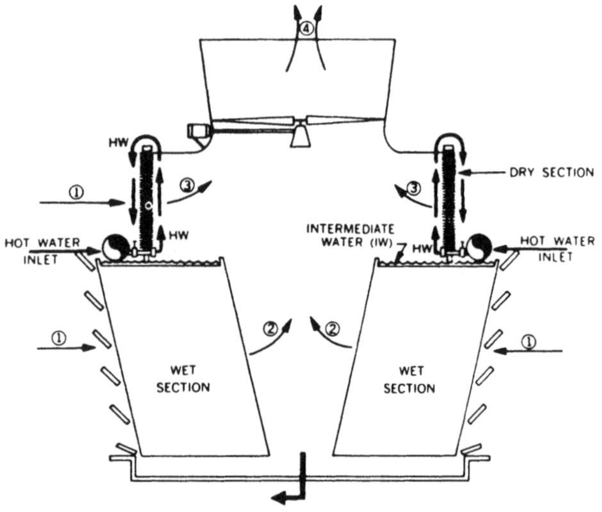

FIG. 12-8g Parallel-path cooling-tower arrangement for plume abatement. (*Marley Co.*)

New Technologies The cooling tower business is constantly changing in an attempt to improve efficiencies of evaporative cooling products. A significant thermal performance improvement over the splash-type fills, covered extensively in the writings above, can be achieved by using film-type fill. Film fills are formed plastic sheets separated by spacing knobs that allow water and air to flow easily between paired plastic surfaces. Fully wetted water flow over these panels creates an extensive "film" of evaporative surface on the plastic. Film fill is more sensitive to water quality than are splash-type fills.

These film fills are not sized via the graphical methods illustrated above for splash fills. They are selected by using manufacturers' proprietary sizing programs, which are based on extensive testing data. Such programs can be obtained by contacting manufacturers and/or industry trade organizations.

Applications for Evaporative Cooling Towers Cooling towers are commonly used in many commercial and industrial processes including

- Power generation (fossil fuel, nuclear)
- Industrial process (refinery, chemical production, plastic molding)
- Comfort cooling (HVAC)

Natural Draft Towers, Cooling Ponds, Spray Ponds Natural draft towers are primarily suited to very large cooling water quantities, and the reinforced concrete structures used are as large as 80 m in diameter and 105 m high.

When large ground areas are available, large cooling ponds offer a satisfactory method of removing heat from water. A pond may be constructed at a relatively small investment by pushing up earth in an earth dike 2 to 3 m high.

Spray ponds provide an arrangement for lowering the temperature of water by evaporative cooling and in so doing greatly reduce the cooling area required in comparison with a cooling pond.

Natural draft towers, cooling ponds, and spray ponds are infrequently used in new construction today in the chemical processing industry. Additional information may be found in previous Perry's editions.

WET SURFACE AIR COOLER (WSAC)

GENERAL REFERENCES: Kals, "Wet Surface Aircoolers," *Chem. Engg.* July 1971; Kals, "Wet Surface Aircoolers: Characteristics and Usefulness," AIChE-ASME Heat Transfer Conference, Denver, Colo., August 6–9, 1972; Elliott and Kals, "Air Cooled Condensers," *Power,* January 1990; Kals, "Air Cooled Heat Exchangers: Conventional and Unconventional," *Hydrocarbon Processing,* August 1994; Hutton, "Properly Apply Closed Circuit Evaporative Cooling," *Chem. Engg. Progress,* October 1996; Hutton, "Improved Plant Performance through Evaporative Steam Condensing," ASME 1998 International Joint Power Conference, Baltimore, Md., August 23–26, 1998; http://www.niagarablower.com/wsac.htm; http://www.baltimoreaircoil.com.

Principles Rejection of waste process heat through a cooling tower (CT) requires transferring the heat in two devices in series, using two different methods of heat transfer. This requires two temperature driving forces in series: first, sensible heat transfer, from the process stream across the heat exchanger (HX) into the cooling water, and, second, sensible and latent heat transfer, from the cooling water to atmosphere across the CT. Rejecting process heat with a wet surface air cooler transfers the waste heat in a single device by using a single-unit operation. The single required temperature driving force is lower, because the WSAC does not require the use of cooling water sensible heat to transfer heat from the process stream to the atmosphere. A WSAC tube cross section (Fig. 12-8*h*) shows the characteristic external tube surface having a continuous flowing film of evaporating water, which cascades through the WSAC tube bundle. Consequently, process streams can be economically cooled to temperatures much closer to the ambient wet-bulb temperature (WBT), as low as to within 2.2°C (4°F), depending on the process requirements and economics for the specific application.

Wet Surface Air Cooler Basics The theory and principles for the design of WSACs are a combination of those known for evaporative cooling tower design and HX design. However, the design practices for engineering WSAC equipment remain a largely proprietary, technical

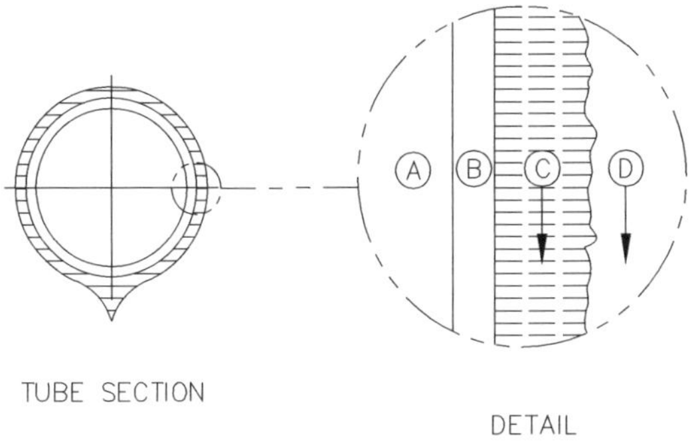

FIG. 12-8*h* WSAC tube cross-section. Using a small T, heat flows from (A) the process stream, through (B) the tube, through (C) the flowing film of evaporating water, into (D) flowing ambient air.

art, and the details are not presented here. Any evaluation of the specifics and economics for any particular application requires direct consultation with a reputable vendor.

Because ambient air is contacted with evaporating water within a WSAC, from a distance a WSAC has a similar appearance to a CT (Fig. 12-8*i*). Economically optimal plot plan locations for WSACs can vary: integrated into, or with, the process structure, remote to it, in a pipe rack, etc.

In the WSAC the evaporative cooling occurs on the wetted surface of the tube bundle. The wetting of the tube bundle is performed by recirculating water the short vertical distance from the WSAC collection basin, through the spray nozzles, and onto the top of the bundle (Fig. 12-8*j*). The tube bundle is completely deluged with this cascading flow of water. Using water application rates between 12 and 24 $(m^3/h)/m^2$ (5 and 10 gpm/ft^2), the tubes have a continuous, flowing external water film, minimizing the potential for water-side biological fouling, sediment deposition, etc. Process inlet temperatures are limited to a maximum of about 85°C (185°F), to prevent external water-side mineral scaling. However, higher process inlet temperatures can be accepted, by incorporating bundles of dry, air-cooled finned tubing within the WSAC unit, to reduce the temperature of the process stream to an acceptable level before it enters the wetted evaporative tube bundles.

The WSAC combines within one piece of equipment the functions of cooling tower, circulated cooling water system, and water-cooled HX. In the basic WSAC configuration (Fig. 12-8*k*), ambient air is drawn in and

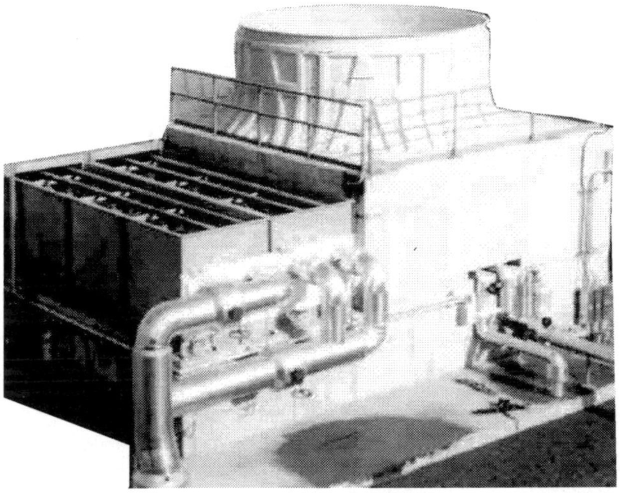

FIG. 12-8*i* Overhead view of a single-cell WSAC.

FIG. 12-8*j* Nozzles spraying onto wetted tube bundle in a WSAC unit.

down through the tube bundle. This airflow is cocurrent with the evaporating water flow, recirculated from the WSAC collection basin sump to be sprayed over the tube bundles. This downward cocurrent flow pattern minimizes the generation of water mist (drift). At the bottom of the WSAC, the air changes direction through 180°, disengaging entrained fine water droplets. Drift eliminators can be added to meet very low drift requirements. Because heat is extracted from the tube surfaces by water latent heat (and not sensible heat), only about 75 percent as much circulating water is required in comparison to an equivalent CT-cooling water-HX application.

The differential head of the circulation water pump is relatively small, since dynamic losses are modest (short vertical pipe and a low ΔP spray nozzle) and the hydraulic head is small, only about 6 m (20 ft) from the basin to the elevation of the spray header. Combined, the pumping energy demand is about 35 percent that for an equivalent CT application. The capital cost for this complete water system is also relatively small. The pumps and motors are smaller, the piping has a smaller diameter and is much shorter, and the required piping structural support is almost negligible, compared to an equivalent CT application. WSAC fan horsepower is typically about 25 percent less than that for an equivalent CT.

A WSAC is **inherently less sensitive to water-side fouling**. This is due to the fact that the deluge rate prevents the adhesion of waterborne material which can cause fouling within a HX. A **WSAC**

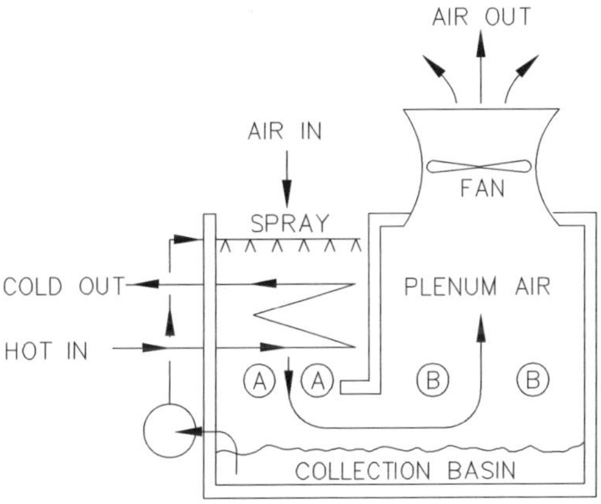

FIG. 12-8*k* Basic WSAC configuration.

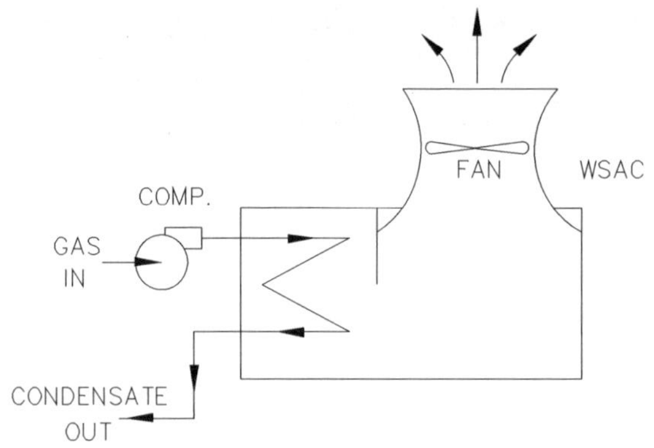

FIG. 12-8*l* WSAC configuration for condensing a compressed gas. A lower condensing pressure reduces compressor operating horsepower.

can accept relatively contaminated makeup water, such as CT blowdown, treated sewage plant effluent, etc. WSACs can endure **more cycles of concentration without fouling than can a CT** application. This higher practical operating concentration reduces the relative volume for the evaporative cooling blowdown, and therefore also reduces the relative volume of required makeup water. For facilities designed for **zero liquid discharge**, the higher practical WSAC blowdown concentration reduces the size and the operating costs for the downstream water treatment system. Since a hot process stream provides the unit with a heat source, a **WSAC has intrinsic freeze protection** while operating.

Common WSAC Applications and Configurations Employment of a WSAC can reduce process system operating costs that are not specific to the WSAC unit itself. A common WSAC application is **condensation of compressed gas** (Fig. 12-8*l*). A compressed gas can be condensed in a WSAC at a lower pressure, by condensing at a temperature closer to the ambient WBT, typically 5.5°C (10°F) above the WBT. This reduced condensation pressure reduces costs, by reducing gas compressor motor operating horsepower. Consequently, WSACs are widely applied for **condensing refrigerant gases**, for HVAC, process chillers, ice makers, gas-turbine inlet air cooling, chillers, etc. WSACs are also used directly to **condense lower-molecular-weight hydrocarbon streams**, such as ethane, ethylene, propylene, and LPG. A related WSAC application is the **cooling of compressed gases** (CO_2, N_2, methane, LNG, etc.), which directly reduces gas compressor operating costs (inlet and interstage cooling) and indirectly reduces downstream condensing costs (aftercooling the compressed gas to reduce the downstream refrigeration load).

For combined cycle electric power generation, employment of a **WSAC increases steam turbine efficiency**. Steam turbine exhaust can be condensed at a lower pressure (higher vacuum) by condensing at a temperature closer to the ambient WBT, typically 15°C (27°F) above the WBT. This reduced condensation pressure results in a lower turbine discharge pressure, **increasing electricity generation** by increasing output shaft power (Fig. 12-8*m*). Due to standard WSAC configurations, a second cost advantage is gained at the turbine itself. The **steam turbine can be placed at grade**, rather than being mounted on an elevated platform, by venting horizontally into the WSAC, rather than venting downward to condensers located below the platform elevation, as is common for conventional water-cooled vacuum steam condensers.

A WSAC can **eliminate chilled water use**, for process cooling applications with required temperatures close to and just above the ambient WBT, typically about 3.0 to 5.5°C (5 to 10°F) above the WBT. This WSAC application can eliminate both chiller capital and operating costs. In such an application, either the necessary process temperature is below the practical CT water supply temperature, or they are so close to it that the use of CT water is uneconomical (a low-HX LMDT).

WSACs can be designed to **simultaneously cool several process streams** in parallel separate tube bundles **within a single cell** of a

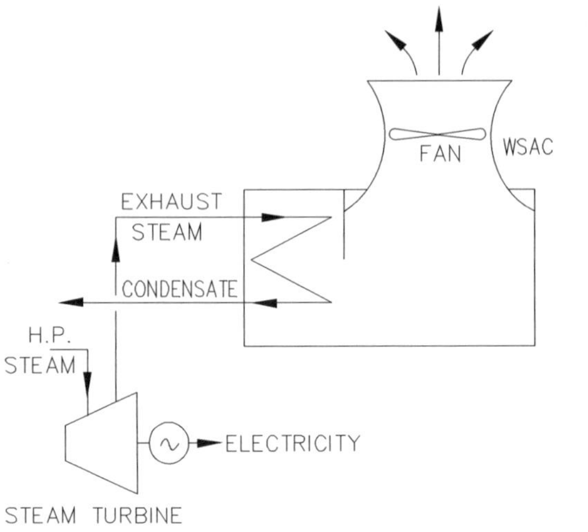

FIG. 12-8*m* WSAC configuration with electricity generation. A lower steam condensing pressure increases the turbine horsepower extracted.

WSAC (Fig. 12-8*n*). Often one of the streams is closed-circuit cooling water to be used for remote cooling applications. These might be applications not compatible with a WSAC (rotating seals, bearings, cooling jackets, internal reactor cooling coils, etc.) or merely numerous, small process streams in small HXs.

WSAC for Closed-Circuit Cooling Systems A closed-circuit cooling system as defined by the Cooling Technology Institute (CTI) employs a closed loop of circulated fluid (typically water) remotely as a cooling medium. By definition, this medium is cooled by **water evaporation involving no direct fluid contact** between the air and the enclosed circulated cooling medium. Applied in this manner, a WSAC can be used as the evaporative device to cool the circulated cooling medium, used remotely to cool process streams. This configuration completely isolates the cooling water (and the hot process streams) from the environment (Fig. 12-8*o*).

The closed circuit permits complete control of the cooling water chemistry, which permits **minimizing** the cost for **water-side materials of construction** and eliminating water-side fouling of, and fouling heat-transfer resistance in, the HXs (or jackets, reactor coils, etc.). **Elimination of water-side fouling** is particularly helpful for high-temperature cooling applications, especially where heat recovery may otherwise be impractical (quench oils, low-density polyethylene reactor cooling, etc.).

Closed-circuit cooling **minimizes circulation pumping horsepower**, which must overcome only dynamic pumping losses. This results through recovery of the returning circulated cooling water hydraulic head. A closed-circuit system can be designed for **operation at elevated pressures**, to guarantee that any process HX leak will be into the

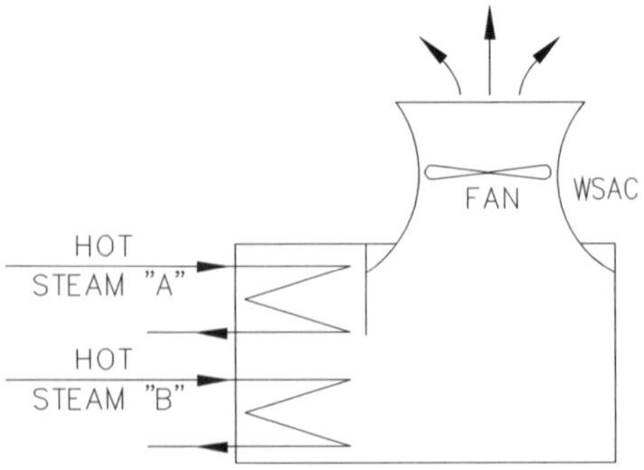

FIG. 12-8*n* WSAC configuration with parallel streams.

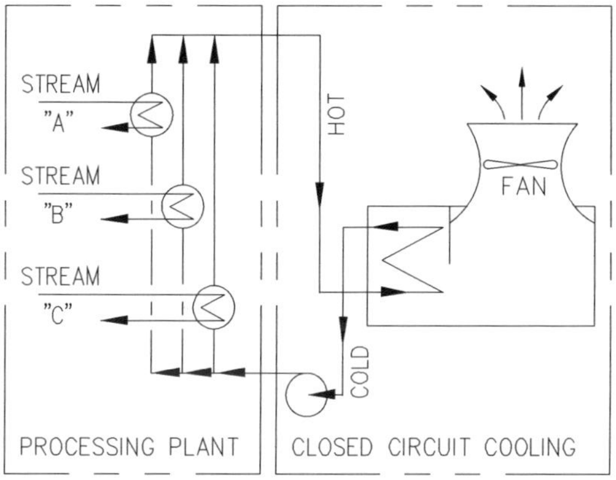

FIG. 12-8*o* WSAC configuration with no direct fluid contact.

process. Such high-pressure operation is economical, since the system overpressure is not lost during return flow to the circulation pump.

Closed-circuit cooling **splits the water chemistry needs** into two isolated systems: the **evaporating section**, exposed to the environment, and the **circulated cooling section**, isolated from the environment. Typically, this split reduces total water chemistry costs and water-related operations and maintenance problems. On the other hand, the split permits the effective **use of a low-quality or contaminated makeup water** for evaporative cooling, or a water source having severe seasonal quality problems, such as high sediment loadings.

If highly saline water is used for the evaporative cooling, a **reduced flow of makeup saline water** would need to be supplied to the WSAC. This reduction results from using latent cooling rather than sensible cooling to reject the waste heat. This consequence reduces the substantial capital investment required for the saline water supply and return systems (canal structures) and pump stations, and the saline supply pumping horsepower. (When saline water is used as the evaporative medium, special attention is paid to materials of construction and spray water chemical treatment due to the aggravated corrosion and scaling tendencies of this water.)

Water Conservation Applications—"Wet-Dry" Cooling A modified and hybridized form of a WSAC can be used to provide what is called "wet-dry" cooling for water conservation applications (Fig. 12-8*p*). A hybridized combination of air-cooled dry finned tubes, standard wetted bare tubes, and wet deck surface area permits the WSAC to **operate without water in cold weather**, reducing water consumption by about 75 percent of the total for an equivalent CT application.

Under design conditions of maximum summer WBT, the unit operates with spray water deluging the wetted tube bundle. The exiting water then flows down into and through the wet deck surface, where the water is cooled adiabatically to about the WBT, and then to the sump.

As the WBT drops, the process load is shifted from the wetted tubes to the dry finned tubes. By bypassing the process stream around the wetted tubes, cooling water evaporation (consumption) is proportionally reduced.

When the WBT drops to the "switch point," the process bypassing has reached 100 percent. This switch point WBT is at or above 5°C (41°F). As the ambient temperature drops further, adiabatic evaporative cooling continues to be used, to lower the dry-bulb temperature

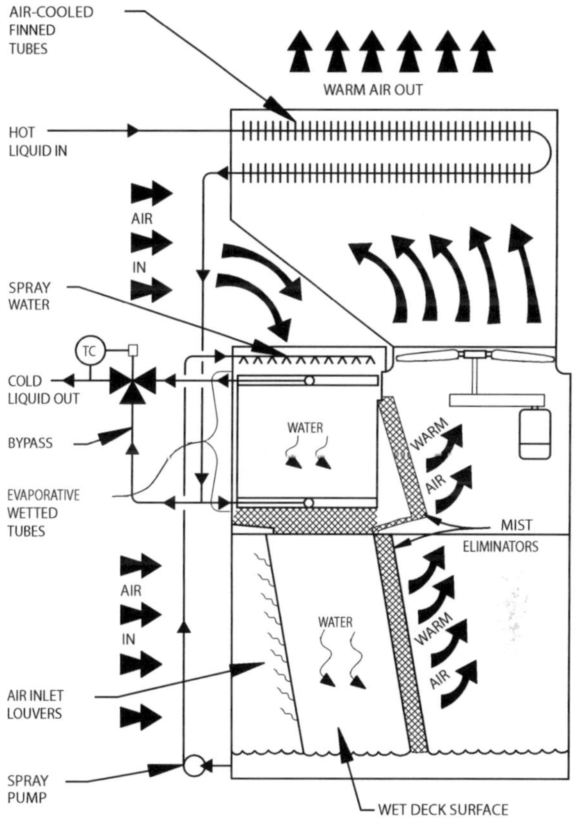

FIG. 12-8*p* As seasonal ambient temperatures drop, the "wet-dry" configuration for a WSAC progressively shifts the cooling load from evaporative to convective cooling.

to below the switch point temperature. This guarantees that the entire cooling load can be cooled in the dry finned tube bundle.

The use of water is discontinued after ambient dry-bulb temperatures fall below the switch point temperature, since the entire process load can be cooled using only cold fresh ambient air. By using this three-step load-shifting practice, total wet-dry cooling water consumption is about 25 percent of that consumption total experienced with an equivalent CT application.

Wet-dry cooling permits **significant reduction of water consumption**, which is useful where makeup water supplies are limited or where water treatment costs for blowdown are high. Because a WSAC (unlike a CT) has a heat source (the hot process stream), wet-dry cooling avoids various cold-weather-related CT problems. **Fogging and persistent plume formation can be minimized or eliminated** during colder weather. **Freezing and icing problems can be eliminated** by designing a wet-dry system for water-free operation during freezing weather, typically below 5°C (41°F). **In the arctic**, or regions of extreme cold, **elimination of freezing fog conditions** is realized by not evaporating any water during freezing weather.

SOLIDS-DRYING FUNDAMENTALS

General References: Cook and DuMont, *Process Drying Practice,* McGraw-Hill, New York, 1991. *Drying Technology—An International Journal,* Taylor and Francis, New York. Hall, *Dictionary of Drying,* Marcel Dekker, New York, 1979. Keey, *Introduction to Industrial Drying Operations,* Pergamon, New York, 1978. Keey, *Drying of Loose and Particulate Materials,* Hemisphere, New York, 1992. Masters, *Spray Drying Handbook,* Wiley, New York, 1990. Mujumdar, *Handbook of Industrial Drying,* Marcel Dekker, New York, 1987. Nonhebel and Moss, *Drying of Solids in the Chemical Industry,* CRC Press, Cleveland, Ohio, 1971. Strumillo and Kudra, *Drying: Principles, Application and Design,* Gordon and Breach, New York, 1986. van't Land, *Industrial Drying Equipment,* Marcel Dekker, New York, 1991.

INTRODUCTION

Drying is the process by which volatile materials, usually water, are evaporated from a material to yield a solid product. Drying is a heat- and mass-transfer process. Heat is necessary to evaporate water. The latent heat of vaporization of water is about 2500 J/g, which means that the drying process requires a significant amount of energy. Simultaneously, the evaporating material must leave the drying material by diffusion and/or convection.

Heat transfer and mass transfer are not the only concerns when one is designing or operating a dryer. The product quality (color, particle density, hardness, texture, flavor, etc.) is also very strongly dependent on the drying conditions and the physical and chemical transformations occurring in the dryer.

Understanding and designing a drying process involves measurement and/or calculation of the following:

1. Mass and energy balances
2. Thermodynamics
3. Mass- and heat-transfer rates
4. Product quality considerations

The section below explains how these factors are measured and calculated and how the information is used in engineering practice.

TERMINOLOGY

Generally accepted terminology and definitions are given alphabetically in the following paragraphs.

Absolute humidity is the mass ratio of water vapor (or other solvent mass) to dry air.

Activity is the ratio of the fugacity of a component in a system relative to the standard-state fugacity. In a drying system, it is the ratio of the vapor pressure of a solvent (e.g., water) in a mixture to the pure solvent vapor pressure at the same temperature. Boiling occurs when the vapor pressure of a component in a liquid exceeds the ambient total pressure.

Bound moisture in a solid is that liquid which exerts a vapor pressure less than that of the pure liquid at the given temperature. Liquid may become bound by retention in small capillaries, by solution in cell or fiber walls, by homogeneous solution throughout the solid, by chemical or physical adsorption on solid surfaces, and by hydration of solids.

Capillary flow is the flow of liquid through the interstices and over the surface of a solid, caused by liquid-solid molecular attraction.

Constant-rate period (unhindered) is that drying period during which the rate of water removal per unit of drying surface is constant, assuming the driving force is also constant.

Convection is heat or mass transport by bulk flow.

Critical moisture content is the average moisture content when the constant-rate period ends, assuming the driving force is also constant.

Diffusion is the molecular process by which molecules, moving randomly due to thermal energy, migrate from regions of high chemical potential (usually concentration) to regions of lower chemical potential.

Dry basis expresses the moisture content of wet solid as kilograms of water per kilogram of bone-dry solid.

Equilibrium moisture content is the limiting moisture to which a given material can be dried under specific conditions of air temperature and humidity.

Evaporation is the transformation of material from a liquid state to a vapor state.

Falling-rate period (hindered drying) is a drying period during which the instantaneous drying rate continually decreases.

Fiber saturation point is the moisture content of cellular materials (e.g., wood) at which the cell walls are completely saturated while the cavities are liquid-free. It may be defined as the equilibrium moisture content as the humidity of the surrounding atmosphere approaches saturation.

Free moisture content is that liquid which is removable at a given temperature and humidity. It may include bound and unbound moisture.

Funicular state is that condition in drying a porous body when capillary suction results in air being sucked into the pores.

Hygroscopic material is material that may contain bound moisture.

Initial moisture distribution refers to the moisture distribution throughout a solid at the start of drying.

Internal diffusion may be defined as the movement of liquid or vapor through a solid as the result of a concentration difference.

Latent heat of vaporization is the specific enthalpy change associated with evaporation.

Moisture content of a solid is usually expressed as moisture quantity per unit weight of the dry or wet solid.

Moisture gradient refers to the distribution of water in a solid at a given moment in the drying process.

Nonhygroscopic material is material that can contain no bound moisture.

Pendular state is that state of a liquid in a porous solid when a continuous film of liquid no longer exists around and between discrete particles so that flow by capillary cannot occur. This state succeeds the funicular state.

Permeability is the resistance of a material to bulk or convective, pressure-driven flow of a fluid through it.

Relative humidity is the partial pressure of water vapor divided by the vapor pressure of pure water at a given temperature. In other words, the relative humidity describes how close the air is to saturation.

Sensible heat is the energy required to increase the temperature of a material without changing the phase.

Unaccomplished moisture change is the ratio of the free moisture present at any time to that *initially* present.

Unbound moisture in a hygroscopic material is that moisture in excess of the equilibrium moisture content corresponding to saturation humidity. All water in a nonhygroscopic material is unbound water.

Vapor pressure is the partial pressure of a substance in the gas phase that is in equilibrium with a liquid or solid phase of the pure component.

Wet basis expresses the moisture in a material as a percentage of the weight of the wet solid. Use of a dry-weight basis is recommended since the percentage change of moisture is constant for all moisture levels. When the wet-weight basis is used to express moisture content, a 2 or 3 percent change at high moisture contents (above 70 percent) actually represents a 15 to 20 percent change in evaporative load. See Fig. 12-9 for the relationship between the dry- and wet-weight bases.

MASS AND ENERGY BALANCES

The most basic type of calculation for a dryer is a mass and energy balance. This calculation only quantifies the conservation of mass and energy in the system; by itself it does not answer important questions of rate and quality.

Some examples here illustrate the calculations. Experimental determination of the values used in these calculations is discussed in a later section.

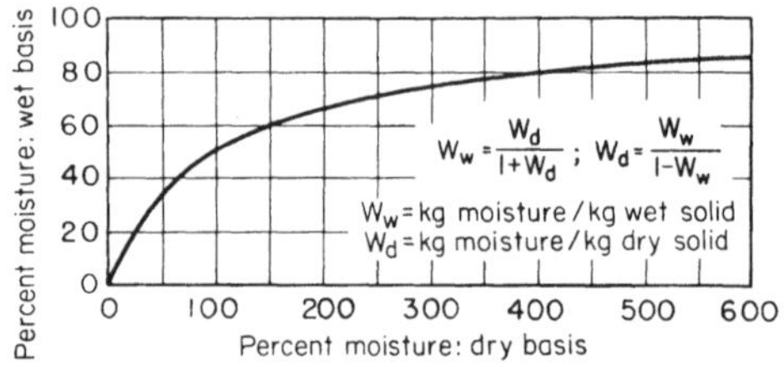

FIG. 12-9 Relationship between wet-weight and dry-weight bases.

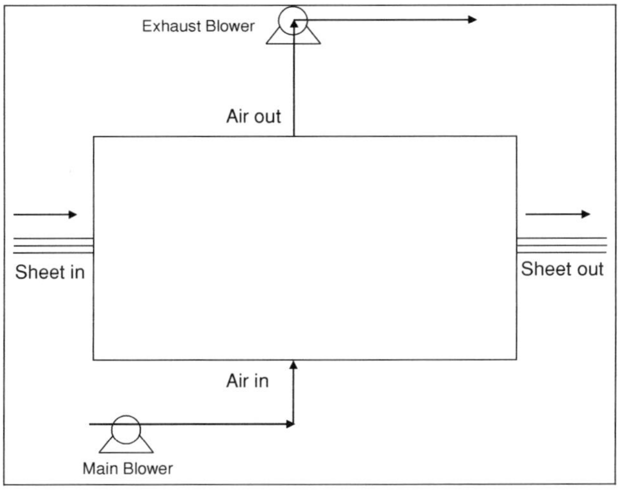

FIG. 12-10 Overall mass and energy balance diagram.

Example 15 illustrates a generic mass and energy balance. Other examples are given in the sections on fluidized bed dryers and rotary dryers.

Example 15: Overall Mass and Energy Balance on a Sheet Dryer Figure 12-10 shows a simple sheet drying system. Hot air enters the dryer and contacts a wet sheet. The sheet leaves a dryer with a lower moisture content, and the air leaves the dryer with a higher humidity.

Given: Incoming wet sheet mass flow rate is 100 kg/h. It enters with 20 percent water on a wet basis and leaves at 1 percent water on a wet basis. The airflow rate is 1000 kg/h, with an absolute humidity of 0.01 g water/g dry air. The incoming air temperature is 170°C. The sheet enters at 20°C and leaves at 90°C.

Relevant physical constants: $C_{p,\text{ air}} = 1$ kJ/(kg·°C), $C_{p,\text{ sheet}} = 2.5$ kJ/(kg·°C), $C_{p,\text{ liquid water}} = 4.184$ kJ/(kg·°C), $C_{p,\text{ water vapor}} = 2$ kJ/(kg·°C) (for superheated steam at low partial pressures). Latent heat of vaporization of water at 20°C = $\lambda_w = 2454$ J/g

Find the following:

1. The absolute humidity of the exiting airstream
2. The exit air temperature

Solution: Answering the questions above involves an overall mass and energy balance. Only the mass and enthalpy of the streams need to be considered to answer the two questions above. Only the streams entering the overall process need to be considered. In this example, wet-basis moisture content (and therefore total mass flow rate including moisture) will be used. Since the same mass of air flows in and out of the dryer, there are no equations to solve for the dry air.

The mass balance is given by the following equations:

$$F_{\text{dry sheet in}} = F_{\text{dry sheet out}} \tag{12-15}$$

$$F_{\text{liquid water in}} = F_{\text{liquid water out}} + F_{\text{evaporated}} \tag{12-16}$$

$$G_{\text{dry air in}} = G_{\text{dry air out}} \tag{12-17}$$

$$G_{\text{water vapor in}} + F_{\text{evaporated}} = G_{\text{water vapor out}} \tag{12-18}$$

The wet-basis moisture contents of the incoming and outgoing sheet are given by

$$w_{\text{in}} = \frac{F_{\text{liquid water in}}}{F_{\text{liquid water in}} + F_{\text{dry sheet in}}} \tag{12-19}$$

$$w_{\text{out}} = \frac{F_{\text{liquid water out}}}{F_{\text{liquid water out}} + F_{\text{dry sheet out}}} \tag{12-20}$$

The relationship between the total airflow, the dry airflow, and the absolute humidity is given by

$$G_{\text{dry air}} = G_{\text{air}}\frac{1}{1+Y} = 1000 \text{ kg/h}\,\frac{1}{1+0.01} = 990 \text{ kg/h}$$

The absolute humidity of each airstream is given by

$$Y_{\text{in}} = \frac{G_{\text{water vapor in}}}{G_{\text{dry air in}}} \tag{12-21}$$

$$Y_{\text{out}} = \frac{G_{\text{water vapor out}}}{G_{\text{dry air out}}} \tag{12-22}$$

The mass flow rates of the dry sheet and the liquid water in can be calculated from the overall sheet flow rate and the incoming moisture content:

$$G_{\text{liquid water in}} = G_{\text{sheet}}w_{\text{in}} = (100 \text{ kg/h})(0.2) = 20 \text{ kg/h} \tag{12-23}$$

$$F_{\text{dry sheet}} = F_{\text{sheet}}(1 - w_{\text{in}}) = (100 \text{ kg/h})(0.8) = 80 \text{ kg/h} \tag{12-24}$$

The mass flow rates of the dry air and incoming water vapor can be calculated from the overall airflow rate and the incoming absolute humidity:

$$G_{\text{water vapor in}} = G_{\text{dry air}}Y_{\text{in}} = (990 \text{ kg/h})(0.01) = 9.9 \text{ kg/h} \tag{12-25}$$

To calculate the exiting absolute humidity, Eq. (12-22) is used. But the evaporation rate $G_{\text{evaporated}}$ is needed. This is calculated from Eqs. (12-16) and (12-20).

$$F_{\text{liquid water out}} = \frac{w_{\text{out}}}{1 - w_{\text{out}}}F_{\text{dry sheet out}} = \frac{0.01}{0.99}\cdot 80 \text{ kg/h} = 0.8 \text{ kg/h} \quad \text{(12-20, rearranged)}$$

$$G_{\text{evaporated}} = F_{\text{liquid water in}} - F_{\text{liquid water out}} = 20 - 1 \text{ kg/h} = 19.2 \text{ kg/h} \tag{12-26}$$

Equation (12-18) is now used to calculate the mass flow of water vapor out of the dryer:

$$G_{\text{water vapor out}} = 9.9 \text{ kg/h} + 19.2 \text{ kg/h} = 29.1 \text{ kg/h} \tag{12-27}$$

Now the absolute humidity of the exiting air is readily calculated from Eq. (12-22):

$$Y_{\text{out}} = \frac{G_{\text{water vapor out}}}{G_{\text{dry air}}} = \frac{29}{990} = 0.0294 \tag{12-28}$$

Next an energy balance must be used to estimate the outgoing air temperature. The following general equation is used:

$$H_{\text{dry air,in}} + H_{\text{water vapor,in}} + H_{\text{dry sheet in}} + H_{\text{liquid water in}} = H_{\text{dry air, out}} + H_{\text{water vapor, out}} + H_{\text{dry sheet out}} + H_{\text{liquid water out}} + \text{heat loss to surroundings} \tag{12-29}$$

Heat losses to the environment are often difficult to quantify, but they can be neglected for a first approximation. This assumption is more valid for large systems than small systems. It is neglected in this example.

Evaluation of the energy balance terms can be done in a couple of ways. Values of the enthalpies above can be calculated by using a consistent reference, or the equation can be rearranged in terms of enthalpy differences. The latter approach will be used here, as shown by Eq. (12-30).

$$\Delta H_{\text{dry air}} + \Delta H_{\text{water vapor}} + \Delta H_{\text{evaporation}} + \Delta H_{\text{liquid water}} + \Delta H_{\text{dry sheet}} = 0 \tag{12-30}$$

The enthalpy change due to evaporation $\Delta H_{\text{evaporation}}$ is given by $F_{\text{evaporated}}\,\lambda_w$. To evaluate λ_w rigorously, a decision has to be made on the calculational path of the evaporating water since this water is both heating and evaporating. Typically, a two-step path is used—isothermal evaporation and heating of either phase. The incoming liquid water can all be heated to the outlet temperature of the sheet, and then the heat of vaporization at the outlet temperature can be used; or the evaporation can be calculated as occurring at the inlet temperature, and the water vapor is heated from the inlet temperature to the outlet temperature. Alternatively a three-step path based on latent heat at the datum (0°C) may be used. All these methods of calculation are equivalent, since the enthalpy is a state function; but in this case, the second method is preferred since the outlet temperature is unknown. In the calculation, the water will be evaporated at 20°C, heated to the air inlet temperature 170°C, and then cooled to the outlet temperature. Alternatively, this enthalpy change can be calculated directly by using tabular enthalpy values available on the psychrometric chart or Mollier diagram.

The terms in these equations can be evaluated by using

$$\Delta H_{\text{dry air}} = G_{\text{dry air in}}C_{p,\text{air}}(T_{\text{air in}} - T_{\text{air, out}}) = (990.1 \text{ kg/h})[1 \text{ kJ/(kg·°C)}][(170 - T_{\text{air,out}}) \text{ kJ/h}] \tag{12-31}$$

$$\Delta H_{\text{water vapor}} = G_{\text{water vapor out}}C_{p,\text{water vapor}}(T_{\text{air in}} - T_{\text{air,out}}) = (29.1 \text{ kg/h})\,[2 \text{ kJ/(kg·°C)}]\,[(170 - T_{\text{air,out}}) \text{ kJ/h}] \tag{12-32}$$

From steam tables, ΔH_{vap} at 20°C = 2454 kJ/kg, h_l = 84 kJ/kg, and h_g at 170°C (superheated, low pressure) = 2820 kJ/kg.

$$-\Delta H_{\text{evaporation}} = -G_{\text{evaporated}} \cdot \Delta H_{\text{vap}}$$
$$= (-19.2 \text{ kg/h})$$
$$[2736 \text{ kJ/kg (from steam table)}]$$
$$= -52{,}530 \text{ kJ/h} \quad (12\text{-}33)$$

$$\Delta H_{\text{liquid water}} = F_{\text{liquid water out}} C_{p,\text{liquid water}} (T_{\text{sheet,in}} - T_{\text{sheet,out}})$$
$$= (0.8 \text{ kg/h})[4.18 \text{ kJ/(kg·°C)}][(20°\text{C} - 90°\text{C})] = -234 \text{ kJ/h} \quad (12\text{-}34)$$

$$\Delta H_{\text{dry sheet}} = F_{\text{dry sheet}} C_{p,\text{sheet}} (T_{\text{sheet,in}} - T_{\text{sheet,out}})$$
$$= (80 \text{ kg/h})[2.5 \text{ kJ/(kg·°C)}][(20°\text{C} - 90°\text{C})] = -14{,}000 \text{ kJ/h}$$

Putting this together gives

$$(990.1)(1)(170 - T_{\text{air, out}}) + (29.1)(2)(170 - T_{\text{air, out}}) - 52{,}530 - 293 - 14{,}000 = 0$$
$$T_{\text{air, out}} = 106°\text{C}$$

THERMODYNAMICS

The thermodynamic driving force for evaporation is the difference in chemical potential or water activity between the drying material and the gas phase. Although drying of water is discussed in this section, the same concepts apply analogously for solvent drying.

For a pure water drop, the driving force for drying is the difference between the vapor pressure of water and the partial pressure of water in the gas phase. The rate of drying is proportional to this driving force; please see the discussion on drying kinetics later in this chapter.

$$\text{Rate} \propto (p^{\text{sat}}_{\text{pure}} - p_{w,\text{air}})$$

The activity of water in the gas phase is defined as the ratio of the partial pressure of water to the vapor pressure of pure water, which is also related to the definition of relative humidity.

$$a_w^{\text{vapor}} = \frac{p_w}{p^{\text{sat}}_{\text{pure}}} = \frac{\%\text{RH}}{100}$$

The activity of water in a mixture or solid is defined as the ratio of the vapor pressure of water in the mixture to that of a reference, usually the vapor pressure of pure water. In solids drying or drying of solutions, the vapor pressure (or water activity) is lower than that for pure water. Therefore, the water activity value equals 1 for pure water and < 1 when binding is occurring. This is caused by thermodynamic interactions between the water and the drying material. In many standard drying references, this is called *bound water.*

$$a_w^{\text{solid}} = \frac{p^{\text{sat}}_{\text{mixture}}}{p^{\text{sat}}_{\text{pure}}}$$

When a solid sample is placed into a humid environment, water will transfer from the solid to the air or vice versa until equilibrium is established. At thermodynamic equilibrium, the water activity is equal in both phases:

$$a_w^{\text{vapor}} = a_w^{\text{solid}} = a_w$$

Sorption isotherms quantify how tightly water is bound to a solid. The goal of obtaining a sorption isotherm for a given solid is to measure the equilibrium relationship between the percentage of water in the sample and the vapor pressure of the mixture. The sorption isotherm describes how dry a product can get if contacted with humid air for an infinite amount of time. An example of a sorption isotherm is shown in Fig. 12-11. In the sample isotherm, a feed material dried with 50 percent relative humidity air (a_w = 0.5) will approach a moisture content of 10 percent on a dry basis. Likewise, a material kept in a sealed container will create a headspace humidity according to the isotherm; a 7 percent moisture sample in the example below will create a 20 percent relative humidity (a_w = 0.2) headspace in a sample jar or package.

Strictly speaking, the equilibrium moisture content of the sample in a given environment should be independent of the initial condition of the sample. However, there are cases where the sorption isotherm of an initially wet sample (sometimes called a desorption isotherm) is different from that of an identical, but initially dry sample. This is called hysteresis and can be caused by irreversible changes in the sample during wetting or drying, micropore geometry in the sample, and other factors. Paper products are notorious for isotherm hysteresis. Most materials show little or no hysteresis.

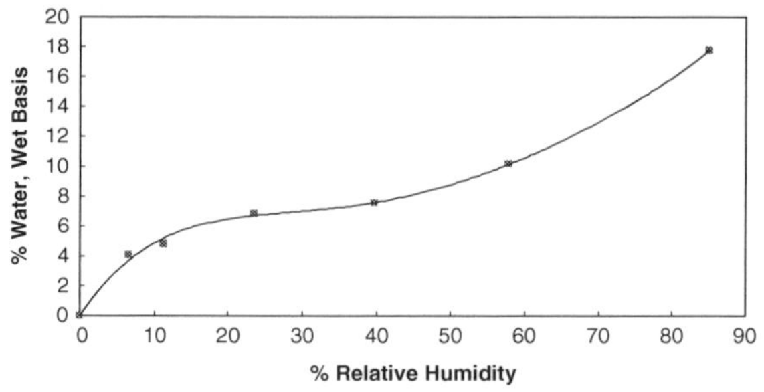

FIG. 12-11 Example of a sorption isotherm (coffee at 22°C).

Sorption isotherms cannot generally be predicted from theory. They need to be measured experimentally. The simplest method of measuring a sorption isotherm is to generate a series of controlled-humidity environments by using saturated salt solutions, allow a solid sample to equilibrate in each environment, and then analyze the solid for moisture content.

The basic apparatus is shown in Fig. 12-12, and a table of salts is shown in Table 12-7. It is important to keep each chamber sealed and to be sure that crystals are visible in the salt solution to ensure that the liquid is saturated. Additionally, the solid should be ground into a powder to facilitate mass transfer. Equilibration can take 2 to 3 weeks. Successive moisture measurements should be used to ensure that the sample has equilibrated, i.e., achieved a steady value. Care must be taken when measuring the moisture content of a sample; this is described later in the chapter.

Another common method of measuring a sorption isotherm is to use a dynamic vapor sorption device. This machine measures the weight change of a sample when exposed to humidity-controlled air. A series of humidity points are programmed into the unit, and it automatically delivers the proper humidity to the sample and monitors the weight. When the weight is stable, an equilibrium point is noted and the air humidity is changed to reflect the next setting in the series. When one is using this device, it is critical to measure and record the starting moisture of the sample, since the results are often reported as a percent of change rather than a percent of moisture.

There are several advantages to the dynamic vapor sorption device. First, any humidity value can be dialed in, whereas salt solutions are not available for every humidity value and some are quite toxic. Second, since the weight is monitored as a function of time, it is clear when equilibrium is reached. The dynamic devices also give the sorption/desorption rates, although these can easily be misused (see the drying kinetics section later). The salt solution method, on

FIG. 12-12 Sorption isotherm apparatus. A saturated salt solution is in the bottom of the sealed chamber; samples sit on a tray in the headspace.

TABLE 12-7 Maintenance of Constant Humidity

Solid phase	Max. temp., °C	% Humidity
$H_3PO_4 \cdot aH_2O$	24.5	9
$ZnCl_2 \cdot aH_2O$	20	10
$KC_2H_3O_2$	168	13
$LiCl \cdot H_2O$	20	15
$KC_2H_3O_2$	20	20
KF	100	22.9
NaBr	100	22.9
$CaCl_2 \cdot 6H_2O$	24.5	31
$CaCl_2 \cdot 6H_2O$	20	32.3
$CaCl_2 \cdot 6H_2O$	18.5	35
CrO_3	20	35
$CaCl_2 \cdot 6H_2O$	10	38
$CaCl_2 \cdot 6H_2O$	5	39.8
$K_2CO_3 \cdot 2H_2O$	24.5	43
$K_2CO_3 \cdot 2H_2O$	18.5	44
$Ca(NO_3)_2 \cdot 4H_2O$	24.5	51
$NaHSO_4 \cdot H_2O$	20	52
$Mg(NO_3)_2 \cdot 6H_2O$	24.5	52
$NaClO_3$	100	54
$Ca(NO_3)_2 \cdot 4H_2O$	18.5	56
$Mg(NO_3)_2 \cdot 6H_2O$	18.5	56
$NaBr \cdot 2H_2O$	20	58
$Mg(C_2H_3O_2) \cdot 4H_2O$	20	65
$NaNO_2$	20	66
$(NH_4)_2SO_4$	108.2	75
$(NH_4)_2SO_4$	20	81
$NaC_2H_3O_2 \cdot 3H_2O$	20	76
$Na_2S_2O_3 \cdot 5H_2O$	20	78
NH_4Cl	20	79.5
NH_4Cl	25	79.3
NH_4Cl	30	77.5
KBr	20	84
Tl_2SO_4	104.7	84.8
$KHSO_4$	20	86
$Na_2CO_3 \cdot 10H_2O$	24.5	87
K_2CrO_4	20	88
$NaBrO_3$	20	92
$Na_2CO_3 \cdot 10H_2O$	18.5	92
$Na_2SO_4 \cdot 10H_2O$	20	93
$Na_2HPO_4 \cdot 12H_2O$	20	95
NaF	100	96.6
$Pb(NO_3)_2$	20	98
$TlNO_3$	100.3	98.7
TLCl	100.1	99.7

For a more complete list of salts, and for references to the literature, see *International Critical Tables,* vol. 1, p. 68.

the other hand, is significantly less expensive to buy and maintain. Numerous samples can be placed in humidity chambers and run in parallel while a dynamic sorption device can process only one sample at a time.

An excellent reference on all aspects of sorption isotherms is by Bell and Labuza, *Moisture Sorption,* 2d ed., American Associated of Cereal Chemists, 2000.

MECHANISMS OF MOISTURE TRANSPORT WITHIN SOLIDS

Drying requires moisture to travel to the surface of a material. There are several mechanisms by which this can occur:

1. *Diffusion of moisture through solids.* Diffusion is a molecular process, brought about by random wanderings of individual molecules. If all the water molecules in a material are free to migrate, they tend to diffuse from a region of high moisture concentration to one of lower moisture concentration, thereby reducing the moisture gradient and equalizing the concentration of moisture.

2. *Convection of moisture within a liquid or slurry.* If a flowable solution is drying into a solid, then liquid motion within the material brings wetter material to the surface.

3. *Evaporation of moisture within a solid and gas transport out of the solid by diffusion and/or convection.* Evaporation can occur within a solid if it is boiling or porous. Subsequently vapor must move out of the sample.

4. *Capillary flow of moisture in porous media.* The reduction of liquid pressure within small pores due to surface tension forces causes liquid to flow in porous media by capillary action.

DRYING KINETICS

This section discusses the rate of drying. The kinetics of drying dictates the size of industrial drying equipment, which directly affects the capital and operating costs of a process involving drying. The rate of drying can also influence the quality of a dried product since other simultaneous phenomena can be occurring, such as heat transfer and shrinkage due to moisture loss.

Drying Curves and Periods of Drying The most basic and essential kinetic information on drying is a *drying curve.* A drying curve describes the drying kinetics and how they change during drying. The drying curve is affected by the material properties, size or thickness of the drying material, and drying conditions. In this section, the general characteristics of drying curves and their uses are described. Experimental techniques to obtain drying curves are discussed in the "Experimental Methods" section and uses of drying curves for scale-up are discussed in "Dryer Modeling Design and Scale-up."

Several representations of a typical drying curve are shown in Fig. 12-13. The top plot, Fig. 12-13*a,* is the moisture content (dry basis) as a function of time. The middle plot, Fig. 12-13*b,* is the drying rate as a function of time, the derivative of the top plot. The bottom plot,

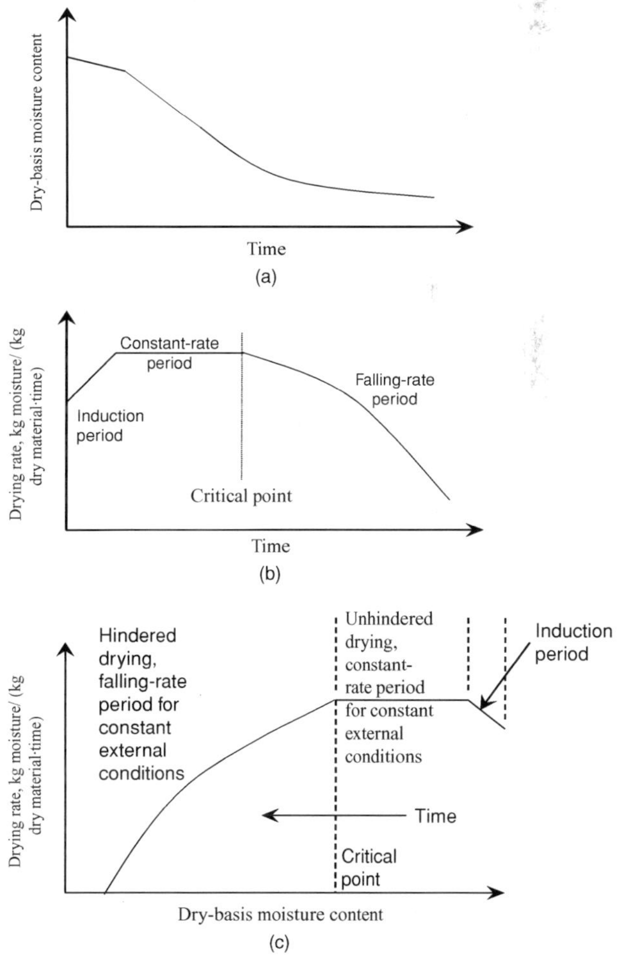

FIG. 12-13 Several common representations of a typical drying curve.

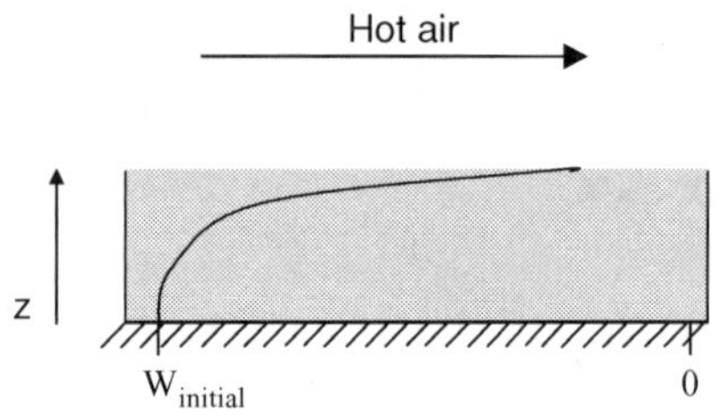

FIG. 12-14 Drying of a slab.

Fig. 12-13*c*, is the drying rate as affected by the average moisture content of the drying material. Since the material loses moisture as time passes, the progression of time in this bottom plot is from right to left.

Some salient features of the drying curve show the different periods of drying. These are common periods, but not all occur in every drying process. The first period of drying is called the induction period. This period occurs when material is being heated early in drying. The second period of drying is called the constant-rate period. During this period, the surface remains wet enough to maintain the vapor pressure of water on the surface. Once the surface dries sufficiently, the drying rate decreases and the falling-rate period occurs. This period can also be referred to as *hindered drying*.

Figure 12-13 shows the transition between constant- and falling-rate periods of drying occurring at the critical point. The *critical point* refers to the average moisture content of a material at this transition.

The sections below show examples of drying curves and the phenomena that give rise to common shapes.

Introduction to Internal and External Mass-Transfer Control—Drying of a Slab The concepts in drying kinetics are best illustrated with a simple example—air drying of a slab. Consider a thick slab of homogeneous wet material, as shown in Fig. 12-14. In this particular example, the slab is dried on an insulating surface under constant conditions. The heat for drying is carried to the surface with hot air, and air carries water vapor from the surface. At the same time, a moisture gradient forms within the slab, with a dry surface and a wet interior. The curved line is the representation of the gradient. At the bottom the slab ($z = 0$), the material is wet and the moisture content is drier at the surface.

The following processes must occur to dry the slab:

1. Heat transfer from the air to the surface of the slab
2. Mass transfer of water vapor from the surface of the slab to the bulk air
3. Mass transfer of moisture from the interior of the slab to the surface of the slab

Depending on the drying conditions, thickness, and physical properties of the slab, any of the above steps can be rate-limiting. Figure 12-15 shows two examples of rate-limiting cases.

The top example shows the situation of *external rate control*. In this situation, the heat transfer to the surface and/or the mass transfer from the surface in the vapor phase is slower than mass transfer to the surface from the bulk of the drying material. In this limiting case, the moisture gradient in the material is minimal, and the rate of drying will be constant as long as the average moisture content remains high enough to maintain a high water activity (see the section on thermodynamics for a discussion of the relationship between moisture content and water vapor pressure). External rate control leads to the observation of a constant-rate period drying curve.

The bottom example shows the opposite situation: *internal rate control*. In the case of heating from the top, internal control refers to a slow rate of mass transfer from the bulk of the material to the surface of the material. Diffusion, convection, and capillary action (in the case of porous media) are possible mechanisms for mass transfer of moisture to the surface of the slab. In the internal rate control situation, moisture is removed from the surface by the air faster than moisture is transported to the surface. This regime is caused by relatively thick layers or high values of the mass- and heat-transfer coefficients in the air. Internal rate control leads to the observation of a falling-rate period drying curve.

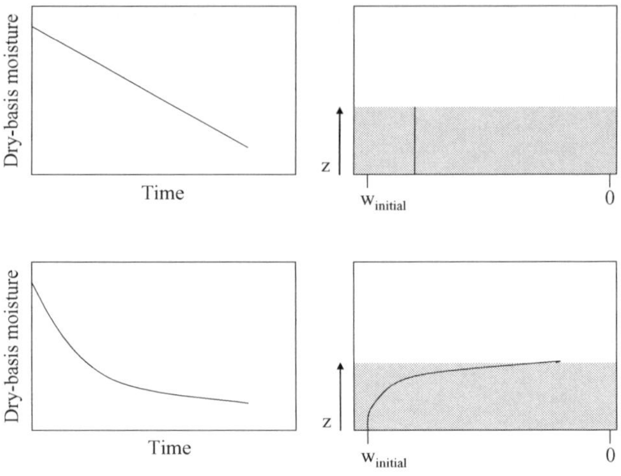

FIG. 12-15 Drying curves and corresponding moisture gradients for situations involving external heat and mass-transfer control and internal mass-transfer control.

Generally speaking, drying curves show both behaviors. When drying begins, the surface is often wet enough to maintain a constant-rate period and is therefore externally controlled. But as the material dries, the mass-transfer rate of moisture to the surface often slows, causing the rate to decrease since the lower moisture content on the surface causes a lower water vapor pressure. However, some materials begin dry enough that there is no observable constant-rate period.

MATHEMATICAL MODELING OF DRYING

Mathematical models can be powerful tools to help engineers understand drying processes. Models can be either purchased or homemade. Several companies offer software packages to select dryers, perform scale-up calculations, and simulate dryers.

Homemade models are often mass and energy balance spreadsheets, simplified kinetic models, or the simultaneous solution of the convection diffusion and heat equations together with nonlinear isotherms. All levels of models have their place.

This section begins with the most rigorous and numerical models. These models are potentially the most accurate, but require physical property data and simultaneous solution of differential and algebraic equations. Generally speaking, simpler models are more accessible to engineers and easier to implement. They can be very useful as long as the inherent limitations are understood.

Numerical Modeling of Drying Kinetics This section summarizes a numerical approach toward modeling drying from a fundamental standpoint. In other words, predictions are made from the appropriate sets of differential and algebraic equations, together with physical properties of the drying medium and drying material. Statistical methods of data analysis, e.g., design of experiments, are not covered.

The approach in this section is lagrangian; i.e., the model is for a drying object (particle, drop, sheet, etc.) as it moves through the drying process in time. More complicated models can use a eulerian frame of reference by simulating the dryer with material moving into and out of the dryer.

The approach taken in this example also assumes that the mechanism of mass transport is by diffusion. This is not always the case and can be significantly incorrect, especially in the case of drying of porous materials.

Any fundamental mathematical model of drying contains mass and energy balances, constituative equations for mass- and heat-transfer rates, and physical properties. Table 12-8 shows the differential mass balance equations that can be used for common geometries. Note there are two sets of differential mass balances—one including shrinkage and one not including shrinkage. When moisture leaves a drying material, the material can either shrink, or develop porosity, or both.

TABLE 12-8 Mass-Balance Equations for Drying Modeling When Diffusion Is Mass-Transfer Mechanism of Moisture Transport

Case	Mass balance without shrinkage	Mass balance with shrinkage
Slab geometry	$\frac{\partial C_w}{\partial t} = \frac{\partial}{\partial z}\left[\mathcal{D}(w)\frac{\partial C_w}{\partial z}\right]$	$\frac{\partial u}{\partial t} = \frac{\partial}{\partial s}\left[\mathcal{D}(w)\frac{\partial u}{\partial s}\right] \quad \frac{\partial s}{\partial z} = \rho_s$
	At top surface, $-\mathcal{D}(w)\frac{\partial C_w}{\partial z}_{\text{top surface}} = k_p \frac{p_w^{\text{bulk}} - p_w^{\text{surface}}}{P - p_w^{\text{surface}}}$	At top surface, $-\mathcal{D}(w)\frac{\partial u}{\partial s}_{\text{top surface}} = k_p \frac{p_w^{\text{bulk}} - p_w^{\text{surface}}}{P - p_w^{\text{surface}}}$
	At bottom surface, $\frac{\partial C_w}{\partial z}_{\text{bottom surface}} = 0$	At bottom surface, $\frac{\partial u}{\partial s}_{\text{bottom surface}} = 0$
Cylindrical geometry	$\frac{\partial C_w}{\partial t} = \frac{1}{r}\frac{\partial}{\partial r}\left[r\mathcal{D}(w)\frac{\partial C_w}{\partial r}\right]$	$\frac{\partial u}{\partial t} = \frac{\partial}{\partial s}\left[\rho_s^2\mathcal{D}(w)\frac{\partial u}{\partial s}\right] \quad \frac{\partial s}{\partial z} = r\rho_s$
	At surface, $-\mathcal{D}(w)\frac{\partial C_w}{\partial r}_{\text{surface}} = k_p \frac{p_w^{\text{bulk}} - p_w^{\text{surface}}}{P - p_w^{\text{surface}}}$	At surface, $-\mathcal{D}(w)r\frac{\partial u}{\partial s}_{\text{surface}} = k_p \frac{p_w^{\text{bulk}} - p_w^{\text{surface}}}{P - p_w^{\text{surface}}}$
	At center, $\frac{\partial C_w}{\partial r}_{\text{center}} = 0$	At center, $\frac{\partial u}{\partial s}_{\text{surface}} = 0$
Spherical geometry	$\frac{\partial C_w}{\partial t} = \frac{1}{r^2}\frac{\partial}{\partial r}\left[r^2\mathcal{D}(w)\frac{\partial C_w}{\partial r}\right]$	$\frac{\partial u}{\partial t} = \frac{\partial}{\partial s}\left[\rho_s^4\mathcal{D}(w)\frac{\partial u}{\partial s}\right] \quad \frac{\partial s}{\partial z} = r^2\rho_s$
	At surface, $-\mathcal{D}(w)\frac{\partial C_w}{\partial r}_{\text{surface}} = k_p \frac{p_w^{\text{bulk}} - p_w^{\text{surface}}}{P - p_w^{\text{surface}}}$	At surface, $-\mathcal{D}(w)r^2\frac{\partial u}{\partial s}_{\text{surface}} = k_p \frac{p_w^{\text{bulk}} - p_w^{\text{surface}}}{P - p_w^{\text{surface}}}$
	At center, $\frac{\partial C_w}{\partial r}_{\text{center}} = 0$	At center, $\frac{\partial u}{\partial s}_{\text{bottom surface}} = 0$

The variable u is the dry-basis moisture content. The equations that include shrinkage are taken from Van der Lijn, doctoral thesis, Wageningen (1976).

The equations in Table 12-8 are insufficient on their own. Some algebraic relationships are needed to formulate a complete problem, as illustrated in Example 16. Equations for the mass- and heat-transfer coefficients are also needed for the boundary conditions presented in Table 12-8. These require the physical properties of the air, the object geometry, and Reynolds number. Example 16 shows the solution for a problem using numerical modeling. This example shows some of the important qualitative characteristics of drying.

Example 16: Air Drying of a Thin Layer of Paste Simulate the drying kinetics of 100 μm of paste initially containing 50 percent moisture (wet-basis) with dry air at 60°C, 0 percent relative humidity air at velocities of 1, 10, or 1000 m/s (limiting case) and at 60°C, 0 percent relative humidity air at 1 m/s. The diffusion coefficient of water in the material is constant at 1×10^{-10} m²/s. The length of the layer in the airflow direction is 2.54 cm.

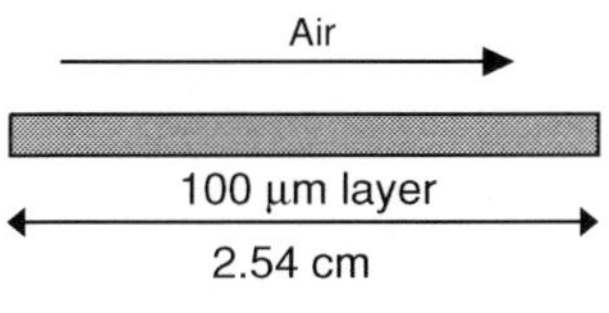

Physical property data: Sorption isotherm data fit well to the following equation:

$$w = 3.10\left(\frac{\%\text{RH}}{100}\right)^5 - 6.21\left(\frac{\%\text{RH}}{100}\right)^4 + 4.74\left(\frac{\%\text{RH}}{100}\right)^3 - 1.70\left(\frac{\%\text{RH}}{100}\right)^2 + 0.378\left(\frac{\%\text{RH}}{100}\right)$$

Solid density = 1150 kg/m³

Heat of vaporization = 2450 J/g

Solid heat capacity: 2.5 J/(g·K)

Water heat capacity: 4.184 J/(g·K)

Solution: The full numerical model needs to include shrinkage since the material is 50 percent water initially and the thickness will decrease from 100 to 46.5 μm during drying. Assuming the layer is viscous enough to resist convection in the liquid, diffusion is the dominant liquid-phase transport mechanism.

Table 12-8 gives the mass balance equation:

$$\frac{\partial u}{\partial t} = \frac{\partial}{\partial s}\left[\mathcal{D}(w)\frac{\partial u}{\partial s}\right] \qquad \frac{\partial s}{\partial z} = \rho_s$$

At top surface,

$$-\mathcal{D}(w)\frac{\partial u}{\partial s}_{\text{top surface}} = k_c \frac{p_w^{\text{bulk}} - p_w^{\text{surface}}}{P - p_w^{\text{surface}}}$$

At bottom surface,

$$\frac{\partial u}{\partial s}_{\text{bottom surface}} = 0$$

The temperature is assumed to be uniform through the thickness of the layer.

$$(1 + w_{\text{avg,dry-basis}}) \cdot m_{\text{solids}}\, C_p \cdot \frac{dT_{\text{layer}}}{dt} = [h(T_{\text{air}} - T_{\text{layer}}) - F \cdot \Delta H_{\text{vap}}]A$$

Mass- and heat-transfer coefficients are given by

$$\text{Nu} = \frac{hL}{k_{\text{air}}} = 0.664 \cdot \text{Re}^{0.5} \cdot \text{Pr}^{0.333}$$

$$\text{Sh} = \frac{k_c L}{\mathcal{D}_{\text{air/water}}} = 0.664 \cdot \text{Re}^{0.5} \cdot \text{Sc}^{0.333}$$

$$k_p = k_c \cdot \rho_{\text{air}}$$

The Reynolds number uses the length of the layer L in the airflow direction:

$$\text{Re} = \frac{VL\rho_{\text{air}}}{\mu_{\text{air}}}$$

where V = air velocity.

The Prandtl and Schmidt numbers, Pr and Sc, for air are given by

$$\mathrm{Pr} = \frac{C_{p,\mathrm{air}}\mu_{\mathrm{air}}}{k_{\mathrm{air}}} = 0.70 \qquad \mathrm{Sc} = \frac{\mu_{\mathrm{air}}}{\rho_{\mathrm{air}} D_{\mathrm{air/water}}} = 0.73$$

The following algebraic equations are also needed:

$$\frac{1}{\rho} = \frac{w}{\rho_w^o} + \frac{1-w}{\rho_s^o}$$ density of wet material (assumes volume additivity)

$$C_w = w \cdot \rho$$ concentration of water

$$\rho_s = (1-w)\rho$$ concentration of solids

$$\frac{\%\mathrm{RH}}{100} = \frac{P_{w,\mathrm{surface}}}{P_{w,\mathrm{sat}}}$$ definition of relative humidity

$$P_{w,\mathrm{sat}} = 0.01 \exp\left(16.262 - \frac{3800}{226.3 + T_{\mathrm{liquid/solid}}}\right)$$ Antoine equation for vapor pressure of water

Result: The results of the simulation are shown in Fig. 12-16. The top plot shows the average moisture content of the layer as a function of time, the middle plot shows the drying rate as a function of time, and the bottom plot shows the moisture gradient in each layer after 10 s of drying.

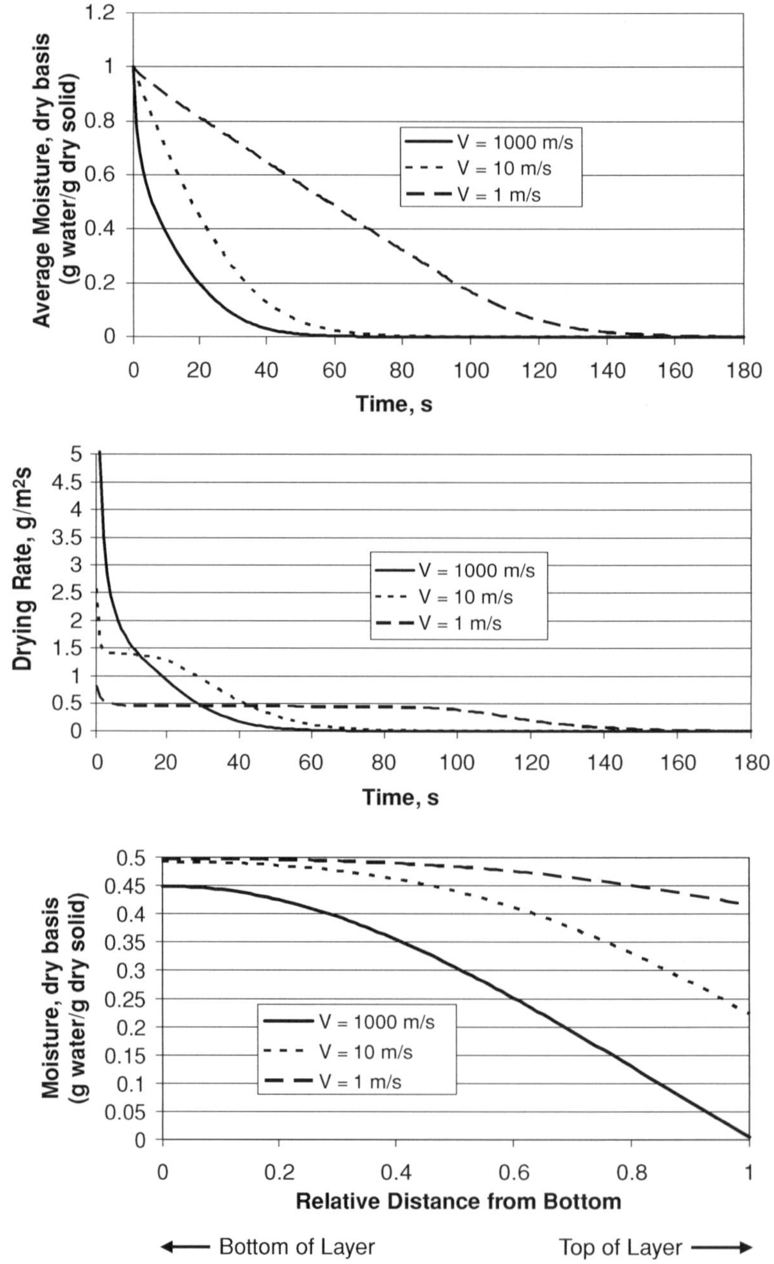

FIG. 12-16 Simulation results for thin layer drying example.

At a velocity of 1 m/s, drying occurs at a constant rate for nearly the entire process; at 10 m/s, drying begins at a high constant rate and then enters a falling-rate period; and at 1000 m/s (limiting case), there is no constant-rate period. These results illustrate the relationships between the external air conditions, drying rate, and moisture gradient. At high air velocity, the drying rate is faster, but becomes limited by internal diffusion and a steep moisture gradient forms. As the air velocity increases, the drying rate becomes less sensitive to air velocity.

The equation set in this example was solved by using a differential-algebraic equation solver called gPROMS from Process Systems Enterprises (www.pse.com). It can also be solved with other software and programming languages such as FORTRAN. Example 16 is too complicated to be done on a spreadsheet.

Simplified Kinetic Models This section presents several examples of simplified kinetic models. A model of the constant-rate period is shown in Example 17. During the constant-rate period, the drying rate is controlled by gas-phase mass and heat transfer. This is easier than modeling the falling-rate period, since the properties of air and water (or other gas-phase molecules) are well understood. Modeling the falling-rate period requires knowledge of and/or assumptions about the physical properties of the drying material.

Example 17: Drying a Pure Water Drop (Marshall, *Atomization & Spray Drying,* 1986.) Calculate the time to dry a drop of water, given the air temperature and relative humidity as a function of drop size.

Solution: Assume that the drop is drying at the wet-bulb temperature. Begin with an energy balance [Eq. (12-35)]

$$\text{Mass flux} = \frac{h(T_{\text{air}} - T_{\text{drop}})}{\Delta H_{\text{vap}}} \tag{12-35}$$

Next, a mass balance is performed on the drop. The change in mass equals the flux times the surface area.

$$\frac{\rho\, dV_{\text{droplet}}}{dt} = -A \cdot \text{mass flux} \tag{12-36}$$

Evaluating the area and volume for a sphere gives

$$\rho \cdot 4\pi R^2 \frac{dR}{dt} = -4\pi R^2 \cdot \text{mass flux} \tag{12-37}$$

Combining Eqs. (12-35) and (12-37) and simplifying gives

$$\rho \frac{dR}{dt} = \frac{-h(T_{\text{air}} - T_{\text{drop}})}{\Delta H_{\text{vap}}} \tag{12-38}$$

A standard correlation for heat transfer to a sphere is given by (Ranz and Marshall, 1952)

$$\text{Nu} = \frac{h(2R)}{k_{\text{air}}} = 2 + 0.6 \cdot \text{Re}^{0.5}\text{Pr}^{0.33} \tag{12-39}$$

For small drop sizes or for stagnant conditions, the Nusselt number has a limiting value of 2.

$$\text{Nu} = \frac{h(2R)}{k_{\text{air}}} = 2 \tag{12-40}$$

$$h = \frac{k_{\text{air}}}{R} \tag{12-41}$$

Inserting into Eq. (12-38) gives

$$R\frac{dR}{dt} = \frac{k_{\text{air}}(T_{\text{air}} - T_{\text{drop}})}{\rho\, \Delta H_{\text{vap}}} \tag{12-42}$$

Integration yields

$$\frac{R^2}{2} - \frac{R_0^2}{2} = \frac{k_{\text{air}}\,(T_{\text{air}} - T_{\text{drop}})t}{\rho\, \Delta H_{\text{vap}}} \tag{12-43}$$

where R_0 = initial drop radius, m.

Now the total lifetime of a drop can be calculated from Eq. (12-43) by setting $R = 0$:

$$t = \frac{\rho\, \Delta H_{\text{vap}} R_0^2}{2k_{\text{air}}(T_{\text{air}} - T_{\text{drop}})} \tag{12-44}$$

The effects of drop size and air temperature are readily apparent from Eq. (12-44). The temperature of the drop is the wet-bulb temperature and can be obtained from a psychrometric chart, as described in the previous section. Sample results are plotted in Fig. 12-17.

The above solution for drying of a pure water drop cannot be used to predict the drying rates of drops containing solids. Drops containing solids will not shrink uniformly and will develop internal concentration gradients (falling-rate period) in most cases.

Modeling of the falling-rate period is usually done by treating the drying problem as a diffusion problem, where the rate-limiting step is the diffusion of moisture from deep within the solid to the surface.

One of the attractions of treating drying as a diffusion problem is its relative simplicity compared with more complex models for moisture movement. This renders the approach tractable for hand calculations, and these calculations are often appropriate given the wide variability in diffusion coefficients and permeabilities both within and between

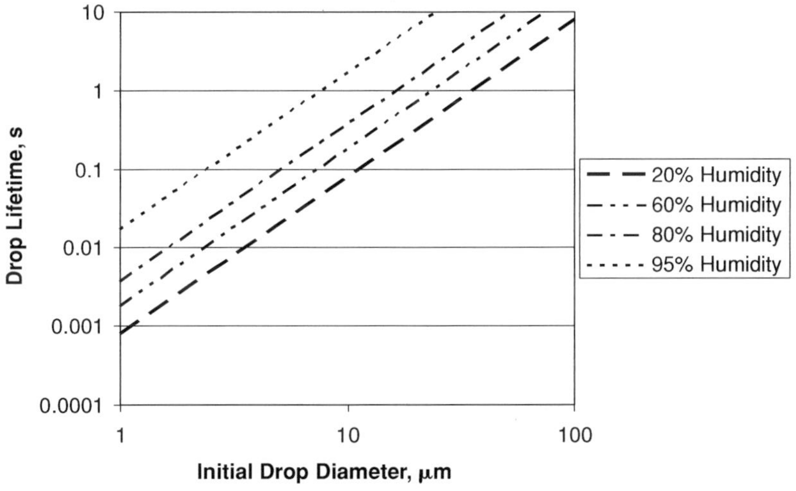

FIG. 12-17 Drying time of pure water drops as function of relative humidity at 25°C.

materials. The simplicity of this approach is also useful when one is optimizing processing conditions, where the number of calculations, even with modern workstations, is considerable. Moreover, this diffusion approach works well for predicting both average moisture contents and moisture-content profiles for some materials.

The three main driving forces which have been used within diffusion models (moisture content, partial pressure of water vapor, and chemical potential) will now be discussed. Attempts to predict diffusion coefficients theoretically will also be reviewed, together with experimental data for fitted diffusion coefficients and their dependence on temperature and moisture content.

Waananen et al. (1993), in their review of drying models, note that most models in their final form express the driving force for moisture movement in terms of a moisture concentration gradient. However, the true potential for transfer may be different, namely, differences in chemical potential, as explored in greater detail by Keey et al. (2000). In theory, the diffusion coefficient will be independent of moisture concentration only if the moisture is unbound, but concentration-independent diffusion coefficients have been successfully used in some cases over a wide range of moisture contents.

Since the true driving force is the chemical potential difference, transfer will occur between two moist bodies in the direction of falling chemical potential rather than decreasing moisture content. Moisture may flow from the drier body to the wetter one.

At low moisture contents, Perré and Turner (1996) suggest that there seems to be little difference between the predictions of drying models with driving forces based on gradients in chemical potential, moisture content, and partial pressure of water vapor, indicating that the simplest approach (a moisture content driving force) might be most practical. The majority of work involving the use of diffusion models has used moisture content driving forces. Hence, there is some empirical support for the use of moisture content driving forces.

In this model, described by Fick's second law, we have

$$\frac{\partial X}{\partial t} = \frac{\partial}{\partial z}\left(\mathcal{D}\frac{\partial X}{\partial z}\right) \tag{12-45}$$

where X is the free moisture content above the equilibrium moisture content, t is time, z is the distance coordinate perpendicular to the airstream, and $\mathcal{D}$ is the diffusion coefficient. Sherwood (1929) was the first to use this approach, and he made the following additional assumptions:

- The diffusion coefficient $\mathcal{D}$ is constant.
- The initial moisture content in the material is uniform.
- Surface material comes into equilibrium with the surrounding air instantaneously, so that the resistance of the boundary layer outside the material is negligible.

Translated to mathematical terms, the last two of these assumptions are

$$t = 0 \qquad X = X_i \tag{12-46}$$

$$z = \delta \qquad X = X_e \quad \text{(at the surface)} \tag{12-47}$$

where δ is the half thickness of the board. This approach allowed Eq. (12-45) to be integrated to yield a predicted moisture content profile. This moisture content profile may be integrated to give average moisture contents, with the characteristic moisture content $\overline{\Phi}$ being defined as before, $\overline{\Phi} = (\overline{X} - X_e)/(X_i - X_e)$, where $\overline{X}$ is the volume-averaged moisture content and X_i and X_e are the initial and equilibrium moisture contents, respectively. The equation for the characteristic moisture content is

$$\overline{\Phi} = \frac{8}{\pi^2}\sum_{n=0}^{\infty}\frac{1}{(2n+1)^2}\exp\left[-\left(\frac{2n+1}{2}\right)^2\pi^2\tau\right] \tag{12-48}$$

With this model, a characteristic parameter which governs the extent of drying is the mass-transfer Fourier number τ, defined as follows:

$$\tau = \frac{\mathcal{D}t}{\delta^2} \tag{12-49}$$

If drying is controlled by diffusion, then for the same drying conditions, doubling the thickness of the material should increase the drying time to the same final moisture content fourfold.

If the diffusion coefficient is constant, the moisture content profile through a material for the steady-state movement of moisture through it would be linear. However, drying is not a steady-state process. When the moisture content change occurs over almost the entire half thickness of the material, in other words when the size of the fully wet region is very small, the moisture content profiles can be shown to be parabolic during drying if the diffusion coefficient is constant.

The surface of the material does not necessarily come instantly to equilibrium. The surface of the material is only at equilibrium with the drying air during the falling-rate period. Although dry patches have been seen and photographed on the surface of moist granular beds as they dry out (Oliver and Clarke, 1973), fine porous material can have a significant fraction of its exposed surface dry before the evaporation from the whole surface is affected (Suzuki et al., 1972; Schlünder, 1988) due to the buffering effect of the external boundary layer.

Concept of a Characteristic Drying Rate Curve In 1958, van Meel observed that the drying rate curves, during the falling-rate period, for a specific material often show the same shape (Figs. 12-18 and 12-19), so that a single characteristic drying curve can be drawn for the material being dried. Strictly speaking, the concept should only

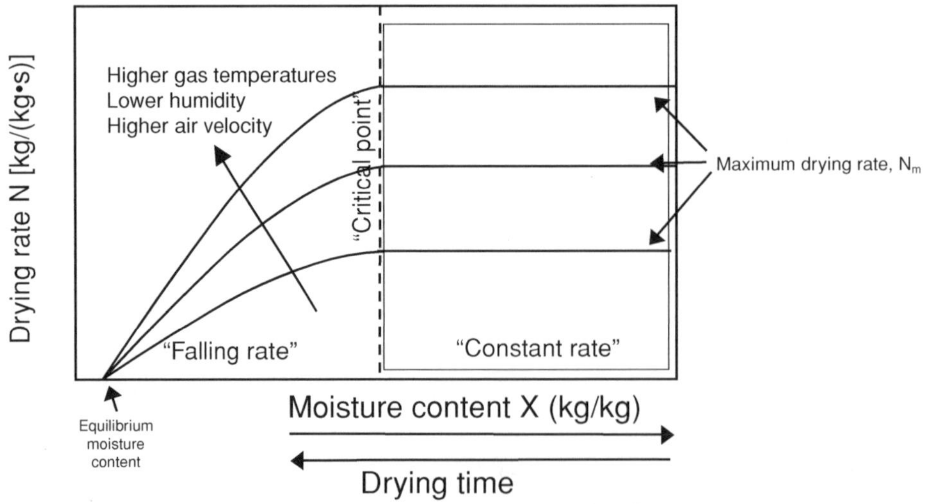

FIG. 12-18 Drying curves for a given material at different constant external conditions.

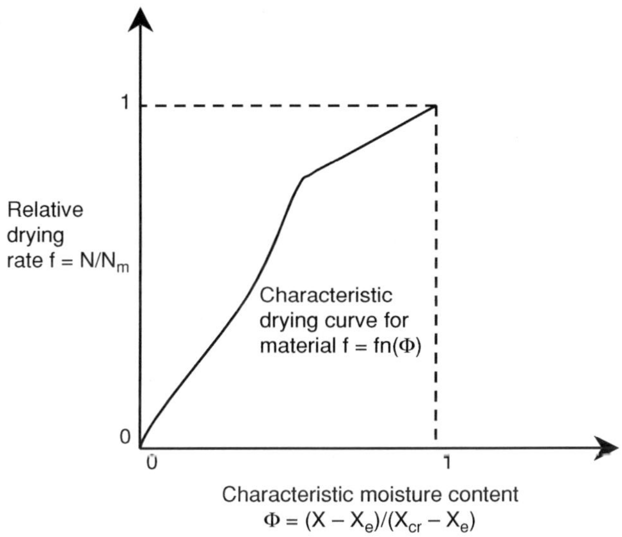

FIG. 12-19 Characteristic drying curve.

apply to materials of the same specific size (surface area to material ratio) and thickness, but Keey (1992) shows evidence that it applies over a somewhat wider range with reasonable accuracy. In the absence of experimental data, a linear falling-rate curve is often a reasonable first guess for the form of the characteristic function (good approximation for milk powder, fair for ion-exchange resin, silica gel). At each volume-averaged, free moisture content, it is assumed that there is a corresponding specific drying rate relative to the unhindered drying rate in the first drying period that is independent of the external drying conditions. *Volume-averaged* means averaging over the volume (distance cubed for a sphere) rather than just the distance. The relative drying rate is defined as

$$f = \frac{N}{N_m} \qquad (12\text{-}50)$$

where N is the drying rate, N_m is the rate in the constant-rate period, and the characteristic moisture content becomes

$$\Phi = \frac{\overline{X} - X_e}{X_{cr} - X_e} \qquad (12\text{-}51)$$

where $\overline{X}$ is the volume-averaged moisture content, X_{cr} is the moisture content at the critical point, and X_e is that at equilibrium. Thus, the drying curve is normalized to pass through the point (1,1) at the critical point of transition in drying behavior and the point (0,0) at equilibrium.

This representation leads to a simple lumped-parameter expression for the drying rate in the falling-rate period, namely,

$$N = fN_m = f[k\phi_m(Y_W - Y_G)] \qquad (12\text{-}52)$$

Here k is the external mass-transfer coefficient, ϕ_m is the humidity-potential coefficient (corrects for the humidity not being a strictly true representation of the driving force; close to unity most of the time), Y_W is the humidity above a fully wetted surface, and Y_G is the bulk-gas humidity. Equation (12-52) has been used extensively as the basis for understanding the behavior of industrial drying plants owing to its simplicity and the separation of the parameters that influence the drying process: the material itself f, the design of the dryer k, and the process conditions $\phi_m(Y_W - Y_G)f$.

For example, suppose (with nonhygroscopic solids, $X_e = 0$ kg/kg) that we have a linear falling-rate curve, with a maximum drying rate N_m of 0.5 kg moisture/(kg dry solids · s) from an initial moisture content of 1 kg moisture/kg dry solids. If the drying conditions around the sample are constant, what is the time required to dry the material to a moisture content of 0.2 kg moisture/kg dry solids?

$$N = \frac{X}{1\text{ kg kg}^{-1}}\,0.5\text{ kg kg}^{-1}\text{ s}^{-1}$$

Given that the drying rate dX/dt is equal to N, we have

$$t = \int_{0.2}^{1}\frac{dX}{N} = \int_{0.2}^{1}\frac{dX}{X(0.5)} = 2\int_{0.2}^{1}\frac{dX}{X} = 2\ln\frac{1}{0.2} = 3.21\text{ s} \qquad (12\text{-}53)$$

The characteristic drying curve, however, is clearly a gross approximation. A common drying curve will be found only if the volume-averaged moisture content reflects the moistness of the surface in some fixed way.

For example, in the drying of impermeable timbers, for which the surface moisture content reaches equilibrium quickly, there is unlikely to be any significant connection between the volume-averaged and the surface moisture contents, so the concept is unlikely to apply. While the concept might not be expected to apply to the same material with different thickness, e.g., Pang finds that it applies for different thicknesses in the drying of softwood timber (Keey, 1992), its applicability appears to be wider than the theory might suggest. A paper by Kemp and Oakley (2002) explains that many of the errors in the assumptions in this method often cancel out, meaning that the concept has wide applicability.

Keey and Suzuki (1974) have explored the conditions for which a characteristic curve might apply, using a simplified analysis based on an evaporative front receding through a porous mass. Their analysis shows that a unique curve pertains only when the material is thinly spread and the permeability to moisture is large. Internal diffusion often controls drying as the material becomes very dry, but the result of Keey and Suzuki suggests that the uniqueness of the curve, in theory, depends on drying not being significantly controlled by internal diffusion. One might expect, then, to find characteristic drying curves for small, microporous particles dried individually, and there is a sufficient body of data to suggest that a characteristic drying curve may be found to describe the drying of discrete particles below 20 mm in diameter over a range of conditions that normally exist within a commercial dryer. Nevertheless, Kemp and Oakley (1992) find that many of the deviations from the assumptions, in practice, cancel out, so that the limitation suggested by Keey and Suzuki (diffusion not controlling) is not as severe as might be expected.

An example of the application of a linear characteristic drying curve is given in the section on rotary dryers.

EXPERIMENTAL METHODS

Lab-, pilot-, and plant-scale experiments all play important roles in drying research. Lab-scale experiments are often necessary to study product characteristics and physical properties; pilot-scale experiments are often used in proof-of-concept process tests and to generate larger quantities of sample material; and plant-scale experiments are often needed to diagnose processing problems and to start or change a full-scale process.

Measurement of Drying Curves Measuring and using experimental drying curves can be difficult. Typically, this is a three-step process. The first step is to collect samples at different times of drying, the second step is to analyze each sample for moisture, and the third step is to interpret the data to make process decisions.

Solid sample collection techniques depend on the type of dryer. Since a drying curve is the moisture content as a function of time, it must be possible to obtain material before the drying process is complete. There are several important considerations when sampling material for a drying curve:

1. The sampling process needs to be fast relative to the drying process. Drying occurring during or after sampling can produce misleading results. Samples must be sealed prior to analysis. Plastic bags do not provide a sufficient seal.

2. In heterogeneous samples, the sample must be large enough to accurately represent the composition of the mixture.

Table 12-9 outlines some sampling techniques for various dryer types.

Moisture measurement techniques are critical to the successful collection and interpretation of drying data. The key message of this section is that the moisture value almost certainly depends on the measurement technique and that it is essential to have a *consistent*

TABLE 12-9 Sample Techniques for Various Dryer Types

Dryer type	Sampling method
Fluid bed dryer	Sampling cup (see Fig. 12-20)
Sheet dryer	Collect at end of dryer. Increase speed to change the drying time.
Tray dryer	Record initial moisture and mass of tray with time.
Indirect dryer	Decrease residence time with higher flow rate and sample at exit.
Spray dryer	Residence time of product is difficult to determine and change. Special probes have been developed to sample partially dried powder in different places within the dryer (ref. Langrish).

technique when measuring moisture. Table 12-10 compares and contrasts some different techniques for moisture measurement.

The most common method is gravimetric ("loss-on-drying"). A sample is weighed in a sample pan or tray and placed into an oven or heater at some high temperature for a given length of time. The sample is weighed again after drying. The difference in weight is then assumed to be due to the complete evaporation of water from the sample. The sample size, temperature, and drying time are all important factors. A very large or thick sample may not dry completely in the given time; a very small sample may not accurately represent the composition of a heterogeneous sample. A low temperature can fail to completely dry the sample, and a temperature that is too high can burn the sample, causing an artificially high loss of mass.

Usually, solid samples are collected as described, but in some experiments, it is more convenient to measure the change in humidity of the air due to drying. This technique requires a good mass balance of the system and is more common in lab-scale equipment than pilot- or plant-scale equipment.

Performing a Mass and Energy Balance on a Large Industrial Dryer Measuring a mass and energy balance on a large dryer is often necessary to understand how well the system is operating and how much additional capacity may be available. This exercise can also be used to detect and debug gross problems, such as leaks and product buildup.

There are several steps to this process.

1. Draw a sketch of the overall process including all the flows of mass into and out of the system. Look for places where air can leak into or out of the system. There is no substitute for physically walking around the equipment to get this information.
2. Decide on the envelope for the mass and energy balance. Some dryer systems have hot-air recycle loops and/or combustion or steam heating systems. It is not always necessary to include these to understand the dryer operation.
3. Decide on places to measure airflows and temperatures and to take feed and product samples. Drying systems and other process equipment are frequently not equipped for such measurements; the system may need minor modification, such as the installation of ports into pipes for pitot tubes or humidity probes. These ports must not leak when a probe is in place.
4. Take the appropriate measurements and calculate the mass and energy balances.

The measurements are inlet and outlet temperatures, humidities, and flow rates of the air inlets and outlets as well as the moisture and temperature of the feed and dry solids. The following are methods for each of the measurements:

Airflow Rate This is often the most difficult to measure. Fan curves are often available for blowers but are not always reliable. A small pitot tube can be used (see Sec. 22, "Waste Management," in this Handbook) to measure local velocity. The best location to use a pitot tube is in a straight section of pipe. Measurements at multiple positions in the cross section of the pipe or duct are advisable, particularly in laminar flow or near elbows and other flow disruptions.

Air Temperature A simple thermocouple can be used in most cases, but in some cases special care must be taken to ensure that wet or sticky material does not build up on the thermocouple. A wet thermocouple will yield a low temperature from evaporative cooling.

Air Humidity Humidity probes need to be calibrated before use, and the absolute humidity (or both the relative humidity and temperature) needs to be recorded. If the probe temperature is below the dew point of the air in the process, then condensation on the probe will occur until the probe heats.

Feed and Exit Solids Rate These are generally known, particularly for a unit in production. Liquids can be measured by using a bucket and stopwatch. Solids can be measured in a variety of ways.

Feed and Exit Solids Moisture Content These need to be measured using an appropriate technique, as described above. Use the same method for both the feed and exit solids. Don't rely on formula sheets for feed moisture information.

Figure 12-20 shows some common tools used in these measurements.

DRYING OF NONAQUEOUS SOLVENTS

Practical Considerations Removal of nonaqueous solvents from a material presents several practical challenges. First, solvents are often flammable and require drying either in an inert environment, such as superheated steam or nitrogen, or in a gas phase comprised solely of solvent vapor. The latter will occur in indirect or

TABLE 12-10 Moisture Determination Techniques

Method	Principle	Advantages	Disadvantages
Gravimetric (loss on drying)	Water evaporates when sample is held at a high temperature. Difference in mass is recorded.	Simple technique. No extensive calibration methods are needed. Lab equipment is commonly available.	Method is slow. Measurement time is several minutes to overnight (depending on material and accuracy). Generally not suitable for process control. Does not differentiate between water and other volatile substances.
IR/NIR	Absorption of infrared radiation by water is measured.	Fast method. Suitable for very thin layers or small particles.	Only surface moisture is detected. Extensive calibration is needed.
RF/microwave	Absorption of RF or microwave energy is measured.	Fast method. Suitable for large particles.	Extensive calibration is needed.
Equilibrium relative humidity (ERH)	The equilibrium relative humidity headspace above sample in a closed chamber is measured. Sorption isotherm is used to determine moisture.	Relatively quick method. Useful particularly if a final moisture specification is in terms of water activity (to retard microorganism growth).	May give misleading results since the surface of the material will equilibrate with the air. Large particles with moisture gradients can give falsely low readings. Measurement of relative humidity can be imprecise.
Karl Fischer titration	Chemical titration that is water-specific. Material can be either added directly to a solvent or heated in an oven, with the headspace purged and bubbled through solvent.	Specific to water only and very precise. Units can be purchased with an autosampler. Measurement takes only a few minutes.	Equipment is expensive and requires solvents. Minimal calibration required. Sample size is small, which may pose a problem for heterogeneous mixtures.

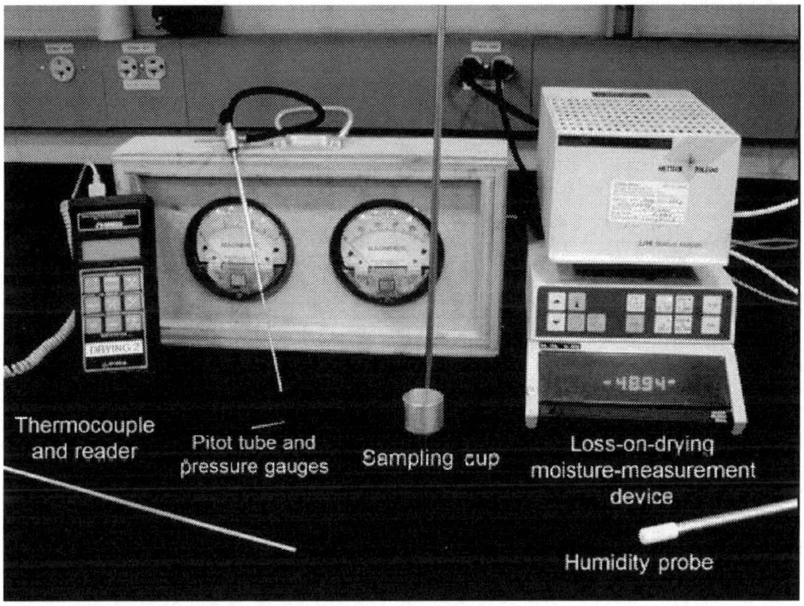

FIG. 12-20 Variety of tools used to measure mass and energy balances on dryers.

vacuum drying equipment. Second, the solvent vapor must be collected in an environmentally acceptable manner.

An additional practical consideration is the remaining solvent content that is acceptable in the final product. Failure to remove all the solvent can lead to problems such as toxicity of the final solid or can cause the headspace of packages, such as drums, to accumulate solvent vapor.

Physical Properties The physical properties that are important in solvent drying are the same as those for an aqueous system. The vapor pressure of a solvent is the most important property since it provides the thermodynamic driving force for drying. Acetone (BP 57°C), for example, can be removed from a solid at atmospheric pressure readily by boiling, but glycerol (BP 200°C) will dry only very slowly. Like water, a solvent may become bound to the solid and have a lower vapor pressure. This effect should be considered when one is designing a solvent-drying process.

Example 18: Preparation of a Psychrometric Chart Make a psychrometric chart for dipropylene glycol. It has a molecular weight of 134.2 g/mol and a normal boiling temperature of 228°C, and the latent heat of vaporization is 65.1 kJ/mol.

The Clausius-Clapeyron equation can be used to estimate the vapor pressure of dipropylene glycol as a function of temperature, with the boiling temperature as a reference.

$$\ln \frac{P_{T_1}^{\text{sat}}}{P_{T_2}^{\text{sat}}} = \frac{-\Delta H_{\text{vap}}}{R}\left(\frac{1}{T_1} - \frac{1}{T_2}\right)$$

where $P_{T_1}^{\text{sat}}, P_{T_2}^{\text{sat}}$ = vapor pressures of the solvent at absolute temperatures T_1 and T_2

ΔH_{vap} = latent heat of vaporization, J/mol

R = gas constant, 8.314 J/(mol·K)

Since the boiling temperature is 228°C, 501.15 K and 1 bar were used as T_2 and P_2. The latent heat value is 65.1 kJ/mol.

Once the vapor pressure of dipropylene glycol is known at a given temperature, the mass of dipropylene glycol/mass of dry air can be calculated. Since dipropylene glycol is the only liquid, the partial pressure of dipropylene glycol equals the vapor pressure.

$$P_{\text{dipropylene glycol}} = P_{\text{dipropylene glycol}}^{\text{Sat}}$$

The mole fraction of dipropylene glycol is the partial pressure divided by the total system pressure, taken to equal 1 bar.

$$y_{\text{dipropylene glycol}} = \frac{P_{\text{dipropylene glycol}}}{P}$$

The saturation mole ratio of dipropylene glycol to air is given by the following.

$$\text{Mole ratio} = \frac{y_{\text{dipropylene glycol}}}{1 - y_{\text{dipropylene glycol}}}$$

The saturation mass ratio of dipropylene glycol to air is calculated by multiplying by the molecular weights. The mass ratio as a function of temperature gives the saturation curve, as shown in Fig. 12-21.

$$\text{Saturation mass ratio} = \frac{\text{g dipropylene glycol}}{\text{g dry air}}$$

$$= \frac{\text{molecular weight of dipropylene glycol}}{\text{molecular weight of dry air}} \cdot \frac{y_{\text{dipropyleneglycol}}}{1 - y_{\text{dipropyleneglycol}}}$$

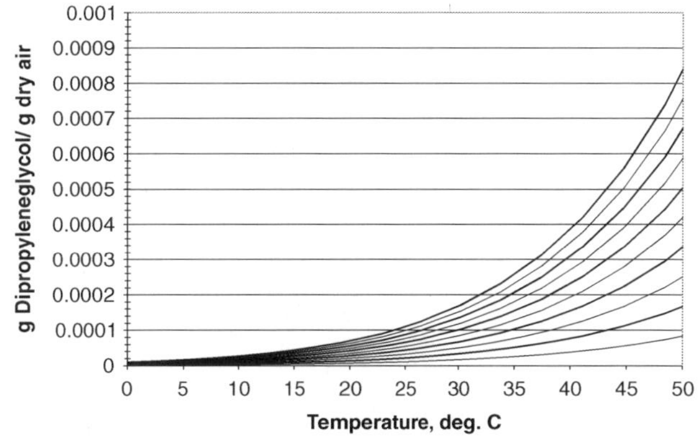

FIG. 12-21 An example of a solvent psychrometric chart.

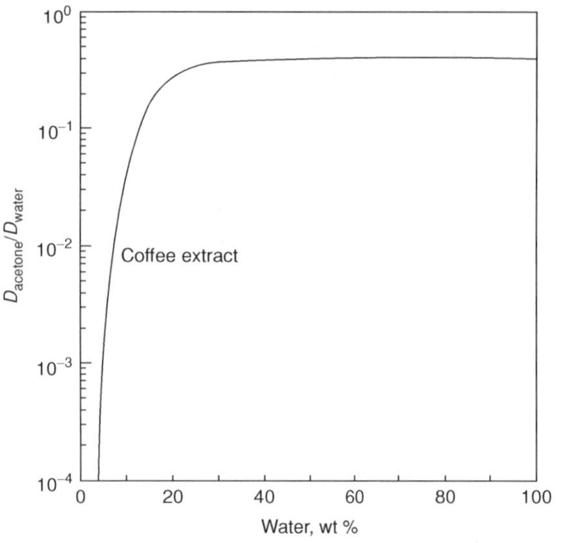

FIG. 12-22 The ratio of the diffusion coefficients of acetone to water in instant coffee as a function of moisture content (taken from Thijssen et al., *De Ingenieur*, JRG, 80, Nr. 47 (1968)]. Acetone has a much higher vapor pressure than water, but is selectively retained in coffee during drying.

Each relative humidity curve is proportional to the saturation value.

$$\text{Mass ratio} = \left(\frac{\%\text{RH}}{100}\right) \cdot \text{saturation mass ratio}$$

Diffusion of nonaqueous solvents through a material can be slow. The diffusion coefficient is directly related to the size of the diffusing molecule, so molecules larger than water typically have diffusion coefficients that have a much lower value. This phenomenon is known as *selective diffusion*. Large diffusing molecules can become kinetically trapped in the solid matrix. Solvents with a lower molecular weight will often evaporate from a material faster than a solvent with a higher molecular weight, even if the vapor pressure of the larger molecule is higher. Some encapsulation methods rely on selective diffusion; an example is instant coffee production using spray drying, where volatile flavor and aroma components are retained in particles more than water, even though they are more volatile than water, as shown in Fig. 12-22.

PRODUCT QUALITY CONSIDERATIONS

Overview The drying operation usually has a very strong influence on final product quality and product performance measures. And the final product quality strongly influences the value of the product. Generally, a specific particle or unit size, a specific density, a specific color, and a specific target moisture are desired. Naturally every product is somewhat different, but these are usually the first things we need to get right.

Target Moisture This seems obvious, but it's very important to determine the right moisture target before we address other drying basics. Does biological activity determine the target, flowability of the powder, shelf life, etc.? Sometimes a very small (1 to 2 percent) change in the target moisture will have a very big impact on the size of the dryer required. This is especially true for difficult-to-dry products with flat falling-rate drying characteristics. Therefore, spend the time necessary to get clear on what really determines the moisture target. And as noted earlier in this subsection, care should be taken to define a moisture measurement method since results are often sensitive to the method.

Particle Size Generally a customer or consumer wants a very specific particle size—and the narrower the distribution, the better. No one wants lumps or dust. The problem is that some attrition and sometimes agglomeration occur during the drying operation. We may start out with the right particle size, but we must be sure the dryer we've selected will not adversely affect particle size to the extent that it becomes a problem. And some dryers will treat particles more gently than others. Particle size is also important from a segregation standpoint. See Sec. 18, "Solid-Solid Operations and Equipment." And of course fine particles can also increase the risk of fire or explosion.

Density Customers and consumers are generally also very interested in getting the product density they have specified or expect. If the product is a consumer product and going into a box, then the density needs to be correct to fill the box to the appropriate level. If density is important, then product shrinkage during drying can be an important harmful transformation to consider. This is particularly important for biological products for which shrinkage can be very high. This is why freeze drying can be the preferred dryer for many of these materials.

Solubility Many dried products are rewet either during use by the consumer or by a customer during subsequent processing. Shrinkage can again be a very harmful transformation. Many times shrinkage is a virtually irreversible transformation which creates an unacceptable product morphology. Case hardening is a phenomenon that occurs when the outside of the particle or product initially shrinks to form a very hard and dense skin that does not easily rewet. A common cause is capillary collapse, discussed along with shrinkage below.

Flowability If we're considering particles, powders, and other products that are intended to flow, then this is a very important consideration. These materials need to easily flow from bins, hoppers, and out of boxes for consumer products. Powder flowability is a measureable characteristic using rotational shear cells (Peschl) or translational shear cells (Jenike) in which the powder is consolidated under various normal loads, and then the shear force is measured, enabling a complete yield locus curve to be constructed. This can be done at various powder moistures to create a curve of flowability versus moisture content. Some minimal value is necessary to ensure free flow. Additional information on these devices and this measure can be found in Sec. 21, "Solid-Solid Operations and Processing."

Color Product color is usually a very important product quality attribute, and a change in color can be caused by several different transformations.

Transformations Affecting Product Quality Drying, as with any other unit operation, has both productive and harmful transformations that occur. The primary productive transformation is water removal of course, but there are many harmful transformations that can occur and adversely affect product quality. The most common of these harmful transformations includes product shrinkage; attrition or agglomeration; loss of flavor, aroma, and nutritional value; browning reactions; discoloration; stickiness; and flowability problems, These were discussed briefly above, but are worth a more in-depth review.

Shrinkage Shrinkage is a particularly important transformation with several possible mechanisms to consider. It's usually especially problematic with food and other biological materials, but is a very broadly occurring phenomenon. Shrinkage generally affects solubility, wettability, texture and morphology, and absorbency. It can be observed when drying lumber when it induces stress cracking and during the drying of coffee beans prior to roasting. Tissue, towel, and other paper products undergo some shrinkage during drying. And many chemical products shrink as water evaporates, creating voids and capillaries prone to collapse as additional water evaporates. As we consider capillary collapse, there are several mechanisms worth mentioning.

Surface tension—the capillary suction created by a receding liquid meniscus can be extremely high.

Plasticization—an evaporating solvent which is also a plasticizer of polymer solute product will lead to greater levels of collapse and shrinkage.

Electric charge effects—the van der Waals and electrostatic forces can also be a strong driver of collapse and shrinkage.

Surface Tension These effects are very common and worth a few more comments. Capillary suction created by a receding liquid meniscus can create very high pressures for collapse. The quantitative expression for the pressure differential across a liquid-fluid interface was first derived by Laplace in 1806. The meniscus, which reflects the differential, is affected by the surface tension of the fluid. Higher surface tensions create greater forces for collapse. These strong capillary suction pressures can easily collapse a pore. We can reduce these suction pressures by using low-surface-tension fluids or by adding surfactants, in the case of water, which will also significantly reduce surface tension (from 72 to 30 dyn/cm).

The collapse can also be reduced with some dryer types. Freeze drying and heat pump drying can substantially reduce collapse, but of course, the capital cost of these dryers sometimes makes them prohibitive. At the other extreme, dryers which rapidly flash off the moisture can reduce collapse. This mechanism can also be affected by particle size such that the drying is primarily boundary-layer-controlled. When the particle size becomes sufficiently small, moisture can diffuse to the surface at a rate sufficient to keep the surface wetted. This has been observed in a gel-forming food material when the particle size reached 150 to 200 μm (Genskow, "Considerations in Drying Consumer Products," *Proceedings International Drying Symposium,* Versailles, France, 1988).

Biochemical Degradation Biochemical degradation is another harmful transformation that occurs with most biological products. There are four key reactions to consider: lipid oxidation, Maillard browning, protein denaturation, and various enzyme reactions. These reactions are both heat- and moisture-dependent such that control of temperature and moisture profiles can be very important during drying.

Lipid oxidation. Lipid oxidation is normally observed as a product discoloration and can be exacerbated with excess levels of bleach. It is catalyzed by metal ions, enzymes, and pigments. Acidic compounds can be used to complex the metal ions. Synthetic antioxidants, such as butylated hydroxtoluene (BHT) and butylated hydroxyanisole (BHA) can be added to the product, but are limited and coming under increased scrutiny due to toxicology concerns. It may be preferable to use natural antioxidants such as lecithin or vitamin E or to dry under vacuum or in an inert (nitrogen, steam) atmosphere.

Protein denaturation. Protein denaturation is normally observed as an increase in viscosity and a decrease in wettability. It is temperature-sensitive, generally occurring between 40 and 80°C. A common drying process scheme is to dry thermally and under wet-bulb drying conditions without overheating and then vacuum, heat-pump, or freeze-dry to the target moisture.

Enzyme reactions. Enzymatic browning is caused by the enzyme polyphenal oxidase which causes phenals to oxidize to orthoquinones. The enzyme is active between pH 5 to 7. A viable process scheme again is to dry under vacuum or in an inert (nitrogen, steam) atmosphere.

Maillard browning reaction. This nonenzymatic reaction is observed as a product discoloration, which in some products creates an attractive coloration. The reaction is temperature-sensitive, and normally the rate passes through a maximum and then falls as the product becomes drier. The reaction can be minimized by minimizing the drying temperature, reducing the pH to acidic, or adding an inhibitor such as sulfur dioxide or metabisulfate. A viable process scheme again is to dry thermally and under wet-bulb drying conditions without overheating and then vacuum, heat-pump, or freeze-dry to the target moisture.

Some of the above reactions can be minimized by reducing the particle size and using a monodisperse particle size distribution. The small particle size will better enable wet-bulb drying, and the monodisperse size will reduce overheating of the smallest particles.

Stickiness, Lumping, and Caking These are not characteristics we generally want in our products. They generally connote poor product quality, but can be a desirable transformation if we are trying to enlarge particle size through agglomeration. Stickiness, lumping, and caking are phenomena which are dependent on product moisture and product temperature. The most general description of this phenomenon can be described by measuring the cohesion (particle to particle) of powders as described below. A related measure is adhesion—particle-to-wall interactions. Finally, sticky point is a special case for materials which undergo glass transitions.

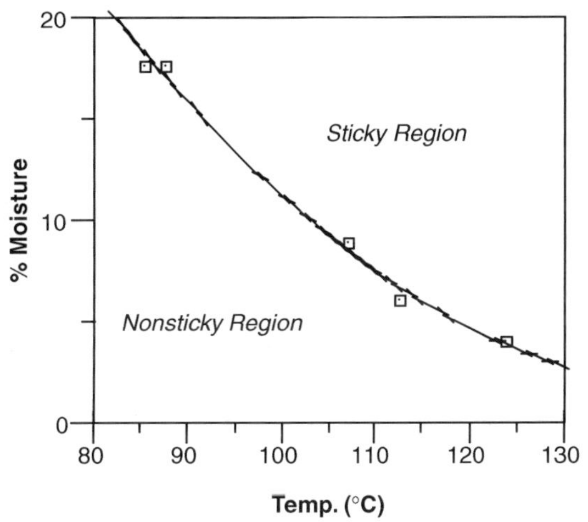

FIG. 12-23 Detergent stickiness curve.

The sticky point can be determined by using a method developed by Lazar and later by Downton [Downton, Flores-Luna, and King, "Mechanism of Stickiness in Hygroscopic, Amorphous Powders," *I&EC Fundamentals* **21:** 447 (1982)]. In the simplest method, a sample of the product, at a specific moisture, is placed in a closed tube which is suspended in a water bath. A small stirrer is used to monitor the torque needed to "stir" the product. The water bath temperature is slowly increased until the torque increases. This torque increase indicates a sticky point temperature for that specific moisture. The test is repeated with other product moistures until the entire stickiness curve is determined. A typical curve is shown in Fig. 12-23.

As noted, a sticky point mechanism is a glass transition—the transition when a material changes from the glassy state to the rubbery liquid state. Glass transitions are well documented in food science (Levine and Slade). Roos and Karel [Roos and Karel, "Plasticizing Effect of Water on Thermal Behavior and Crystallization of Amorphous Food Models," *J. Food Sci.* **56**(1): 38–43 (1991)] have demonstrated that for these types of products, the glass transition temperature follows the sticky point curve within about 2°C. This makes it straightforward to measure the stickiness curve by using a differential scanning calorimeter (DSC). Somewhat surprisingly, even materials which are not undergoing glass transitions exhibit this behavior, as demonstrated with the detergent stickiness curve above.

Lumping and caking can be measured by using the rotational shear cells (Peschl) or translational shear cells (Jenike) noted above for measuring flowability. The powder is consolidated under various normal loads, and then the shear force is measured, enabling a complete yield locus curve to be constructed. This can be done at various powder moistures to create a curve of "cake strength" versus moisture content. Slurries and dry solids are free-flowing, and there is a cohesion/adhesion peak at an intermediate moisture content, typically when voids between particles are largely full of liquid. A variety of other test methods for handling properties and flowability are available.

Product quality was addressed quite comprehensively by Evangelos Tsotsas at the 2d Nordic Drying Conference [Tsotsos, "Product Quality in Drying—Luck, Trial, Experience, or Science?" 2d Nordic Drying Conference, Copenhagen, Denmark, 2003]. Tsotsos notes that 31 percent of the papers at the 12th International Drying Symposium refer to product quality. The top 5 were color (12 percent), absence of chemical degradation (10 percent), absence of mechanical damage (9 percent), bulk density (8 percent), and mechanical properties (7 percent). These are all properties that are reasonably straightforward to measure. They are physical properties, and we are familiar with them for the most part. However, down the list at a rank of 20 with only 2 percent of the papers dealing with it, we have sensory properties.

This is the dilemma—sensory properties should rank very high, but they don't because we lack the tools to measure them effectively. For the most part, these quality measures are subjective rather than objective, and frequently they require direct testing with consumers to determine efficacy of a particular product attribute. So the issue is really a lack of physical measurement tools that directly assess the performance measures important to the consumer of the product. The lack of objective performance measures and unknown mechanistic equations also makes mathematical modeling very difficult for addressing quality problems.

The good news is that there has been a shift from the macro to the meso and now to the microscale in drying science. We have some very powerful analytical tools to help us understand the transformations that are occurring at the meso and microscale.

ADDITIONAL READING

Keey, *Drying of Loose and Particulate Materials.* Hemisphere, New York, 1992.

Keey, Langrish, and Walker, *Kiln Drying of Lumber,* Springer-Verlag, Heidelberg, 2000.

Keey and Suzuki, "On the Characteristic Drying Curve," *Int. J. Heat Mass Transfer* **17:**1455–1464 (1974).

Kemp and Oakley, "Modeling of Particulate Drying in Theory and Practice," *Drying Technol.* **20**(9):1699–1750 (2002).

Kock et al., "Design, Numerical Simulation and Experimental Testing of a Modified Probe for Measuring Temperatures and Humidities in Two-Phase Flow," *Chem. Eng. J.* **76**(1):49–60 (2000).

Liou and Bruin, "An Approximate Method for the Nonlinear Diffusion Problem with a Power Relation between the Diffusion Coefficient and Concentration. 1. Computation of Desorption Times," *Int. J. Heat Mass Transfer* **25:**1209–1220 (1982a).

Liou and Bruin, "An Approximate Method for the Nonlinear Diffusion Problem with a Power Relation between the Diffusion Coefficient and Concentration. 2. Computation of the Concentration Profile," *Int. J. Heat Mass Transfer* **25:** 1221–1229 (1982b).

Marshall, "Atomization and Spray Drying," *AICHE Symposium Series*, No. 2, p. 89 (1986).

Oliver and Clarke, "Some Experiments in Packed-Bed Drying," *Proc. Inst. Mech. Engrs.* **187:**515–521 (1973).

Perré and Turner, "The Use of Macroscopic Equations to Simulate Heat and Mass Transfer in Porous Media," in Turner and Mujumdar (eds.), *Mathematical Modeling and Numerical Techniques in Drying Technology*, Marcel Dekker, New York, 1996, pp. 83–156.

Ranz and Marshall, "Evaporation from Drops," *Chem. Eng. Prog.* **48**(3):141–146 and **48**(4):173–180 (1952).

Schoeber and Thijssen, "A Short-cut Method for the Calculation of Drying Rates for Slabs with Concentration-Dependent Diffusion Coefficient," *AIChE. Symposium Series*, **73**(163):12–24 (1975).

Schlünder, "On the Mechanism of the Constant Drying Rate Period and Its Relevance to Diffusion Controlled Catalytic Gas Phase Reactions," *Chem. Eng. Sci.* **43:**2685–2688 (1988).

Sherwood, "The Drying of Solids," *Ind. and Eng. Chem.* **21**(1):12–16 (1929).

Suzuki et al., "Mass Transfer from a Discontinuous Source," *Proc. PACHEC '72*, Kyoto, Japan, **3:**267–276 (1972).

Thijssen et al., *De Ingenieur, JRG*, **80**(47) (1968).

Thijssen and Coumans, "Short-cut Calculation of Non-isothermal Drying Rates of Shrinking and Non-shrinking Particles Containing an Expanding Gas Phase," *Proc. 4th Int. Drying Symp., IDS '84*, Kyoto, Japan, **1:**22–30 (1984).

Van der Lijn, doctoral thesis, Wageningen, 1976.

van Meel, "Adiabatic Convection Batch Drying with Recirculation of Air," *Chem. Eng. Sci.* **9:**36–44 (1958).

Viollez and Suarez, "Drying of Shrinking Bodies," *AIChE J.* **31:**1566–1568 (1985).

Waananan, Litchfield, and Okos, "Classification of Drying Models for Porous Solids," *Drying Technol.* **11**(1):1–40 (1993).

SOLIDS-DRYING EQUIPMENT—GENERAL ASPECTS

GENERAL REFERENCES: *Aspen Process Manual (Internet knowledge base)*, Aspen Technology, 2000 onward. Cook and DuMont, *Process Drying Practice*, McGraw-Hill, New York, 1991. *Drying Technology—An International Journal*, Taylor and Francis, New York, 1982 onward. Hall, *Dictionary of Drying*, Marcel Dekker, New York, 1979. Keey, *Introduction to Industrial Drying Operations*, Pergamon, New York, 1978. Mujumdar (ed.), *Handbook of Industrial Drying*, Marcel Dekker, New York, 1995. van't Land, *Industrial Drying Equipment*, Marcel Dekker, New York, 1991.

CLASSIFICATION OF DRYERS

Drying equipment may be classified in several ways. Effective classification is vital in selection of the most appropriate dryer for the task and in understanding the key principles on which it operates. The main categories are as follows:

1. Form of feed and product—particulate (solid or liquid feed), sheet, slab
2. Mode of operation—batch or continuous
3. Mode of heat transfer—convective (direct), conductive (indirect), radiative, or dielectric
4. Condition of solids—static bed, moving bed, fluidized or dispersed
5. Gas-solids contacting—parallel flow, perpendicular flow, or through-circulation
6. Gas flow pattern—cross-flow, cocurrent, or countercurrent

Other important features of the drying system are the type of carrier gas (air, inert gas, or superheated steam/solvent), use of gas or solids recycle, type of heating (indirect or direct-fired), and operating pressure (atmospheric or vacuum). However, these are primarily related to the choice of the overall system and operating conditions, not to the individual dryer used, and are discussed briefly at the end of this section. The relative importance of the different categories depends on the purpose of the classification. For distinguishing differences in dryer design, construction, and operation, categories 2 and 3 are particularly useful. A classification chart of drying equipment on this basis is shown in Table 12-11, and the grouping in "Solids-Drying Equipment—Specific Types" follows this pattern. Simplified diagrams for batch and continuous dryers are shown in Figs. 12-24 and 12-25, respectively. However, in the selection of a group of dryers for preliminary consideration in a given drying problem, the most important factor is often category 1, the form, handling characteristics, and physical properties of the wet material. (See Table 12-12.)

In Table 12-11, dryers in round brackets are semicontinuous forms of batch dryers, not commonly used. Dryers in square brackets are semibatch forms of continuous dryers, also fairly rare.

The feed type is a very basic description; *particulate* can also include powders, granules, pastes, pellets, performs, etc.; *liquid/slurry* also includes solutions and sludges. Table 12-12 gives a more comprehensive classification based on particle size and handling properties.

Description of Dryer Classification Criteria

1. Form of Feed and Product Dryers are specifically designed for particular feed and product forms; dryers handling films, sheets, slabs, and bulky artifacts form a clear subset. Most dryers are for particulate products, but the feed may range from a solution or slurry (free-flowing liquid) through a sticky paste to wet filter cakes, powders, or granules (again relatively free-flowing). The ability to successfully mechanically handle the feed and product is a key factor in dryer selection (see Table 12-12).

The drying kinetics (rate of drying, and hence required drying time) also depend strongly on solids properties, particularly particle size and porosity. The surface area/mass ratio and the internal pore structure control the extent to which an operation is diffusion-limited, i.e., diffusion into and out of the pores of a given solids particle, not through the voids among separate particles.

2. Mode of Operation Batch dryers are typically used for low throughputs (averaging under 50 kg/h), long drying times, or where the overall process is predominantly batch. Continuous dryers dominate for high throughputs (over 1 ton/h), high evaporation rates, and where the rest of the process is continuous. Often, there are batch and continuous dryers working on similar principles, but one batch dryer has two or more continuous equivalents, using different methods to move the solids through the dryer. For example, batch tray dryers (nonagitated solids) are equivalent to turbo-tray and plate dryers (vertical gravity transport) and to band dryers (horizontal mechanical transport). Also, dryers which are inherently continuous can be operated in semibatch mode (e.g., small-scale spray dryers) and vice versa.

3. Mode of Heat Transfer

Direct (convective) dryers The general operating characteristics of direct dryers are these:

TABLE 12-11 Classification of Drying Equipment

Dryer group	Feed type	Dryer type	Heating mode	Synonyms and variants
Batch tray	Particulate	Cross-circulated tray	Cross-circulation	Atmospheric tray
Nonagitated		Perforated tray	Through-circulation	Through-circulation, drying room
		Contact/vacuum tray	Conduction	Vacuum oven, vacuum shelf
Batch agitated	Particulate	Vertical pan	Conduction	Vertical agitated
Mechanical agitation		Conical	Conduction	Sidescrew, Nauta
		Spherical	Conduction	Turbosphere
		Horizontal pan	Conduction	Batch paddle, ploughshare
Continuous tray	Particulate	Turbo-tray	Cross-circulation	Rotating tray/shelf, Wyssmont Turbo-dryer
Nonagitated		Plate	Conduction	Krauss-Maffei
		Cascade	Through-circulation	Wenger
		Moving bed	Through-circulation	Tower, silo, gravity
Continuous band/tunnel	Particulate	Tunnel	Cross-circulation	Moving truck/trolleys
Nonagitated		Perforated band	Through-circulation	Atmospheric band/belt, vibrated bed
		Contact/vacuum band	Conduction	Vacuum belt, vibrated tray
Continuous agitated	Particulate	Paddle, low-speed	Conduction	Horizontal agitated, Disc, Porcupine, Nara
Mechanical agitation		Paddle, high-speed	Conduction	Solidaire
		High-speed convective paddle	Through-circulation	Rapid, Forberg
Continuous rotary	Particulate	Indirect rotary	Conduction	Steam-tube, Louisville
Rotational agitation		Rotary louvre	Through-circulation	Rotolouvre
		Cascading rotary	Dispersion	Direct rotary, rotary drum
Continuous dispersion	Particulate	Fluidized bed	Dispersion	Well-mixed/plug-flow fluid bed
Airborne transport		Vibrofluidized bed	Dispersion	Vibrated fluid bed
		Pneumatic conveying	Dispersion	Flash, ring, swept mill
		Spin-flash	Dispersion	Swirl fluidizer
		Spouted bed	Dispersion	Circulating fluid bed
Continuous special	Particulate	(Freeze)	Conduction	Continuous freeze
		Radiofrequency/microwave	Radiation	Dielectric
Continuous liquid feed	Liquid/slurry	Spray	Dispersion	Atomizing
		Spray/fluidized bed	Dispersion	Spray/belt
		Fluid bed granulator	Dispersion	Recirculating inert balls
		Thin-film	Conduction	Evaporator-dryer, wiped-film, LUWA
		Drum	Conduction	Film-drum
		(Filter-dryer)	Conduction	Nutsche, Rosenmund
		Centrifuge-dryer	Through-circulation	Henkel
Continuous sheet/film	Film/sheet	Cylinder	Conduction	Paper machine, roller
		Yankee	Conduction	Impingement
		Rotary through	Through-circulation	
		Stenter	Through-circulation	Tenter, range (textiles)
		Flotation	Through-circulation	Coanda, floating web
		Continuous oven	Conduction	Festoon, Spooner oven
		Infrared	Radiation	Curing

a. Direct contacting of hot gases with the solids is employed for solids heating and vapor removal.

b. Drying temperatures may range up to 1000 K, the limiting temperature for most common structural metals. At higher temperatures, radiation becomes an important heat-transfer mechanism.

c. At gas temperatures below the boiling point, the vapor content of gas influences the rate of drying and the final moisture content of the solid. With gas temperatures above the boiling point throughout, the vapor content of the gas has only a slight retarding effect on the drying rate and final moisture content. Thus, superheated vapors of the liquid being removed (e.g., steam) can be used for drying.

d. For low-temperature drying, dehumidification of the drying air may be required when atmospheric humidities are excessively high.

e. The lower the final moisture content, the more fuel per pound of water evaporated, that a direct dryer consumes.

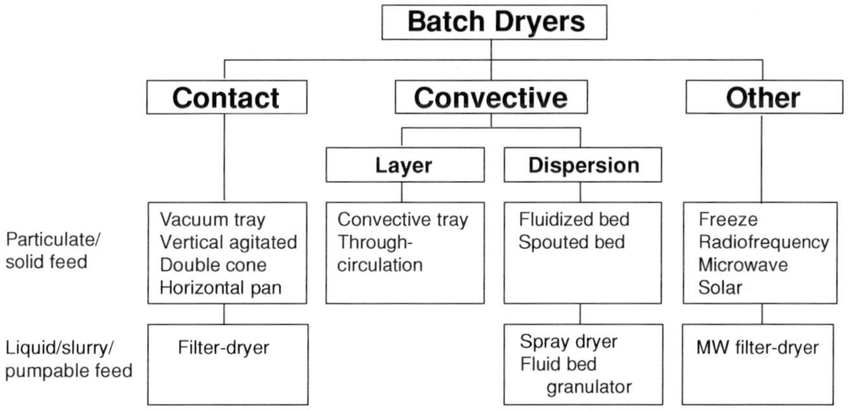

FIG. 12-24 Classification of batch dryers.

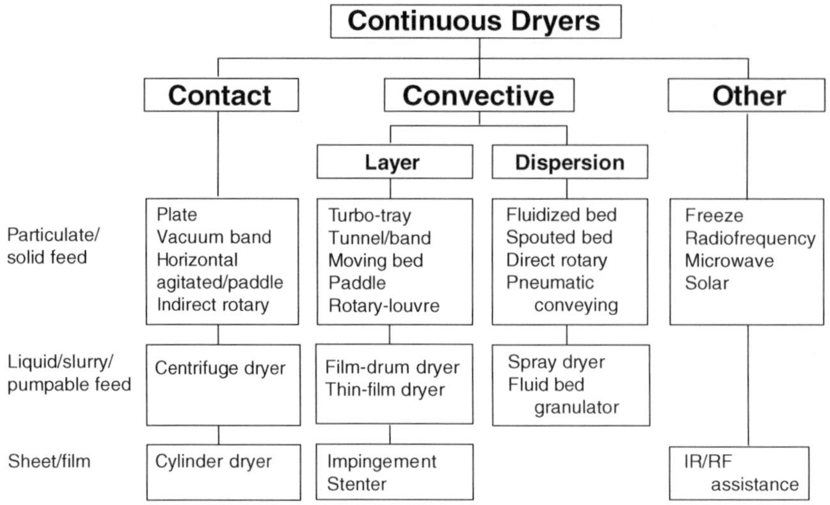

FIG. 12-25 Classification of continuous dryers.

f. Efficiency increases with an increase in the inlet gas temperature for a constant exhaust temperature.

g. Because large amounts of gas are required to supply all the heat for drying, dust recovery equipment may be very large and expensive, especially when drying very small particles.

Indirect (contact or conductive) dryers These differ from direct dryers with respect to heat transfer and vapor removal:

a. Heat is transferred to the wet material by conduction through a solid retaining wall, usually metallic.

b. Surface temperatures may range from below freezing in the case of freeze dryers to above 800 K in the case of indirect dryers heated by combustion products.

c. Indirect dryers are suited to drying under reduced pressures and inert atmospheres, to permit the recovery of solvents and to prevent the occurrence of explosive mixtures or the oxidation of easily decomposed materials.

d. Indirect dryers using condensing fluids as the heating medium are generally economical from the standpoint of heat consumption, since they furnish heat only in accordance with the demand made by the material being dried.

e. Dust recovery and dusty or hazardous materials can be handled more satisfactorily in indirect dryers than in direct dryers.

Miscellaneous dryers

a. **Infrared dryers** depend on the transfer of radiant energy to evaporate moisture. The radiant energy is supplied electrically by infrared lamps, by electric resistance elements, or by incandescent refractories heated by gas. The last method has the added advantage of convection heating. Infrared heating is not widely used in the chemical industries for the removal of moisture. Its principal use is in baking or drying paint films (curing) and in heating thin layers of materials. It is sometimes used to give supplementary heating on the initial rolls of paper machines (cylinder dryers).

b. **Dielectric dryers (radio-frequency or microwave)** have not as yet found a wide field of application, but are increasingly used. Their fundamental characteristic of generating heat within the solid indicates potentialities for drying massive geometric objects such as wood, sponge-rubber shapes, and ceramics, and for evening out moisture gradients in layers of solids. Power costs are generally much higher than the fuel costs of conventional methods; a small amount of dielectric heating (2 to 5 percent) may be combined with thermal heating to maximize the benefit at minimum operating cost. The high capital costs of these dryers must be balanced against product and process improvements.

4. Condition of Solids In solids-gas contacting equipment, the solids bed can exist in any of the following four conditions.

Static This is a dense bed of solids in which each particle rests upon another at essentially the settled bulk density of the solids phase. Specifically, *there is no relative motion among solids particles* (Fig. 12-26).

Moving This is a slightly expanded bed of solids in which the particles are separated only enough to flow one over another. Usually the flow is downward under the force of gravity (Fig. 12-27*a*), but upward motion by mechanical lifting or agitation may also occur within the process vessel (Fig.12-27*b*). In some cases, lifting of the solids is accomplished in separate equipment, and solids flow in the presence of the gas phase is downward only. The latter is a moving bed as usually defined in the petroleum industry. In this definition, *solids motion is achieved by either mechanical agitation or gravity force.*

Fluidized This is an expanded condition in which the solids particles are supported by drag forces caused by the gas phase passing through the interstices among the particles at some critical velocity. The superficial gas velocity upward is less than the terminal setting velocity of the solids particles; the gas velocity is not sufficient to entrain and convey continuously all the solids. Specifically, the solids phase and the gas phase are intermixed and *together behave as a boiling fluid* (Fig. 12-28). The gas forms the continuous phase, but the bulk density is not much lower than a continuous packed bed of solids.

Dispersed or dilute. This is a fully expanded condition in which the solids particles are so widely separated that they exert essentially no influence upon one another. Specifically, the solids phase is so fully dispersed in the gas that the *density of the suspension is essentially that of the gas phase alone* (Fig. 12-29). Commonly, this situation exists when the gas velocity at all points in the system exceeds the terminal settling velocity of the solids and the particles can be lifted and continuously conveyed by the gas; however, this is not always true. Cascading rotary dryers, countercurrent-flow spray dryers, and gravity settling chambers such as prilling towers are three exceptions in which gas velocity is insufficient to entrain the solids completely.

Cascading (direct) rotary dryers with lifters illustrate all four types of flow in a single dryer. Particles sitting in the lifters (flights) are a static bed. When they are in the rolling bed at the bottom of the dryer, or rolling off the top of the lifters, they form a moving bed. They form a falling curtain which is initially dense (fluidized) but then spreads out and becomes dispersed.

Dryers where the solid forms the continuous phase (static and moving beds) are called **layer dryers**, while those where the gas forms the continuous phase (fluidized and dispersed solids) are classified as **dispersion dryers**. Gas-particle heat and mass transfer is much faster in dispersion dryers, and these are therefore often favored where high drying rates, short drying times, or high solids throughput is required.

TABLE 12-12 Classification of Commercial Dryers Based on Feed Materials Handled

Type of dryer	Liquids	Slurries	Pastes and sludges	Free-flowing powders	Granular, crystalline, or fibrous solids	Large solids, special forms and shapes	Continuous sheets	Discontinuous sheets
	True and colloidal solutions; emulsions. Examples: inorganic salt solutions, extracts, milk, blood, waste liquors, rubber latex	Pumpable suspensions. Examples: pigment slurries, soap and detergents, calcium carbonate, bentonite, clay slip, lead concentrates	Examples: filter-press cakes, sedimentation sludges, centrifuged solids, starch	100-mesh (150 μm) or less. Relatively free-flowing in wet state. Dusty when dry. Examples: centrifuged precipitates	Larger than 100-mesh (150 μm). Examples: rayon staple, salt crystals, sand, ores, potato strips, synthetic rubber	Examples: pottery, brick, rayon cakes, shotgun shells, hats, painted objects, rayon skeins, lumber	Examples: paper, impregnated fabrics, cloth, cellophane, plastic sheets	Examples: veneer, wallboard, photograph prints, leather, foam rubber sheets
Vacuum freeze. Indirect type, batch or continuous operation	Expensive. Usually used only for high-value products such as pharmaceuticals; products, which are heat-sensitive and readily oxidized.	See comments under Liquids.	See comments under Liquids.	See comments under Liquids.	Expensive. Usually used on pharmaceuticals and related products which cannot be dried successfully by other means. Applicable to fine chemicals	See comments under Granular solids.	Applicable in special cases such as emulsion-coated films	See comments under Granular solids.
Vacuum tray/shelf. Indirect type, batch operation	Not applicable	Relatively expensive. Applicable for small-batch production	Relatively expensive. Suitable for batch operation, small capacities. Useful for heat-sensitive or readily oxidizable materials. Solvents can be recovered.	See comments under Pastes and Sludges.	Suitable for batch operation, small capacities. Useful for heat-sensitive or readily oxidizable materials. Solvents can be recovered.	See comments under Granular solids.	Not applicable	See comments under Granular solids.
Pan. Indirect type, batch operation including vertical agitated pan, spherical, conical, filter-dryer, double-cone tumbler	Atmospheric or vacuum. Suitable for small batches. Easily cleaned. Solvents can be recovered. Material agitated while dried.	See comments under Liquids.	See comments under Liquids.	See comments under Liquids.	Suitable for small batches. Easily cleaned. Material is agitated during drying, causing some degradation and/or balling up.	Not applicable	Not applicable	Not applicable
Vacuum horizontal agitated and rotary. indirect type, batch operation. Includes indirect rotary, horizontal pan	Not applicable, except when pumping slowly on dry "heel"	May have application in special cases when pumping onto dry "heel"	Material usually cakes to dryer walls and agitator. Special precautions needed, e.g., cleaning hooks, twin screws. Solvents can be recovered.	Suitable for nonsticking materials. Useful for large batches of heat-sensitive materials and for solvent recovery.	Useful for large batches of heat-sensitive materials or where solvent is to be recovered. Product will suffer some grinding action, or may ball up. Dust collectors may be required.	Not applicable	Not applicable	Not applicable
Screw conveyor and indirect rotary. Indirect type, continuous operation. Includes paddle, horizontal agitated and steam-tube dryers, rotary kilns.	Applicable with dry-product recirculation	Applicable with dry-product recirculation	Generally requires recirculation of dry product. Little dusting occurs.	Chief advantage is low dust loss. Well suited to most materials and capacities, particularly those requiring drying at steam temperature	Low dust loss. Material must not stick or be temperature-sensitive	Not applicable	Not applicable	Not applicable

TABLE 12-12 Classification of Commercial Dryers Based on Feed Materials Handled *(Concluded)*

Type of dryer	Liquids	Slurries	Pastes and sludges	Free-flowing powders	Granular, crystalline, or fibrous solids	Large solids, special forms and shapes	Continuous sheets	Discontinuous sheets
Vibrating tray, vacuum band. Indirect type, continuous operation	Not applicable	Not usually applicable. Belt with raised edges possible, but rare	Not usually applicable due to feed and discharge problems	Suitable for free-flowing materials	Suitable for free-flowing materials that can be conveyed on a vibrating tray or belt	Not applicable	Not applicable	Not applicable
Drum. Indirect type, continuous operation	Single, double, or twin. Atmospheric or vacuum operation. Product flaky and usually dusty. Maintenance costs may be high.	See comments under Liquids. Twin-drum dryers are widely used.	Can be used only when paste or sludge can be made to flow. See comments under Liquids.	Not applicable	Not applicable	Not applicable	Not applicable	Not applicable
Cylinder. Indirect type, continuous operation	Not applicable	Not applicable	Not applicable	Not applicable	Not applicable	Not applicable	Suitable for thin or mechanically weak sheets which can be dried in contact with a heated surface. Special surface effects obtainable	Suitable for materials which need not be dried flat and which will not be injured by contact with hot drum
Tray and compartment. Direct type, batch operation. Includes cross-circulated tray	Not applicable	For very small batch production. Laboratory drying	Suited to batch operation. At large capacities, investment and operating costs are high. Long drying times	Dusting may be a problem. See comments under Pastes and Sludges.	Suited to batch operation. At large capacities, investment and operating costs are high. Long drying times	See comments under Granular solids.	Not applicable	See comments under Granular solids.
Batch through-circulation. Direct type, batch operation includes perforated tray, drying room	Not applicable	Not applicable	Suitable only if material can be preformed. Suited to batch operation. Shorter drying time than tray dryers	Not applicable	Usually not suited for materials smaller than 30-mesh (0.5 mm). Suited to small capacities and batch operation	Primarily useful for small objects	Not applicable	Not applicable
Tunnel/continuous tray. Direct type, continuous operation. Includes tunnel, turbo-tray	Not applicable	Not applicable	Suitable for small- and large-scale production	See comments under Pastes and Sludges. Vertical turbo-tray applicable	Essentially large-scale, semi-continuous tray drying	Suited to a wide variety of shapes and forms, especially tunnel type. Operation can be made continuous. Widely used	Not applicable	Suited for leather, wallboard, veneer
Continuous through-circulation (nonagitated). Direct type, continuous operation. Includes perforated band, moving bed, centrifuge-dryer.	Not applicable	Only crystal filter dryer or centrifuge dryer may be suitable.	Suitable for materials that can be preformed. Will handle large capacities	Not applicable	Usually not suited for materials smaller than 30-mesh (0.5 mm). Material does not tumble or mix.	Suited to smaller objects that can be loaded on each other. Can be used to convey materials through heated zones	Not applicable	Special designs are required. Suited to veneers
Continuous through-circulation (agitated/rotary). Direct type, continuous operation. Includes high-speed convective paddle, rotary-louvre.	Not applicable	Applicable with special high-speed fountain-type dryers, e.g., Hazemag Rapid, Forberg	Suitable for materials that can be preformed. Will handle large capacities. Rotary-louvre requires dry-product recirculation.	Not generally applicable, except rotary-louvre in certain cases	Usually not suited for materials smaller than 30-mesh (0.5 mm). Material is tumbled and mixed, may suffer attrition in paddle dryers.	Not applicable	Not applicable	Not applicable

Direct rotary. Direct-type, continuous operation	Applicable with dry-product recirculation	Applicable with dry-product recirculation	Suitable only if product does not stick to walls and does not dust. Recirculation of product may prevent sticking.	Suitable for most materials and especially for high capacities, provided dusting is not too severe	Suitable for most materials especially for high capacities. Dusting or crystal abrasion will limit its use.	Not applicable	Not applicable	Not applicable
Fluid beds. Direct type, batch or continuous	Applicable only as fluid bed granulator with inert bed or dry-solids recirculator	See comments under Liquids.	See comments under Liquids.	Suitable, if not too dusty. Internal coils can supplement heating, especially for fine powders. Suitable for high capacities.	Suitable for crystals, granules, and very short fibers. Suitable for high capacities	Not applicable	Use hot inert particles for contacting; rare	Use hot inert particles for contacting; rare
Spouted beds. Direct type, batch or continuous	Applicable only with inert bed or dry-solids recirculator. Usable to grow large particles by layering	See comments under Liquids.	See comments under Liquids.	Not applicable	Suitable for large particles and granules over 20-mesh (800 μm) which are spoutable	Not applicable unless objects are spoutable (conveyable in gas stream)	Use hot inert particles for contacting; rare	Use hot inert particles for contacting; rare
Pneumatic conveying. Direct-type, continuous operation includes flash, spin-flash, and ring dryers.	See comments under Slurries.	Can be used only if product is recirculated (backmixed) to make feed suitable for handling	Usually requires recirculation of dry product to make suitable feed. Well suited to high capacities. Disintegration usually required	Suitable for materials that are easily suspended in a gas stream and lose moisture readily. Well suited to high capacities	Suitable for materials conveyable in a gas stream. Well suited to high capacities. Only surface moisture usually removed. Product may suffer physical degradation.	Not applicable	Not applicable	Not applicable
Spray. Direct type, continuous operation. Rotary atomizer, pressure nozzle, or two-fluid nozzle. Includes combined spray-fluid bed and spray-belt dryers	Suited for large capacities. Product is usually powdery, spherical, and free-flowing. High temperatures can sometimes be used with heat-sensitive materials. Products generally have low bulk density.	See comments under Liquids. Pressure-nozzle atomizers subject to erosion	Requires special pumping equipment to feed the atomizer. See comments under Liquids.	Not applicable unless feed is pumpable	Not applicable	Not applicable	Not applicable	Not applicable
Continuous sheeting. Direct-type, continuous operation includes stenter, Yankee, impingement.	Not applicable	Not applicable	Not applicable	Not applicable	Not applicable	Not applicable	Generally high capacity. Different types are available for different requirements. Suitable for drying without contacting hot surfaces	Not applicable
Infrared. Batch or continuous operation. Electric heating or gas-fired	Only for thin films. Can be used in combination with other dryers such as drum.	See comments under Liquids.	See comments under Liquids (only for thin layers).	Only for thin layers	Primarily suited to drying surface moisture. Not suited for thick layers	Specially suited for drying and baking paint and enamels	Useful when space is limited. Usually used in conjunction with other methods, e.g., in drying paper coatings	Useful for laboratory work or in conjunction with other methods
Dielectric. Batch or continuous operation includes microwave, radio-frequency (RF)	Expensive, may be used in small batch filter-dryers, often as supplement to thermal heating. Sometimes useful in combination with other dryers.	See comments under Liquids.	See comments under Liquids.	Expensive, may be used on small batch dryers, often as supplement to thermal heating	Expensive, can assist thermal drying especially to dry center of large granules/pellets	Rapid drying of large objects suited to this method	Applications for final stages of paper and textile dryers	Successful on foam rubber. Not fully developed on other materials

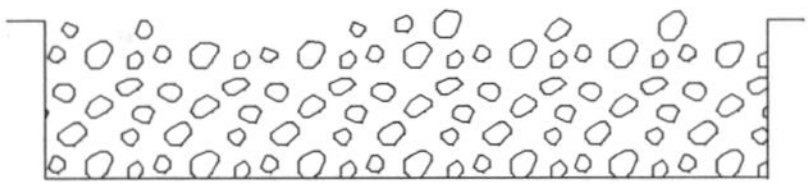

FIG. 12-26 Solids bed in static condition (tray dryer).

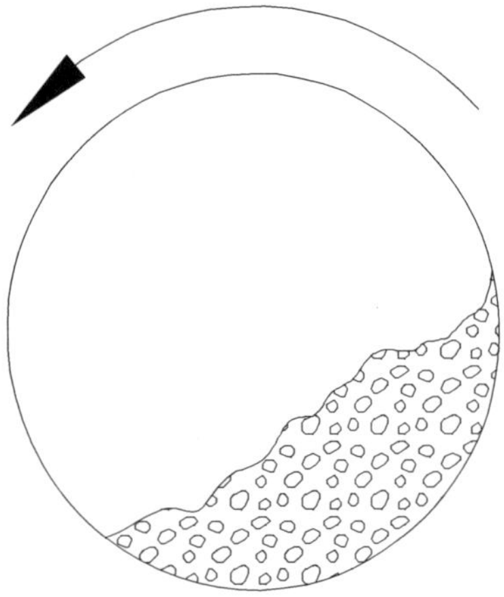

FIG. 12-27*a* Horizontal moving bed.

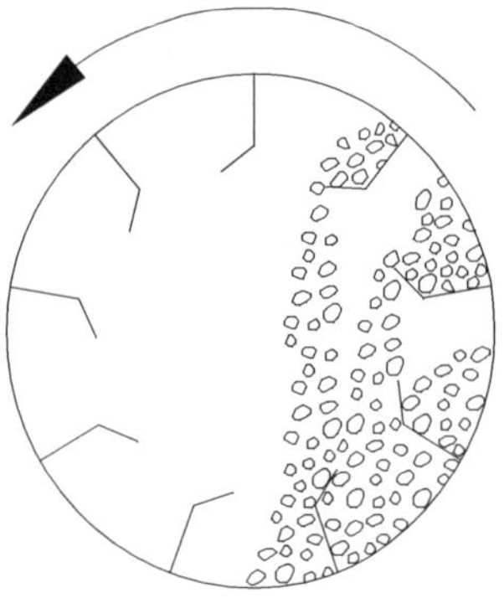

FIG. 12-27*b* Moving solids bed in a rotary dryer with lifters.

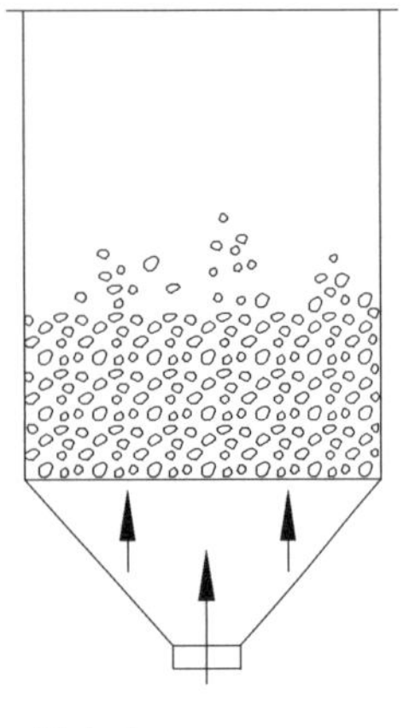

FIG. 12-28 Fluidized solids bed.

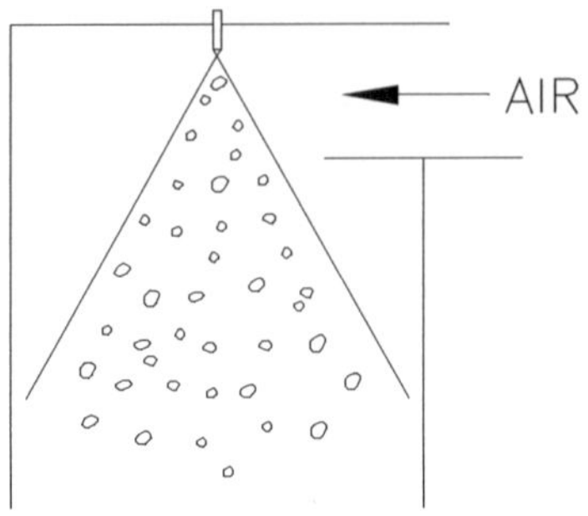

FIG. 12-29 Solids in a dilute condition near the top of a spray dryer.

Layer dryers are very suitable for slow-drying materials requiring a long residence time.

Because in a gas-solids-contacting operation heat transfer and mass transfer take place at the solids' surfaces, maximum process efficiency can be expected with a maximum exposure of solids surface to the gas phase, together with thorough mixing of gas and solids. Both are important. Within any arrangement of particulate solids, gas is present in the voids among the particles and contacts all surfaces except at the points of particle contact. When the solids are fluidized or dispersed, the gas moves past them rapidly, and external heat- and mass-transfer rates are high. When the solids bed is in a static or slightly moving condition, however, gas within the voids is cut off from the main body of the gas phase and can easily become saturated, so that local drying rates fall to low or zero values. Some transfer of energy and mass may occur by diffusion, but it is usually insignificant. The problem can be much reduced by using through-circulation of gas instead of cross-circulation, or by agitating and mixing the solids.

Solids Agitation and Mixing There are four alternatives:

1. No agitation, e.g., tray and band dryers. This is desirable for friable materials. However, drying rates can be extremely low, particularly for cross-circulation and vacuum drying.
2. Mechanical agitation, e.g., vertical pan and paddle dryers. This improves mixing and drying rates, but may give attrition depending on agitator speed; and solids may stick to the agitator, as shown in Fig. 12-30.
3. Vessel rotation, e.g., double-cone and rotary dryers. Mixing and heat transfer are better than for static dryers but may be less than for mechanical agitation. Formation of balls and lumps may be a problem.
4. Airborne mixing, e.g., fluidized beds and flash and spray dryers. Generally there is excellent mixing and mass transfer, but feed must be dispersible and entrainment and gas cleaning are higher.

Mechanical vibration may also be used to assist solids movement in some dryers.

Solids transport In continuous dryers, the solids must be moved through the dryer. The main methods of doing this are

1. Gravity flow (usually vertical), e.g., turbo-tray, plate and moving-bed dryers, and rotary dryers (due to the slope)

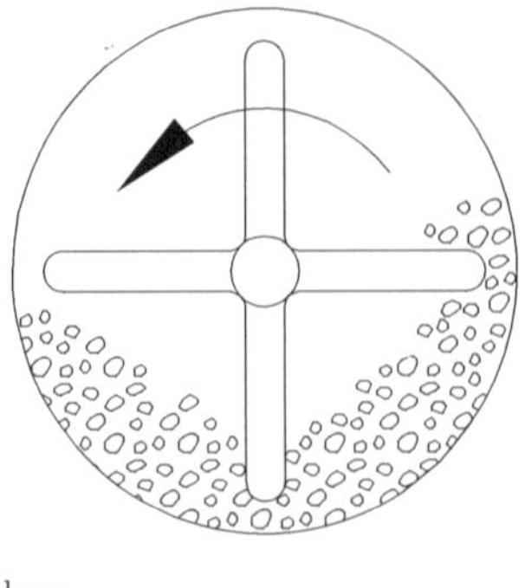

FIG. 12-30 Paddle dryer.

2. Mechanical conveying (usually horizontal), e.g., band, tunnel, and paddle dryers

3. Airborne transport, e.g., fluidized beds and flash and spray dryers

Solids flow pattern For most continuous dryers, the solids are basically in plug-flow; backmixing is low for nonagitated dryers but can be extensive for mechanical, rotary, or airborne agitation. Exceptions are well-mixed fluidized beds, fluid-bed granulators, and spouted beds (well-mixed) and spray and spray/fluidized-bed units (complex flow patterns).

5. *Gas-Solids Contacting* Where there is a significant gas flow, it may contact a bed of solids in the following ways:

a. Parallel flow or *cross-circulation.* The direction of gas flow is parallel to the surface of the solids phase. Contacting is primarily at the interface between phases, with possibly some penetration of gas into the voids among the solids near the surface. The solids bed is usually in a static condition (Fig. 12-31).

b. Perpendicular flow or *impingement.* The direction of gas flow is normal to the phase interface. The gas impinges on the solids bed. Again the solids bed is usually in a static condition (Fig. 12-32). This most commonly occurs when the solids are a continuous sheet, film, or slab.

c. Through circulation. The gas penetrates and flows through interstices among the solids, circulating more or less freely around the individual particles (Fig. 12-33). This may occur when solids are in static, moving, fluidized, or dilute conditions.

6. *Gas Flow Pattern in Dryer* Where there is a significant gas flow, it may be in cross-flow, cocurrent, or countercurrent flow compared with the direction of solids movement.

a. Cocurrent gas flow. The gas phase and solids particles both flow in the same direction (Fig. 12-34).

b. Countercurrent gas flow. The direction of gas flow is exactly opposite to the direction of solids movement.

c. Cross-flow of gas. The direction of gas flow is at a right angle to that of solids movement, across the solids bed (Fig. 12-35).

The difference between these is shown most clearly in the gas and solids temperature profiles along the dryer. For cross-flow dryers, all solids particles are exposed to the same gas temperature, and the solids temperature approaches the gas temperature near the end of drying (Fig. 12-36). In cocurrent dryers, the gas temperature falls throughout the dryer, and the final solids temperature is much lower than that for the cross-flow dryer (Fig. 12-37). Hence cocurrent dryers are very suitable for drying heat-sensitive materials, although it is possible to get a solids temperature peak inside the dryer. Conversely, countercurrent dryers give the most even temperature gradient throughout the dryer, but the exiting solids come into contact with the hottest, driest gas (Fig. 12-38). These can be used to heat-treat the solids or to give low final moisture content (minimizing the local equilibrium moisture content) but are obviously unsuitable for thermally sensitive solids.

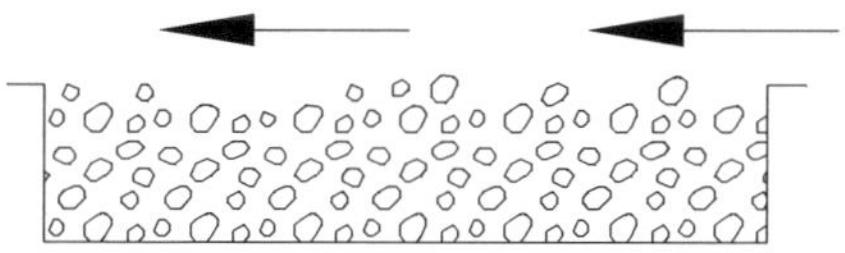

FIG. 12-31 Parallel gas flow over a static bed of solids.

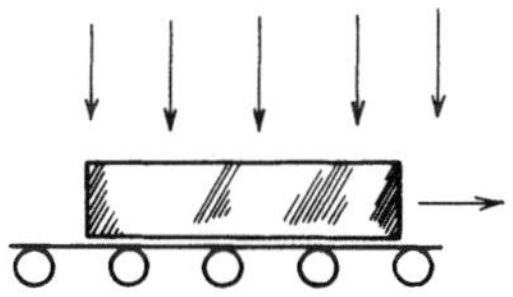

FIG. 12-32 Circulating gas impinging on a large solid object in perpendicular flow, in a roller-conveyor dryer.

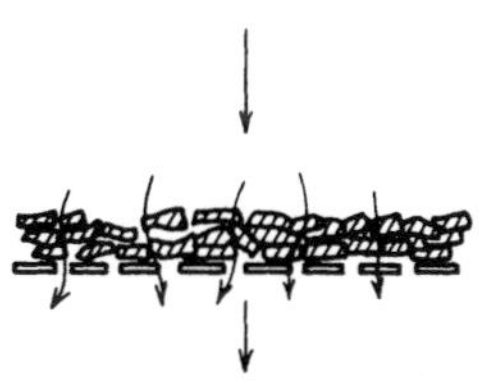

FIG. 12-33 Gas passing through a bed of preformed solids, in through-circulation on a perforated-band dryer.

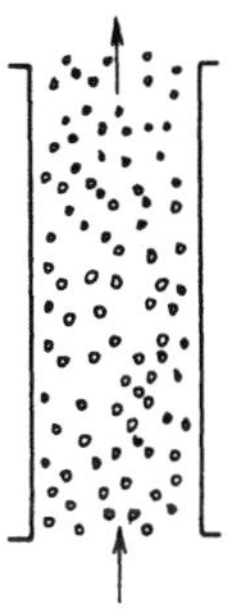

FIG. 12-34 Cocurrent gas-solids flow in a vertical-lift dilute-phase pneumatic conveyor dryer.

Subclassifications

Heater This may be an indirect heat exchanger or a direct-fired burner, or heating may be electrical (including RF/microwave absorption).

Gas Circuit This may be open-cycle (once-through) or closed-cycle (gas recycle, often using inert gas). A closed-cycle system with a direct-fired burner can be operated as a self-inerting system with reduced oxygen concentration.

Solids Feeders These convey the solids into the dryer and may also perform as a metering or sealing dryer. Dry solids may be back-mixed into the wet feed if the latter is sticky and difficult to handle. See Sec. 21.

Gas-Solids Separations After the solids and gas have been brought together and mixed in a gas-solids contactor, it becomes necessary to separate the two phases, particularly for dispersion dryers where the solids loading in the exhaust gas can be very high. If the solids are sufficiently coarse and the gas velocity sufficiently low, it is possible to effect a complete gravitational separation in the primary contactor. Applications of this type are rare, however, and supplementary dust collection equipment is commonly required. The recovery step may even dictate the type of primary contacting device selected. For example, in treating an extremely friable solid material, a deep fluidized-solids contactor might overload the collection system with fines, whereas the more gentle contacting of a traveling-screen contactor

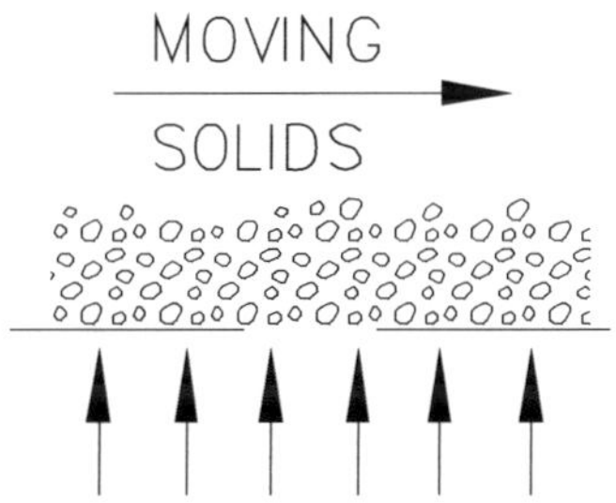

FIG 12-35 Cross-flow of gas and solids in a fluid bed or band dryer.

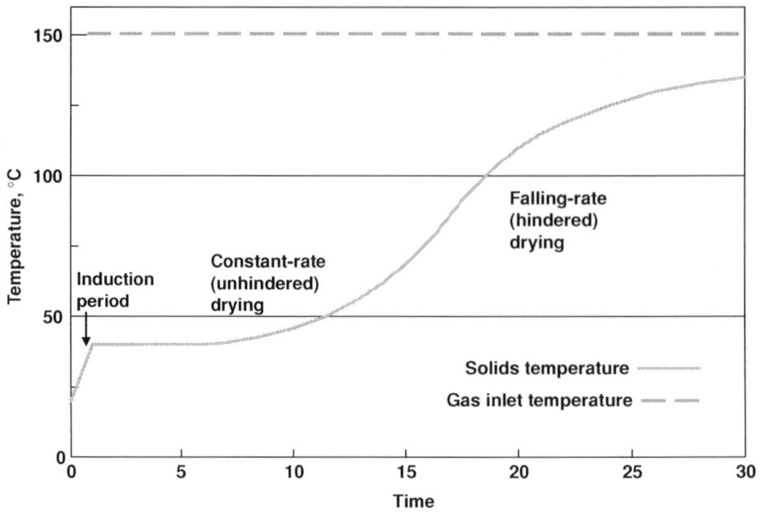

FIG. 12-36 Temperature profiles along a continuous plug-flow dryer for cross-flow of gas and solids. *(Aspen Technology Inc.)*

would be expected to produce a minimum of fines by attrition. Therefore, although gas solids separation is usually considered as separate and distinct from the primary contacting operation, it is usually desirable to evaluate the separation problem at the same time as contacting methods are evaluated. Methods are noted later in "Environmental Considerations." The subject is covered in depth in Sec. 17, "Gas-Solids Operations."

SELECTION OF DRYING EQUIPMENT

Dryer Selection Considerations Dryer selection is a challenging task and rarely clear-cut. For 500-μm particles, there may be several different dryer types which are likely to handle the task well, at similar cost. For 5-μm particles, there may be no dryers that are fully suitable, and the task is to find the "least bad"!

Dryer classification often helps to reveal the broad choices for which equipment is suitable. For instance:

- Batch dryers are almost invariably used for mean throughputs below 50 kg/h and continuous dryers above 1 ton/h; in the intervening range, either may be suitable.
- Liquid or slurry feeds, large artifacts, or continuous sheets and films require completely different equipment to particulate feeds.
- Particles and powders below 1 mm are effectively dried in dispersion or contact dryers, but most through-circulation units are unsuitable. Conversely, for particles of several millimeters or above, through-circulation dryers, rotary dryers, and spouted beds are very suitable.
- Through-circulation and dispersion convective dryers (including fluidized-bed, rotary, and pneumatic types), and agitated or rotary contact dryers, generally give better drying rates than nonagitated cross-circulated or contact tray dryers.
- Nonagitated dryers (including through-circulation) may be preferable for fragile particles where it is desired to avoid attrition.
- For organic solvents, or solids which are highly flammable, are toxic, or decompose easily, contact dryers are often preferable to

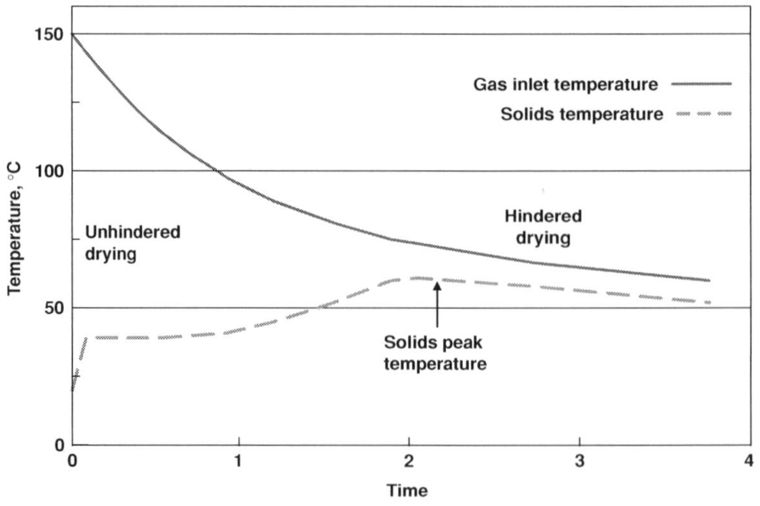

FIG. 12-37 Temperature profiles along a continuous plug-flow dryer for cocurrent flow of gas and solids. *(Aspen Technology Inc.)*

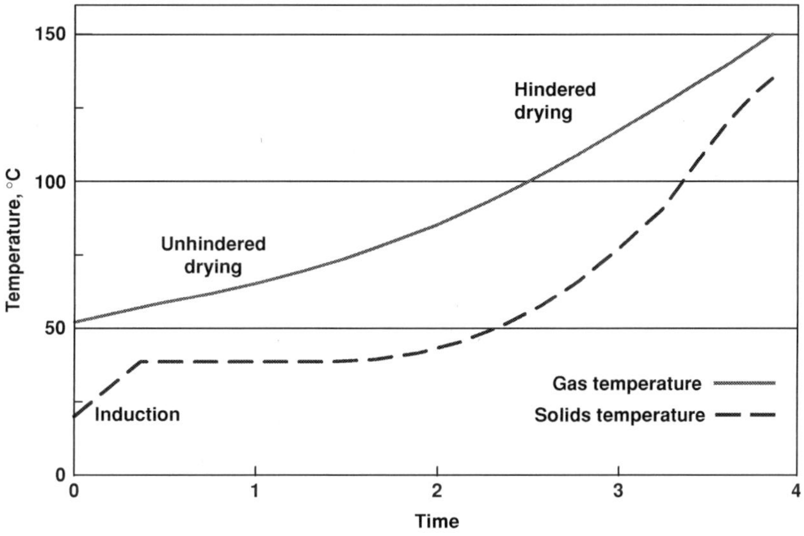

FIG. 12-38 Temperature profiles along a continuous plug-flow dryer for countercurrent flow of gas and solids. *(Aspen Technology Inc.)*

convective, as containment is better and environmental emissions are easier to control. If a convective dryer is used, a closed-cycle system using an inert carrier gas (e.g., nitrogen) is often required.

- Cocurrent, vacuum, and freeze dryers can be particularly suitable for heat-sensitive materials.

A detailed methodology for dryer selection, including the use of a rule-based expert system, has been described by Kemp [*Drying Technol.* **13**(5–7): 1563–1578 (1995) and **17**(7 and 8): 1667–1680 (1999)].

A simpler step-by-step procedure is given here.

1. *Initial selection of dryers.* Select those dryers which appear best suited to handling the wet material and the dry product, which fit into the continuity of the process as a whole, and which will produce a product of the desired physical properties. This preliminary selection can be made with the aid of Table 12-12, which classifies the various types of dryers on the basis of the materials handled.

2. *Initial comparison of dryers.* The dryers so selected should be evaluated approximately from available cost and performance data. From this evaluation, those dryers which appear to be uneconomical or unsuitable from the standpoint of performance should be eliminated from further consideration.

3. *Drying tests.* Drying tests should be conducted in those dryers still under consideration. These tests will determine the optimum operating conditions and the product characteristics and will form the basis for firm quotations from equipment vendors.

4. *Final selection of dryer.* From the results of the drying tests and quotations, the final selection of the most suitable dryer can be made.

The important factors to consider in the preliminary selection of a dryer are the following:

1. Properties of the material being handled
   a. Physical characteristics when wet (stickiness, cohesiveness, adhesiveness, flowability)
   b. Physical characteristics when dry
   c. Corrosiveness
   d. Toxicity
   e. Flammability
   f. Particle size
   g. Abrasiveness
2. Drying characteristics of the material
   a. Type of moisture (bound, unbound, or both)
   b. Initial moisture content (maximum and range)
   c. Final moisture content (maximum and range)
   d. Permissible drying temperature
   e. Probable drying time for different dryers
   f. Level of nonwater volatiles
3. Flow of material to and from the dryer
   a. Quantity to be handled per hour (or batch size and frequency)
   b. Continuous or batch operation
   c. Process prior to drying
   d. Process subsequent to drying
4. Product quality
   a. Shrinkage
   b. Contamination
   c. Uniformity of final moisture content
   d. Decomposition of product
   e. Overdrying
   f. State of subdivision
   g. Product temperature
   h. Bulk density
5. Recovery problems
   a. Dust recovery
   b. Solvent recovery
6. Facilities available at site of proposed installation
   a. Space
   b. Temperature, humidity, and cleanliness of air
   c. Available fuels
   d. Available electric power
   e. Permissible noise, vibration, dust, or heat losses
   f. Source of wet feed
   g. Exhaust-gas outlets

The physical nature of the material to be handled is the primary item for consideration. A slurry will demand a different type of dryer from that required by a coarse crystalline solid, which, in turn, will be different from that required by a sheet material (Table 12-12).

Following preliminary selection of suitable types of dryers, a rough evaluation of the size and cost should be made to eliminate those which are obviously uneconomical. Information for this evaluation can be obtained from material presented under discussion of the various dryer types. When data are inadequate, preliminary cost and performance data can usually be obtained from the equipment manufacturer. In comparing dryer performance, the factors in the preceding list which affect dryer performance should be properly weighed. The possibility of eliminating or simplifying processing steps which precede or follow drying, such as filtration, grinding, or conveying, should be carefully considered.

Drying Tests These tests should establish the optimum operating conditions, the ability of the dryer to handle the material physically, product quality and characteristics, and dryer size. The principal manufacturers of drying equipment are usually prepared to perform the required tests on dryers simulating their equipment. Occasionally, simple laboratory experiments can serve to reduce further the number of dryers under consideration.

Once a given type and size of dryer have been installed, the product characteristics and drying capacity can be changed only within relatively narrow limits. Thus it is more economical and far more satisfactory to experiment in small-scale units than on the dryer that is finally installed.

On the basis of the results of the drying tests that establish size and operating characteristics, formal quotations and guarantees should be obtained from dryer manufacturers. Initial costs, installation costs, operating costs, product quality, dryer operability, and dryer flexibility can then be given proper weight in final evaluation and selection.

Effective scale-up from tests to industrial equipment is obviously very important, and it was covered in a special issue of *Drying Technol.* **12** (1 and 2): 1–452 (1994).

DRYER MODELING, DESIGN, AND SCALE-UP

General Principles Models and calculations on dryers can be categorized in terms of (1) the level of complexity used and (2) the purpose or type of calculation (design, performance rating, or scale-up). A fully structured approach to dryer modeling can be developed from these principles, as described below and in greater detail by Kemp and Oakley (2002).

Levels of Dryer Modeling Modeling can be carried out at four different levels, depending on the amount of data available and the level of detail and precision required in the answer.

Level 1. *Heat and mass balances*. These balances give information on the material and energy flows to and from the dryer, but say nothing about the required equipment size or the performance which a given dryer is capable of.

Level 2. *Scoping* Approximate or scoping calculations give rough sizes and throughputs (mass flow rates) for dryers, using simple data and making some simplifying assumptions. Either heat-transfer control or first-order drying kinetics is assumed.

Level 3. *Scaling* Scaling calculations give overall dimensions and performance figures for dryers by scaling up drying curves from small-scale or pilot-plant experiments.

Level 4. *Detailed* Rigorous or detailed methods aim to track the temperature and drying history of the solids and find local conditions inside the dryer. Naturally, these methods use more complex modeling techniques with many more parameters and require many more input data.

Types of Dryer Calculations The user may wish to either design a new dryer or improve the performance of an existing one. Three types of calculations are possible:

- Design of a new dryer to perform a given duty, using information from the process flowsheet and physical properties databanks
- Performance calculations for an existing dryer at a new set of operating conditions
- Scale-up from laboratory-scale or pilot-plant experiments to a full-scale dryer

Solids drying is very difficult to model reliably, particularly in the falling-rate period which usually has the main effect on determining the overall drying time. Falling-rate drying kinetics depend strongly on the internal moisture transport within a solid. This is highly dependent on the internal structure, which in turn varies with the upstream process, the solids formation step, and often between individual batches. Hence, many key drying parameters within solids (e.g., diffusion coefficients) cannot be predicted from theory alone, or obtained from physical property databanks; practical measurements are required. Because of this, experimental work is almost always necessary to design a dryer accurately, and scale-up calculations are more reliable than design based only on thermodynamic data. The experiments are used to verify the theoretical model and find the difficult-to-measure parameters; the full-scale dryer can then be modeled more realistically.

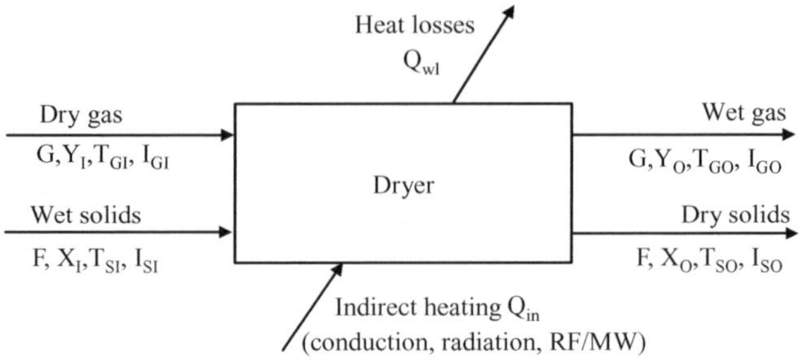

FIG. 12-39 Heat and material flows around a continuous dryer.

Heat and Mass Balance The heat and mass balance on a generic continuous dryer is shown schematically in Fig. 12-39. In this case, mass flows and moisture contents are given on a dry basis.

The **mass balance** is usually performed on the principal solvent and gives the evaporation rate E (kg/s). In a contact or vacuum dryer, this is approximately equal to the exhaust vapor flow, apart from any noncondensibles. In a convective dryer, this gives the increased outlet humidity of the exhaust. For a continuous dryer at steady-state operating conditions,

$$E = F(X_I - X_O) = G(Y_O - Y_I) \tag{12-54}$$

This assumes that the dry gas flow G and dry solids flow F do not change between dryer inlet and outlet. Mass balances can also be performed on the overall gas and solids flows to allow for features such as air leaks and solids entrainment in the exhaust gas stream.

In a design mode calculation (including scale-up), the required solids flow rate, inlet moisture content X_I and outlet moisture X_O are normally specified, and the evaporation rate and outlet gas flow are calculated. In performance mode, the calculation is normally reversed; the evaporation rate under new operating conditions is found, and the new solids throughput or outlet moisture content is back-calculated.

For a batch dryer with a dry mass m of solids, a mass balance only gives a snapshot at one point during the drying cycle and an instantaneous drying rate, given by

$$E = m\left(\frac{-dX}{dt}\right) = G\,(Y_O - Y_I) \tag{12-55}$$

The **heat balance** on a continuous dryer takes the generic form

$$GI_{GI} + FI_{SI} + Q_{\text{in}} = GI_{GO} + FI_{SO} + Q_{wl} \tag{12-56}$$

Here I is the enthalpy (kJ/kg dry material) of the solids or gas plus their associated moisture. Enthalpy of the gas includes the latent heat term for the vapor. Expanding the enthalpy terms gives

$$\begin{aligned} &G(C_{sI}T_{GI} + \lambda Y_I) + F(C_{PS} + X_I C_{PL})T_{SI} + Q_{\text{in}} \\ &\quad = G(C_{sO}T_{GO} + \lambda Y_O) + F(C_{PS} + X_O C_{PL})T_{SO} + Q_{wl} \end{aligned} \tag{12-57}$$

Here C_s is the humid heat $C_{PG} + YC_{PY}$. In convective dryers, the left-hand side is dominated by the sensible heat of the hot inlet gas $GC_{sI}T_{GI}$; in contact dryers, the heat input from the jacket Q_{in} is dominant. In both cases, the largest single term on the right-hand side is the latent heat of the vapor $G\lambda Y_O$. Other terms are normally below 10 percent. *This shows why the operating line of a convective dryer on a psychrometric chart is roughly parallel to a constant-enthalpy line.*

The corresponding equation for a batch dryer is

$$GI_{GI} + Q_{\text{in}} = GI_{GO} + m\,\frac{dI_S}{dt} + Q_{wl} \tag{12-58}$$

Further information on heat and mass balances, including practical challenges on industrial dryers and a worked example, is given in the "Drying Fundamentals" section.

Scoping Design Calculations In scoping calculations, some approximate dryer dimensions and drying times are obtained based mainly on a heat and mass balance, without measuring a drying curve or other experimental drying data. They allow the cross-sectional area of convective dryers and the volume of batch dryers to be estimated quite accurately, but are less effective for other calculations and can give overoptimistic results.

Continuous Convective Dryers In design mode, the required solids throughput F and the inlet and outlet moisture content X_I and X_O are known, as is the ambient humidity Y_I. If the inlet gas temperature T_{GI} is chosen, the outlet gas conditions (temperature T_{GO} and humidity Y_O) can be found, either by calculation or (more simply and quickly) by using the constant-enthalpy lines on a psychrometric chart. However, it may be necessary to allow for heat losses and sensible heating of solids, which typically reduce the useful enthalpy of the inlet gas by 10 to 20 percent. Also, if tightly bound moisture is being removed, the heat of wetting to break the bonds should be allowed for. The gas mass flow rate G can now be calculated, as it is the only unknown in the mass balance on the solvent [Eq. (12-56)]. A typical gas velocity U_G along the dryer is now chosen, for example, 20 m/s for a flash dryer, 0.5 m/s for a fluidized bed, and 3 m/s for a cocurrent rotary dryer. For through-circulation and dispersion dryers, the cross-sectional area A is given by

$$A = \frac{G}{\rho_{GI}U_G} = \frac{F(X_I - X_O)}{\rho_{GI}U_G(Y_O - Y_I)} \tag{12-59}$$

The dryer diameter, or linear dimensions of a rectangular bed, can then be calculated. The result is usually accurate within 10 percent, and can be further improved by better estimates of velocity and heat losses. In performance mode, the equation is reversed to find the gas flow rate from $G = \rho_{GI}U_GA$.

The method gives no information about solids residence time or dryer length. A minimum drying time $t_{\min}$ can be calculated by evaluating the maximum (unhindered) drying rate N_{cr}, assuming gas-phase heat-transfer control and estimating a gas-to-solids heat-transfer coefficient. The simple equation (12-60) then applies:

$$t_{\min} = \frac{X_I - X_O}{N_{cr}} \tag{12-60}$$

Alternatively, it may be assumed that first-order falling-rate kinetics apply throughout the drying process, and scale the estimated drying time by using Eq. (12-63). However, these crude methods can give serious underestimates of the required drying time, and it is much better to measure the drying time experimentally and apply scaling (level 3) methods.

Continuous Contact Dryers The key parameter is the area of the heat-transfer surface A_S. In design mode, this can be found from the equation:

$$A_S = \frac{Q}{h_{WS}\,\Delta T_{WS}} = \frac{E\lambda_{ev}}{h_{WS}\,\Delta T_{WS}} = \frac{F(X_I - X_O)\lambda_{ev}}{h_{WS}\,\Delta\,T_{WS}} \tag{12-61}$$

Here Q is the rate of heat transfer from the heated wall to the solids, and ΔT_{WS} is the temperature driving force. The latent heat of evaporation λ_{ev} should allow for bound moisture and heating of solids and vapor to the final temperature. A typical wall-to-solids heat-transfer coefficient h_{WS} for the given dryer type should be used. The calculation is less accurate than the one for convective dryers. Again, the heat-transfer rate is assumed to be the overall limiting factor.

If the drying process is strongly limited by falling-rate drying kinetics, the calculated size of dryer corresponding to the given heating surface A_S may not give sufficient solids residence time to reach the desired final moisture content. Again, experimental measurement of a drying curve is strongly recommended.

Batch Dryers If the batch size is stipulated, the requirement is simply that the dryer be able to physically contain the volume of the solids, and the dryer volume and dimensions can thus be calculated directly. Solids residence time must then be calculated. Equation (12-61) can be reversed and modified to give

$$t_{CR} = \frac{m_s(X_I - X_O)\lambda_{ev}}{h_{WS}\Delta T_{WS}A_S} \tag{12-62}$$

Heat-transfer control and constant-rate drying are assumed. Again, the calculation will be inaccurate and overoptimistic for falling-rate drying, and it is preferable to measure a drying curve and use a scaling calculation, as outlined in the next section. It is possible to compare the surface area/volume ratios of various types of dryers and deduce how their drying times will compare with each other (see "Drying Equipment—Batch Agitated and Rotary Dryers").

Falling-Rate Kinetics To correct from a calculated constant-rate (unhindered) drying time t_{CR} to first-order falling-rate kinetics, the following equation is used, where X_1 is the initial, X_2 the final, and X_E the equilibrium moisture content (all must be dry-basis):

$$\frac{t_{FR}}{t_{CR}} = \frac{X_1 - X_E}{X_1 - X_2}\ln\left(\frac{X_1 - X_E}{X_2 - X_E}\right) \tag{12-63}$$

Note that $t_{FR} \geq t_{CR}$. Likewise, to convert to a two-stage drying process with constant-rate drying down to X_{cr} and first-order falling-rate drying beyond, the equation is

$$\frac{t_{2S}}{t_{CR}} = \frac{X_1 - X_{cr}}{X_1 - X_2} + \frac{X_{cr} - X_E}{X_1 - X_2}\ln\left(\frac{X_{cr} - X_E}{X_2 - X_E}\right) \tag{12-64}$$

Scale-up Effects As dryers get larger, if the drying rate is either controlled by heat transfer (unhindered or constant-rate drying) or proportional to it (first-order drying kinetics following the characteristic drying curve), then the drying rate N (kg/kg/s) and drying time t will be proportional to the ratio between the area over which heat enters and the mass or volume of solids.

For most types of dryer, it is found that the specific drying rate (SDR), which is a mass flux (evaporation rate per unit area), is constant for a given set of operating conditions. The concept is described by Moyers [*Drying Technol.* **12**(1 and 2):393 (1994)]. For convective layer dryers, both through-circulation and cross-circulation, mass increases proportionately to bed area if layer depth remains constant; hence the drying time should remain the same. This is also true of fluidized beds and of contact dryers where the solids rest as a layer on a heated plate (tray, vacuum band, plate, film-drum, and thin-film dryers). However, for mechanically agitated and rotating contact dryers (vertical pan, conical, double-cone tumbler, and paddle), the heat-transfer surface area increases as the square of dryer diameter and volume as the cube, and hence drying time increases with the cube root of batch size:

$$\frac{t_2}{t_1} = \left(\frac{m_2}{m_1}\right)^{1/3} \tag{12-65}$$

Providing additional internal heating surfaces, such as heated agitators or steam tubes in paddle or rotary dryers, gives a higher area/volume ratio and faster drying, so these will be the preferred contact dryer types for large batches or high throughput. This applies if the drying rate is proportional to the rate of heat supply. For a continuous dryer, heating the agitator allows a smaller dryer for a given solids throughput; for a batch dryer with fixed batch size, a heated agitator shortens the required drying time. However, if a minimum residence time is required to allow removal of tightly bound moisture, there will be little or no gain from providing very large amounts of heat-transfer area.

Again, these methods take no account of the actual drying kinetics of the particle, which are included in the next section.

Example 19: Drying of Particles A convective dryer is to be used to dry 720 kg/h (0.2 kg/s) of particulate material from 0.2 to 0.02 kg/kg moisture content (all flows and moistures on dry basis), using air at 180°C and 0.005 kg/kg humidity. Estimate the required air flow rate and dryer size for a fluidized-bed dryer (0.5 m/s inlet velocity) and a pneumatic conveying dryer (20 m/s inlet velocity). Assume outlet RH is approximately 20 percent. What is the effect of 10 percent heat losses?

Solution: Using a psychrometric chart with T_{GI} = 180°C, the outlet gas temperature is approximately 70°C, Y_O = 0.048 kg/kg with no losses, or 0.040 kg/kg with 10 percent losses.

From the mass balance, Eq. (12-54): $0.2(0.2 - 0.02) = W_G(Y_O - 0.005)$. Hence W_G = 0.837 kg/s (no losses) or 1.03 kg/s (10 percent losses).

The cross-sectional area of the dryer is obtained from Eq. (12-59), by taking ρ_G at 180°C as 0.78 kg/m^3. For the fluidized-bed dryer, assuming 10 percent losses; A_B = 1.03/(0.5 × 0.78) = 2.64 m^2. For a circular bed, D = 1.83 m. For the pneumatic conveying dryer, assuming 10 percent losses, A_{xs} = 1.03/(20 × 0.78) = 0.066 m^2. For a circular duct, D = 0.29 m; for a square duct, D = 0.257 m.

A similar example for batch dryers may be found in the section "Batch Agitated and Rotating Dryers," including constant-rate and falling-rate kinetics and scale-up from an experimental test result.

SCALING MODELS

These models use experimental data from drying kinetics tests in a laboratory, pilot-plant or full-scale dryer, and are thus more accurate and reliable than methods based only on estimated drying kinetics. They treat the dryer as a complete unit, with drying rates and air velocities averaged over the dryer volume, except that, if desired, the dryer can be subdivided into a small number of sections. These methods are used for layer dryers (tray, oven, horizontal-flow band, and vertical-flow plate types) and for a simple estimate of fluidized-bed dryer performance. For batch dryers, they can be used for scale-up by refining the scoping design calculation.

The basic principle is to take an experimental drying curve and perform two transformations: (1) from test operating conditions to full-scale operating conditions and (2) for test dimensions to full-scale dryer dimensions. If the operating conditions of the test (e.g., temperature, gas velocity, agitation rate) are the same as those for the full-scale plant, the first correction is not required.

Scaling models are the main design method traditionally used by dryer manufacturers. Pilot-plant test results are scaled to a new set of conditions on a dryer with greater airflow or surface area by empirical rules, generally based on the external driving forces (temperature, vapor pressure, or humidity driving forces). By implication, therefore, a characteristic drying curve concept is again being used, scaling the external heat and mass transfer and assuming that the internal mass transfer changes in proportion. A good example is the set of rules described under "Fluidized-Bed Dryers," which include the effects of temperature, gas velocity, and bed depth on drying time in the initial test and the full-scale dryer.

The integral model is a development of a simple scale-up model which allows for mixing and residence time effects, first suggested for fluidized beds by Vanecek et al. (1964, 1966). The mean outlet moisture content is given by summing the product of the particle moisture content and the probability that it emerges at time t:

$$\overline{X} = \int_0^\infty E(t)X(t)\,dt \tag{12-66}$$

Here $X(t)$ is the drying curve, corrected as before to the new scale and new operating conditions, and $E(t)$ is the residence time function, which must be known. This approach has been used successfully for well-mixed fluidized beds. For pure plug flow, $E(t)$ is a spike (Green's function) and $\overline{X} = X(t)$.

Scale-up of Batch Dryers We can use the same equations as before but base drying time on an experimental value rather than one obtained from an unhindered drying calculation.

Example 20: Scaling of Data An experimental batch drying curve has been measured at 100°C, and drying time was 2 h. Estimate the drying time at (*a*) 100°C and (*b*) 150°C for (i) a fluidized-bed dryer and (ii) a conical vacuum dryer at 100-mbar absolute pressure, for a batch size 10 times greater than that of the test. Assume for the fluidized bed that temperature driving forces are proportional to $T - T_{wb}$ and batch drying time is proportional to bed depth, and for the conical dryer that the solids temperature is equal to the saturation temperature at 100-mbar pressure (46°C for water vapor).

Solution:

(i) For the fluid bed, T = 100°C and 150°C, T_{wb} = 30°C and 38°C, respectively. The increase in heat transfer and drying rate for case (*b*) is a factor of 112/70 = 1.6.

The bed could be scaled up by increasing the bed area by a factor of 10 and keeping depth z constant, in which case drying time will remain at 2 h for case (*a*) and become 2/1.6 = 1.25 h for case (*b*).

Alternatively, all dimensions could be scaled up proportionately; as $V = \rho D^2 z$, D and z will increase by $\sqrt[3]{10}$ = 2.16. Drying time then becomes 4.32 h for (*a*) and 2.70 h for (*b*).

(ii) For the conical dryer, in case (*b*), temperature driving force increases by a factor of 104/54 = 1.93. From Eq. (12-65), linear dimensions and drying time all scale up by a factor of $\sqrt[3]{10}$. Drying time becomes 4.32 h for (*a*) and $2 \times \sqrt[3]{10}/1.93$) = 2.24 h for (*b*).

DETAILED OR RIGOROUS MODELS

These models aim to predict local conditions within the dryer and the transient condition of the particles and gas in terms of temperature, moisture content, velocity, etc. Naturally, they require much more input data. There are many published models of this type in the academic literature. They give the possibility of more detailed results, but the potential cumulative errors are also greater.

- Incremental models track the local conditions of the gas and particles through the dryer, mainly in one dimension. They are especially suitable for cocurrent and countercurrent dryers, e.g., flash (pneumatic conveying) and rotary dryers. The air conditions are usually treated as uniform across the cross-section and dependent only on axial position. This method can also be used to determine local conditions (e.g., temperature) where a simpler model has been used to find the overall drying rate. A two- or three-dimensional grid can also be used, e.g., modeling vertical and horizontal variations in a band dryer or plug-flow fluidized bed.
- Complex three-dimensional models, e.g., CFD (computational fluid dynamics), aim to solve the gas conditions and particle motion throughout the dryer. They are the only effective models for spray dryers because of the complex swirling flow pattern; they can also be used to find localized conditions in other dryers.

Incremental Model The one-dimensional incremental model is a key analysis tool for several types of dryers. A set of simultaneous equations is solved at a given location (Fig. 12-40), and the simulation moves along the dryer axis in a series of steps or increments—hence the name. The procedure may be attempted by hand if a few large steps (say, 5 to 10) are used; but for an accurate simulation, a computer program is needed and thousands of increments may be used.

Increments may be stated in terms of time (dt), length (dz), or moisture content (dX). A set of six simultaneous equations is then solved, and ancillary calculations are also required, e.g., to give local values of gas and solids properties. The generic set of equations (for a time increment Δt) is as follows:

Heat transfer to particle:
$$Q_P = h_{PG}A_P(T_G - T_S) \tag{12-67}$$

Mass transfer from particle:
$$\frac{-dX}{dt} = \text{function } (X, Y, T_P, T_G, h_{PG}, A_P) \tag{12-68}$$

Mass balance on moisture:
$$G\Delta Y = -F\Delta X = F\frac{-dX}{dt}\Delta t \tag{12-69}$$

Heat balance on particle:
$$\Delta T_S = \frac{Q_P\,\Delta t - \lambda_{ev} m_P\,\Delta X}{m_P(C_{PS} + C_{PL}X)} \tag{12-70}$$

Heat balance for increment:
$$-\Delta T_G = \frac{F(C_{PS} + C_{PL}X)\,\Delta T_S + G(\lambda_0 + C_{PY}T_G)\,\Delta Y + \Delta Q_{Wl}}{G(C_{PG} + C_{PY}Y)} \tag{12-71}$$

Particle transport:
$$\Delta z = U_S\,\Delta t \tag{12-72}$$

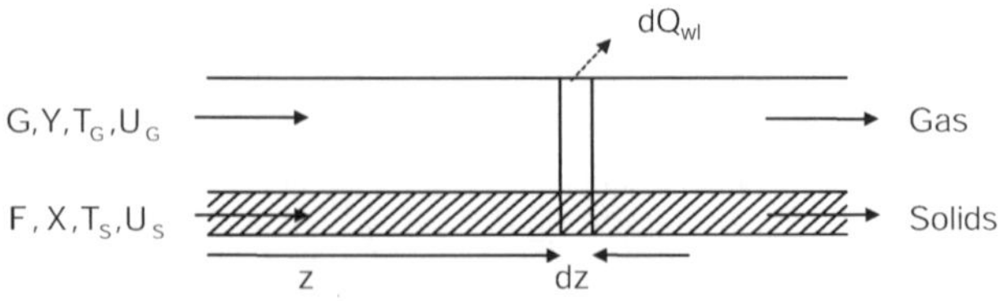

FIG. 12-40 Principle of the incremental model.

The mass and heat balance equations are the same for any type of dryer, but the particle transport equation is completely different, and the heat- and mass-transfer correlations are also somewhat different as they depend on the environment of the particle in the gas (i.e., single isolated particles, agglomerates, clusters, layers, fluidized beds, or packed beds). The mass-transfer rate from the particle is regulated by the drying kinetics and is thus obviously material-dependent (at least in falling-rate drying).

The model is effective and appropriate for dryers where both solids and gas are approximately in axial plug flow, such as pneumatic conveying and cascading rotary dryers. However, it runs into difficulties where there is recirculation or radial flow.

The incremental model is also useful for measuring variations in local conditions such as temperature, solids moisture content, and humidity along the axis of a dryer (e.g., plug-flow fluidized bed), through a vertical layer (e.g., tray or band dryers), or during a batch drying cycle (using time increments, not length). It can be applied in these situations even though the integral model has been used to determine the overall kinetics and drying time.

Example 21: Sizing of a Cascading Rotary Dryer The average gas velocity passing through a cocurrent, adiabatic, cascading rotary dryer is 4 m/s. The particles moving through the dryer have an average diameter of 5 mm, a solids density of 600 kg/m^3, and a shape factor of 0.75. The particles enter with a moisture content of 0.50 kg/kg (dry basis) and leave with a moisture content of 0.15 kg/kg (dry basis). The drying rate may be assumed to decrease linearly with average moisture content, with no unhindered ("constant-rate") drying period. In addition, let us assume that the solids are nonhygroscopic (so that the equilibrium moisture content is zero; hygroscopic means that the equilibrium moisture content is nonzero).

The inlet humidity is 0.10 kg/kg (dry basis) due to the use of a direct-fired burner, and the ratio of the flow rates of dry solids to dry gas is unity ($F/G = 1$). The gas temperature at the inlet to the dryer is 800°C, and the gas may be assumed to behave as a pure water vapor/air mixture.

What is the gas-phase residence time that is required?

Data:

$U = 4$ m/s	$X_I = 0.50$ kg/kg	$F/G = 1$
$d_{PSM} = 0.005$ m	$X_O = 0.15$ kg/kg	$T_{GI} = 800$°C
$\rho_P = 600$ kg/m^3	$X_{cr} = 0.50$ kg/kg	$Y_I = 0.10$ kg/kg
$\alpha_P = 0.75$	$X_e = 0.0$ kg/kg	

Application of concept of characteristic drying curve: A linear-falling rate curve implies the following equation for the drying kinetics:

$$f = \Phi \quad \text{assumption of linear drying kinetics} \tag{12-73}$$

where f is the drying rate relative to the initial drying rate.

$$f = \frac{N}{N_{\text{initial}}}$$

Since the material begins drying in the falling-rate period, the critical moisture content can be taken as the initial moisture content. The equilibrium moisture content is zero since the material is not hygroscopic.

$$\Phi = \frac{\overline{X} - X_{eq}}{X_{cr} - X_{eq}} = \frac{\overline{X}}{0.5}$$

Application of mass balances (theory): A mass balance around the inlet and any section of the dryer is shown in Fig. 12-41.

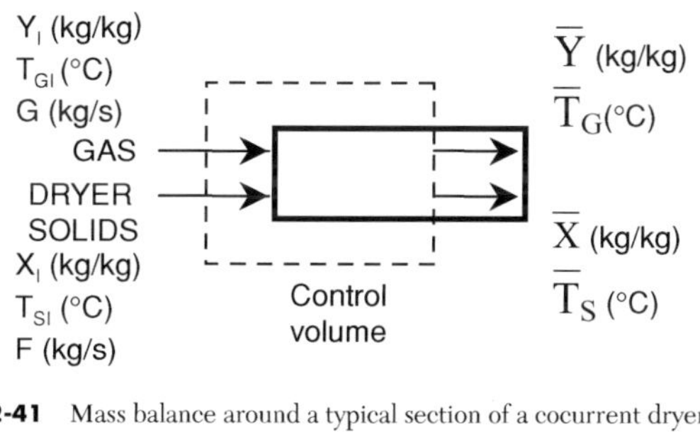

FIG. 12-41 Mass balance around a typical section of a cocurrent dryer.

The essential idea is to calculate the average gas humidity $\overline{Y}$ at each average moisture content $\overline{X}$.

A differential mass balance on the air at any position in the bed is given below.

$$F \cdot dX = -G \cdot dy$$

$$-\frac{dY}{dX} = \frac{F}{G} = \frac{\overline{Y} - Y_I}{X_I - \overline{X}} \tag{12-74}$$

where Y = gas humidity, kg moisture/kg dry gas
X = solids moisture content, kg moisture/kg dry solids
W_G = flow rate of dry gas, kg dry gas/s
W_S = flow rate of dry solids, kg dry solids/s

Application of mass balances: Plugging in the numbers gives the relationship between absolute humidity and moisture in the solids at any position.

$$\frac{\overline{Y} - 0.1}{0.5 - \overline{X}} = 1 = \frac{F}{G}$$

$$\overline{Y} = 0.6 - \overline{X}$$

$$Y_O = 0.6 - 0.15 = 0.45 \text{ kg/kg}$$

From Mollier chart: $T_{wb} = 79$°C $\quad Y_S^{\circ} = 0.48$ kg/kg

For the whole dryer, $\overline{Y} = 0.275$ kg/kg

The mass balance information is important, but not the entire answer to the question. Now the residence time can be calculated from the kinetics.

Application of concept of characteristic drying curve to estimating drying rates in practice (theory): The overall (required) change in moisture content is divided into a number of intervals of size ΔX. The sizes of the intervals need not be the same and should be finer where the fastest moisture content change occurs. For the sake of simplicity, this example will use intervals of uniform size. Then the application of the concept of a characteristic drying curve gives the following outcomes.

$$-\frac{d\overline{X}}{dt} = \text{drying rate in interval}$$

$$= \frac{f \cdot k \cdot \phi}{\rho_P} \frac{A_P}{V_P} (Y_S^{\circ} - \overline{Y}) \tag{12-75}$$

where f = relative drying rate in interval (dimensionless)
$\overline{Y}$ = average humidity in interval, kg/kg
ϕ = humidity potential coefficient, close to unity
ρ_P = density of dry solids, kg/m^3
Y_S° = humidity at saturation, from the adiabatic saturation contour on Mollier chart

$$\frac{A_P}{V_P} = \frac{\text{particle surface area}}{\text{particle volume}} = \frac{6}{\phi_P \cdot d_{PSM}} \tag{12-76}$$

d_{PSM} = Sauter-mean particle diameter for mixture (volume-surface diameter), m

ϕ_P = particle shape factor, unity for spheres (dimensionless)

k = **mass-transfer coefficient,** kg/(m^2·s), obtained from the heat-transfer coefficient (often easier to obtain) using the Chilton-Colburn analogy

$$k \cdot \phi = \frac{\delta}{C_{PY}} h \tag{12-77}$$

δ = psychrometric ratio, close to unity for air/water vapor system
C_{PY} = humid heat capacity = $C_{PG} + YC_{PV}$
C_{PG} = specific heat capacity of dry gas (air), J/(kg·K)
C_{PV} = specific heat capacity of water vapor, J/(kg·K)
h = **heat-transfer coefficient,** W/(m^2·K)

Define

$$Re_P = \text{particle Reynolds number}$$

$$= \frac{U \cdot d_{PSM}}{\nu} \tag{12-78}$$

U = relative velocity between gas and particles; in cascading rotary dryers, this is almost constant throughout the dryer and close to the superficial gas velocity U_{Gsuper}
ν = kinematic viscosity of gas at average T_G in dryer, m^2/s

For cascading rotary dryers:

$$\mathrm{Nu}_P = \text{particle Nusselt number}$$

$$= \min(0.03\ \mathrm{Re}_P^{1.3},\ 2 + 0.6\ \mathrm{Re}_P^{0.5}\mathrm{Pr}^{0.3}) \qquad (12\text{-}79)$$

$$h = \frac{\lambda G}{d_{PSM}}\mathrm{Nu}_P \qquad (12\text{-}80)$$

λ_G = thermal conductivity of gas, W/(m·K)

Application of concept of characteristic drying curve to estimating drying rates in practice

GAS PROPERTIES

$$\nu = 15 \times 10^{-6}\ \mathrm{m^2/s}$$
$$\mathrm{Pr} = 0.7$$
$$\lambda_G = 0.02\ \mathrm{W/(m\cdot K)}$$
$$C_{PG} = 1050\ \mathrm{J/(kg\cdot K)}$$
$$C_{PV} = 2000\ \mathrm{J/(kg\cdot K)}$$
$$C_{PY} = 1050 + 0.275 \cdot 2000 = 1600\ \mathrm{J/(kg\cdot K)}$$

We might do a more accurate calculation by calculating the gas properties at the conditions for each interval.

HEAT AND MASS-TRANSFER COEFFICIENTS

$$\mathrm{Re}_P = \frac{U \cdot d_{PSM}}{\nu} = \frac{4 \cdot 0.005}{15 \times 10^{-6}} = 1333$$

$$\mathrm{Nu} = \min[0.03(1333)^{1.3},\ 2 + 0.6(1333)^{0.5}0.7^{0.3}]$$
$$= 21.5$$

$$h = \frac{0.02}{0.005}21.5 = 86\ \mathrm{W/(m^2\cdot K)}$$

$$k = \frac{86}{1600} = 0.054\ \mathrm{kg/(m^2\cdot s)} \qquad (12\text{-}81)$$

$$\frac{A_P}{V_P} = \frac{6}{0.75 \cdot 0.005} = 1600\ \mathrm{m^{-1}}$$

$$\frac{d\overline{X}}{dt} = f\frac{0.054 \cdot 1600}{600}(0.48 - \overline{Y})$$
$$= 0.14f(0.48 - \overline{Y})$$

The particle temperature is obtained by analogy with the falling-rate expression for the drying rate (Keey, 1978). This procedure assumes that the particles reach a quasi-steady state temperature when they are resting on flights after cascading through the gas. The heat-transfer Fourier number typically approaches unity for the dwell time on the flights, meaning that the temperature distribution in the particles at the end of the dwell time is almost uniform. The heat-transfer Fourier number Fo is defined as $\alpha t/d_P^2$, where d_P is the particle diameter, t is time, and α is the thermal diffusivity. The thermal diffusivity α is the ratio of the thermal conductivity to the product of the density and the specific heat capacity.

The energy accumulation term in the energy balance for a particle is assumed to be zero. As above, this assumption can be justified because of the significant resting time that particles remain on a flight during the time that it is lifted around the drum. This quasi-steady state temperature in the energy balance, neglecting the energy accumulation term, is

$$h(T_G - T_s) = fN_w$$

where N_w is the maximum (unhindered) drying rate. For completely unhindered drying, $f = 1$ and T_s is the wet-bulb temperature T_W, so that

$$f = \frac{T_G - T_s}{T_G - T_W}$$

Under these conditions, the solids temperature may be obtained from the equation

$$T_s = T_G - f(T_G - T_W) \qquad (12\text{-}82)$$

This procedure has been used to calculate the average solids temperature in the eighth column of Table 12-13. This approach to the energy balance has been indicated experimentally for rendered meat solids through the accurate prediction of the maximum particle temperatures.

This procedure gives the particle residence time in the gas (38 s), and a typical variation of process conditions through a cocurrent cascading rotary dryer is shown in Fig. 12-42. We want the total residence time, which is the sum of the time in the gas τ_G and the time soaking on the flights τ_S. Figure 12-43 shows the enthalpy humidity chart used to generate the results in Table 12-13. There is an incomplete final row in Table 12-13 because the first two columns refer to the inlets and the outlets of the control volumes, while the remaining columns refer to the average conditions inside the control volumes, which are assumed to be the average of the inlet and outlet conditions. Hence column 3 ($\overline{X}$) is the average of the inlet and outlet moisture contents for each of the cells in column 2 (X_i). Column 4 ($\overline{Y}$) follows from column 3, using Eq. (12-74). Column 5 ($\overline{T_G}$) follows from column 4, using the enthalpy humidity chart in Fig. 12-43. Column 6 (T_{wb}) is read off the same enthalpy humidity chart. Column 7 (f) follows from the linear falling-rate curve, using the average moisture contents in column 3. Column 8 ($\overline{T_S}$) comes from columns 5 ($\overline{T_G}$), 7 (f), and the energy balance in Eq. (12-82), while column 9 ($d\overline{X}/dt$) comes from columns 4 ($\overline{Y}$), 7 (f), and Eq. (12-81). The final column comes from the difference between inlet and outlet moisture contents in column 2, divided by the average drying rate in column 9.

Computational Fluid Dynamics (CFD) CFD provides a very detailed and accurate model of the gas phase, including three-dimensional effects and swirl. Where localized flow patterns have a major effect on the overall performance of a dryer and the particle history, CFD can give immense improvements in modeling and in understanding of physical phenomena. Conversely, where the system is well mixed or drying is dominated by falling-rate kinetics and local conditions are unimportant, CFD modeling will give little or no advantage over conventional methods, but will incur a vastly greater cost in computing time.

CFD has been extensively applied in recent years to spray dryers (Langrish and Fletcher, 2001), but it has also been useful for other local three-dimensional swirling flows, e.g., around the feed point of pneumatic conveying dryers (Kemp et al., 1991), and for other cases where airflows affect drying significantly, e.g., local overdrying and warping in timber stacks (Langrish, 1999).

Design and Scale-up of Individual Dryer Types

Oven and Tray Dryers Scale up from tests with an oven or single tray at identical conditions (temperature, airflow or pressure, layer thickness, and agitation, if any). The total area of trays required is then proportional to the mass of material to be dried, compared to the small-scale test.

TABLE 12-13 The Variation in Process Conditions for the Example of a Cocurrent Cascading Rotary Dryer

Interval	X_j, kg/kg	$\overline{X}$, kg/kg	$\overline{Y}$, kg/kg	$\overline{T_G}$, °C	T_{wb}, °C	$f = \Phi$	$\overline{T_S}$, °C	$d\overline{X}/dt$, kg/(kg·s)	Δt_p, s
1	0.500	0.478	0.122	720	79.0	0.956	107	0.04656	0.94
2	0.456	0.434	0.166	630	78.5	0.869	151	0.03683	1.19
3	0.412	0.391	0.209	530	78.5	0.781	177	0.02820	1.55
4	0.369	0.347	0.253	430	78.0	0.694	186	0.02067	2.12
5	0.325	0.303	0.297	340	78.0	0.606	181	0.01424	3.07
6	0.281	0.259	0.341	250	78.0	0.519	161	0.00892	4.91
7	0.238	0.216	0.384	200	78.0	0.431	147	0.00470	9.32
8	0.194	0.172	0.428	130	78.0	0.344	112	0.00158	27.73
out	0.150	**Total required gas-phase residence time (s) = 50.82 s (summation of last column)**							

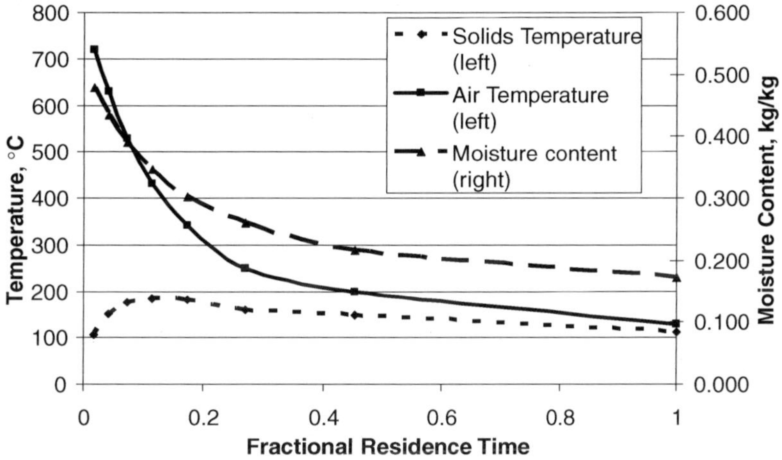

FIG. 12-42 Typical variation of process conditions through a cocurrent cascading rotary dryer.

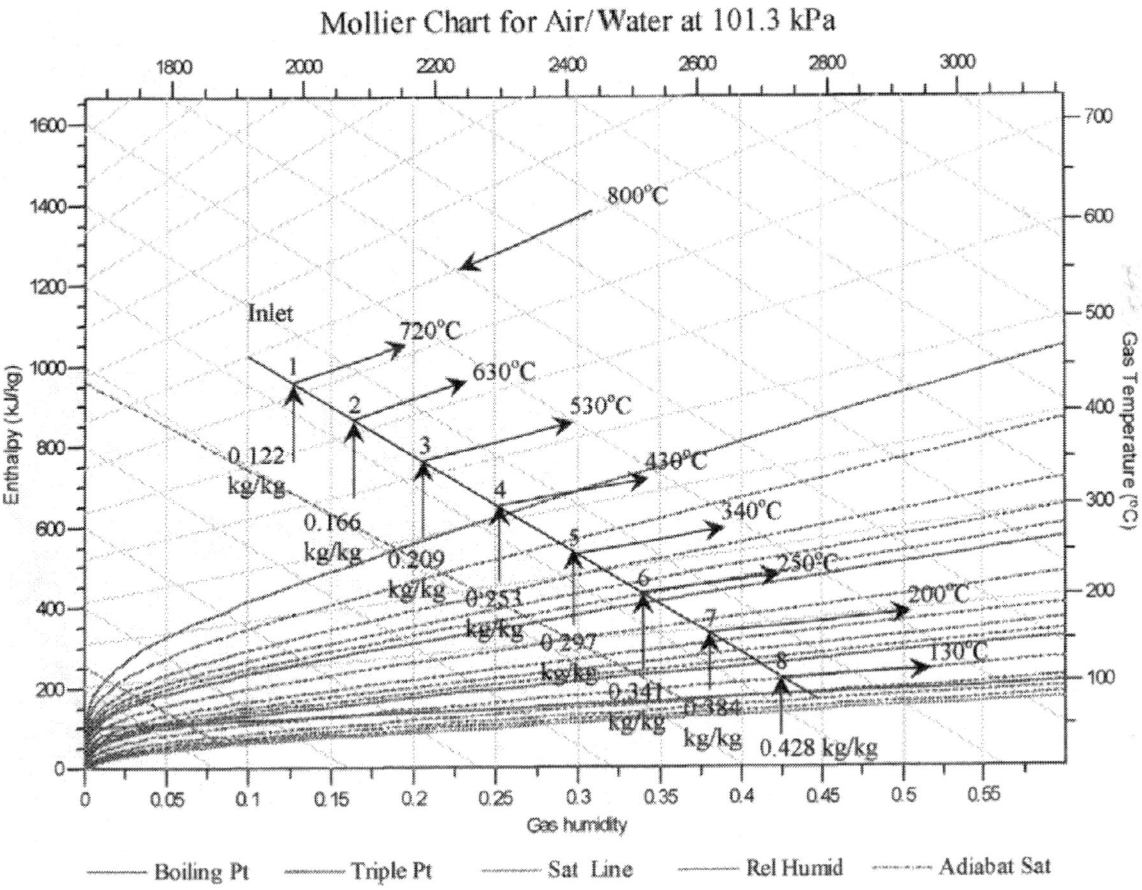

FIG. 12-43 Enthalpy humidity chart used to generate the results in Table 12-13 plots humidity (abscissa) against enthalpy (lines sloping diagonally from top left to bottom right).

Agitated and Rotating Batch Dryers Scale up from pilot-plant tests in a small-scale dryer at the same temperature and pressure and similar agitation conditions. As noted under scoping design, scale-up depends on the surface area/volume ratio, and hence normally to the one-third power of mass. Results from one dryer type may be extrapolated to a different type if assumptions are made on the heat transfer coefficients in both dryers; obviously this is less reliable than measurements on the same dryer type.

Fluidized-Bed Dryers In design mode, the required gas flow rate can be obtained from a heat and mass balance. Bed cross-sectional area is found from the scoping design calculation; the required gas velocity should be found from fluidization tests, but for initial design purposes, a typical value is 0.5 m/s.

For scale-up based on an experimentally recorded batch drying curve, including performance mode calculations and altering operating conditions, Kemp and Oakley (2002) showed that the drying time for a given range of moisture content ΔX scales according to the relationship

$$\frac{\Delta\tau_2}{\Delta\tau_1} = Z = \frac{(m_B/A)_2 G_1 (T_{GI} - T_{wb})_1 (1 - e^{-f \cdot NTU_z})_1}{(m_B/A)_1 G_2 (T_{GI} - T_{wb})_2 (1 - e^{-f \cdot NTU_z})_2} \qquad (12\text{-}83)$$

where 1 denotes experimental or original conditions and 2 denotes full-scale or new conditions; Z is the normalization factor; G is gas mass flux; m_B/A is bed mass per unit area, proportional to bed depth z; NTU is number of transfer units through the bed; and f is falling-rate kinetics factor. This method can be used to scale a batch drying curve section by section. Almost always, one of two simplified limiting cases applies, known as type A and type B normalization. In type A, f(NTU) is high, the exponential term is negligible, and the drying time is proportional to $G(T_{GI} - T_{wb})/(m_B/A)$. This applies to all fast-drying materials and the vast majority of other materials, even well into the falling-rate period. In type B, f(NTU) is low, and expanding the exponential term shows that drying time is simply proportional to $T_{GI} - T_{wb}$. This applies to a few very slow-drying materials, at very low moisture contents or where drying kinetics is completely controlled by internal moisture movement (e.g., wheat and grain, which have thick cell walls).

For a typical pilot-plant experiment, the fluidization velocity and temperature driving forces are similar to those of the full-size bed, but the bed diameter and depth are much less. Hence, for type A normalization, the m_B/A term dominates, Z is much greater than 1, and the drying time in the full-scale bed is typically 5 to 10 times that in the pilot-plant.

The drying time, bed area, solids throughput, and bed depth expressed as m_B/A are linked by

$$W_S = \frac{A_B}{\tau_S}\left(\frac{m_B}{A_B}\right) \qquad (12\text{-}84)$$

The consequence is that increasing gas velocity is beneficial for type A normalization (giving reduced drying time and either a higher throughput or a smaller bed area) but gives no real benefit for type B; likewise, increasing bed depth is beneficial for type B (giving either a higher throughput or a smaller bed area with the same drying time) but not type A. However, using unnecessarily high gas velocity or an unnecessarily deep bed increases pressure drop and operating costs.

Cascading Rotary Dryers In design mode, the required gas flow rate can be obtained from a heat and mass balance. Bed cross-sectional area is found from the scoping design calculation (a typical gas velocity is 3 m/s for cocurrent and 2 m/s for countercurrent units). Length is normally between 5 and 10 times drum diameter (an *L/D* value of 8 can be used for initial estimation) or can be calculated by using an incremental model (see worked example).

Entrainment Dryers In design mode, the required gas flow rate can be obtained from a heat and mass balance. For pneumatic conveying dryers, duct cross-sectional area and diameter are found from the scoping design calculation (if required gas velocity is unknown, a typical value is 20 m/s). Duct length can be estimated by an incremental model, but some parameters are hard to obtain and conditions change rapidly near the feed point, so the model is most effective for scaling up from pilot-plant data; see Kemp and Oakley (2002). Spray dryer chamber design is complex, and sizing should normally be done by manufacturers.

ADDITIONAL READING

Kemp, Bahu, and Oakley, "Modeling Vertical Pneumatic Conveying Dryers," Drying '91 (7th Int. Drying Symp., Prague, Czechoslovakia, Aug. 1990), Mujumdar et al. (eds.), Elsevier, 1991, pp. 217–227.

Kemp and Oakley, "Modeling of Particulate Drying in Theory and Practice," *Drying Technol.* **20**(9): 1699–1750 (2002).

Langrish, "The Significance of the Gaps between Boards in Determining the Moisture Content Profiles in the Drying of Hardwood Timber," *Drying Technol.* **17**(Pt. 7–8): 1481–1494 (1999).

Langrish and Fletcher, "Spray Drying of Food Ingredients and Applications of CFD in Spray Drying," *Chemical Engineering and Processing* **40**(4): 345–354 (2001).

Vanecek, Picka, and Najmr, "Some Basic Information on the Drying of Granulated NPK Fertilisers," *Int. Chem. Eng.* **4** (1): 93–99 (1964).

Vanecek, Picka, and Najmr, *Fluidized Bed Drying*, Leonard Hill, London, 1966.

DRYER DESCRIPTIONS

GENERAL REFERENCES: *Aspen Process Manual (Internet knowledge base)*, Aspen Technology, 2000 onward. Baker (ed.), *Industrial Drying of Foods*, Blaikie, London, 1997; Cook and DuMont, *Process Drying Practice*, McGraw-Hill, 1991. *Drying Technology—An International Journal*, Marcel Dekker, New York, 1982 onward. Keey, *Drying of Loose and Particulate Materials*, Hemisphere, New York, 1992. Masters, *Spray Drying Handbook*, Wiley, New York, 1990. Mujumdar (ed.), *Handbook of Industrial Drying*, Marcel Dekker, New York, 1995. Nonhebel and Moss, *Drying of Solids in the Chemical Industry*, CRC Press, Cleveland Ohio, 1971. van't Land, *Industrial Drying Equipment*, Marcel Dekker, New York, 1991.

Batch Tray Dryers

Description A tray or compartment dryer is an enclosed, insulated housing in which solids are placed upon tiers of trays in the case of particulate solids or stacked in piles or upon shelves in the case of large objects. Heat transfer may be *direct* from gas to solids by circulation of large volumes of hot gas or *indirect* by use of heated shelves, radiator coils, or refractory walls inside the housing. In indirect-heat units, excepting vacuum-shelf equipment, circulation of a small quantity of gas is usually necessary to sweep moisture vapor from the compartment and prevent gas saturation and condensation. Compartment units are employed for the heating and drying of lumber, ceramics, sheet materials (supported on poles), painted and metal objects, and all forms of particulate solids.

Classification Batch; nonagitated; layer; convective (cross-circulation or through-circulation) or contact/conduction.

Field of Application Because of the high labor requirements usually associated with loading or unloading the compartments, batch compartment equipment is rarely economical except in the following situations:

1. A long heating cycle is necessary because the size of the solid objects or permissible heating temperature requires a long holdup for internal diffusion of heat or moisture. This case may apply when the cycle will exceed 12 to 24 h.
2. The production of several different products requires strict batch identity and thorough cleaning of equipment between batches. This is a situation existing in many small multiproduct plants, e.g., for pharmaceuticals or specialty chemicals.
3. The quantity of material to be processed does not justify investment in more expensive, continuous equipment. This case would apply in many pharmaceutical drying operations.

Further, because of the nature of solids-gas contacting, which is usually by parallel flow and rarely by through-circulation, heat transfer and mass transfer are comparatively inefficient. For this reason, use of tray and compartment equipment is restricted primarily to ordinary drying and heat-treating operations. Despite these harsh limitations, when the listed situations do exist, economical alternatives are difficult to develop.

Auxiliary Equipment If noxious gases, fumes, or dust is given off during the operation, dust or fume recovery equipment will be necessary in the exhaust gas system. Wet scrubbers are employed for the recovery of valuable solvents from dryers. To minimize heat losses, thorough insulation of the compartment with brick, asbestos, or other insulating compounds is necessary. Modern fabricated dryer compartment panels usually have 7.5 to 15 cm of blanket insulation placed between the internal and external sheet-metal walls. Doors and other access openings should be gasketed and tight. In the case of tray and truck equipment, it is usually desirable to have available extra trays and trucks so that they can be preloaded for rapid emptying and loading of the compartment between cycles. Air filters and gas dryers are occasionally employed on the inlet air system for direct-heat units.

Vacuum-shelf dryers require auxiliary stream jets or other vacuum-producing devices, intercondensers for vapor removal, and occasionally wet scrubbers or (heated) bag-type dust collectors.

Uniform depth of loading in dryers and furnaces handling particulate solids is essential to consistent operation, minimum heating cycles, or control of final moisture. After a tray has been loaded, the bed should be leveled to a uniform depth. Special preform devices, noodle extruders, pelletizers, etc., are employed occasionally for preparing pastes and filter cakes so that screen bottom trays can be used and the advantages of through-circulation approached.

Control of tray and compartment equipment is usually maintained by control of the circulating air temperature (and humidity) and rarely by the solids temperature. On vacuum units, control of the absolute pressure and heating-medium temperature is utilized. In direct dryers, cycle controllers are frequently employed to vary the air temperature or velocity across the solids during the cycle; e.g., high air temperatures may be employed during a constant-rate drying period while the solids surface remains close to the air wet-bulb temperature. During the falling-rate periods, this temperature may be reduced to prevent case hardening or other degrading effects caused by overheating the solids surfaces. In addition, higher air velocities may be employed during early drying stages to improve heat transfer; however, after surface drying has been completed, this velocity may need to be reduced to prevent dusting. Two-speed circulating fans are employed commonly for this purpose.

Direct-Heat Tray Dryers Satisfactory operation of tray-type dryers depends on maintaining a constant temperature and a uniform air velocity over all the material being dried.

Circulation of air at velocities of 1 to 10 m/s is desirable to improve the surface heat-transfer coefficient and to eliminate stagnant air pockets. Proper airflow in tray dryers depends on sufficient fan capacity, on the design of ductwork to modify sudden changes in direction, and on properly placed baffles. *Nonuniform airflow is one of the most serious problems in the operation of tray dryers.*

Tray dryers may be of the tray-truck or the stationary-tray type. In the former, the trays are loaded on trucks which are pushed into the dryer; in the latter, the trays are loaded directly into stationary racks within the dryer. Trucks may be fitted with flanged wheels to run on tracks or with flat swivel wheels. They may also be suspended from and moved on monorails. Trucks usually contain two tiers of trays, with 18 to 48 trays per tier, depending upon the tray dimensions.

Trays may be square or rectangular, with 0.5 to 1 m^2 per tray, and may be fabricated from any material compatible with corrosion and temperature conditions. When the trays are stacked in the truck, there should be a clearance of not less than 4 cm between the material in one tray and the bottom of the tray immediately above. When material characteristics and handling permit, the trays should have screen bottoms for additional drying area. Metal trays are preferable to nonmetallic trays, since they conduct heat more readily. Tray loadings range usually from 1 to 10 cm deep.

Steam is the usual heating medium, and a standard heater arrangement consists of a main heater before the circulating fan. When steam is not available or the drying load is small, electric heat can be used. For temperatures above 450 K, products of combustion can be used, or indirect-fired air heaters.

Air is circulated by propeller or centrifugal fans; the fan is usually mounted within or directly above the dryer. Above 450 K, external or water-cooled bearings become necessary. Total pressure drop through the trays, heaters, and ductwork is usually in the range of 2.5 to 5 cm of water. Air recirculation is generally in the order of 80 to 95 percent except during the initial drying stage of rapid evaporation. Fresh air is drawn in by the circulating fan, frequently through dust filters. In most installations, air is exhausted by a separate small exhaust fan with a damper to control air recirculation rates.

Prediction of heat- and mass-transfer coefficients in direct heat tray dryers In convection phenomena, heat-transfer coefficients depend on the geometry of the system, the gas velocity past the evaporating surface, and the physical properties of the drying gas. In estimating drying rates, the use of heat-transfer coefficients is preferred because they are usually more reliable than mass-transfer coefficients. In calculating mass-transfer coefficients from drying experiments, the partial pressure at the surface is usually inferred from the measured or calculated temperature of the evaporating surface. Small errors in temperature have negligible effect on the heat-transfer coefficient but introduce relatively large errors in the partial pressure and hence in the mass-transfer coefficient.

For many cases in drying, the heat-transfer coefficient is proportional to U_g^n, where U_g is an appropriate local gas velocity. For flow parallel to plane plates, the exponent n has been reported to range from 0.35 to 0.8. The differences in exponent have been attributed to differences in flow pattern in the space above the evaporating surface, particularly whether it is laminar or turbulent, and whether the length is sufficient to allow fully developed flow. In the absence of applicable specific data, the heat-transfer coefficient for the parallel-flow case can be taken, for estimating purposes, as

$$h = \frac{8.8J^{0.8}}{D_c^{\,0.2}} \tag{12-85}$$

where h is the heat-transfer coefficient, W/(m^2·K) [or J/s·m^2·K); J is the gas mass flux, kg/(m^2·S); and D_c is a characteristic dimension of the system. The experimental data have been weighted in favor of an exponent of 0.8 in conformity with the usual Colburn j factor, and average values of the properties of air at 370 K have been incorporated. Typical values are in the range 10 to 50 W/(m^2·K).

Experimental data for drying from flat surfaces have been correlated by using the equivalent diameter of the flow channel or the length of the evaporating surface as the characteristic length dimension in the Reynolds number. However, the validity of one versus the other has not been established. The proper equivalent diameter probably depends at least on the geometry of the system, the roughness of the surface, and the flow conditions upstream of the evaporating surface. For most tray drying calculations, the equivalent diameter (4 times the cross-sectional area divided by the perimeter of the flow channel) should be used.

For airflow impinging normally to the surface from slots, nozzles, or perforated plates, the heat-transfer coefficient can be obtained from the data of Friedman and Mueller *(Proceedings of the General Discussion on Heat Transfer,* Institution of Mechanical Engineers, London, and American Society of Mechanical Engineers, New York, 1951, pp. 138–142). These investigators give

$$h = \alpha J^{0.78} \tag{12-86}$$

where the gas mass flux J is based on the total heat-transfer area and is dependent on the plate open area, hole or slot size, and spacing between the plate, nozzle, or slot and the heat-transfer surface.

Most efficient performance is obtained with plates having open areas equal to 2 to 3 percent of the total heat-transfer area. The plate should be located at a distance equal to four to six hole (or equivalent) diameters from the heat-transfer surface.

Data from tests employing multiple slots, with a correction calculated for slot width, were reported by Korger and Kizek *[Int. J. Heat Mass Transfer, London,* **9:**337 (1966)].

Another well-known correlation has been used to predict heat- and mass-transfer coefficients for air impinging on a surface from arrays of holes (jets). This correlation uses relevant geometric properties such as the diameter of the holes, the distance between the holes, and the distance between the holes and the sheet [Martin, "Heat and Mass Transfer Between Impinging Gas Jets and Solid

Surfaces," *Advances in Heat Transfer,* vol. 13, Academic Press, 1977, pp. 1–66].

$$\frac{Sh}{Sc^{0.42}} = \frac{Nu}{Pr^{0.42}} = \left[1 + \left(\frac{H/D}{0.6/\sqrt{f}}\right)^6\right]^{-0.05} \times \sqrt{f}\,\frac{1 - 2.2\sqrt{f}}{1 + 0.2(H/D - 6)\sqrt{f}} \times Re^{2/3}$$

where D = diameter of nozzle, m

$$f = \frac{\pi}{2\sqrt{3}}\left(\frac{D}{L_D}\right)^2$$

H = distance from nozzle to sheet, m
L_D = average distance between nozzles, m
Nu = Nusselt number
Pr = Prandtl number
$Re = \dfrac{wD}{\nu}$, Reynolds number
Sc = Schmidt number
Sh = Sherwood number
w = velocity of air at nozzle exit, m/s
ν = kinematic viscosity of air, m^2/s

The heat- and mass-transfer coefficients were then calculated from the definitions of the Nusselt and Sherwood numbers.

$$Nu = \frac{hD}{k_{th}}$$ where k_{th} = thermal conductivity of air, W/(m·K)

D = diameter of holes in air bars, m

$$Sh = \frac{k_m^\circ D}{\text{diff}}$$ where diff = diffusion coefficient of water vapor in air, m^2/s

Air impingement is commonly employed for drying sheets, film, thin slabs, and coatings. The temperature driving force must also be found. When radiation and conduction are negligible, the temperature of the evaporating surface approaches the wet-bulb temperature and is readily obtained from the humidity and dry-bulb temperatures. Frequently, however, radiation and conduction cause the temperature of the evaporating surface to exceed the wet-bulb temperature. When this occurs, the true surface temperature must be estimated. The easiest way is to use a psychrometric chart and to change the slope of the adiabatic saturation line; a typical figure for the additional radiation is about 10 percent. In many cases this is canceled out by heat losses and the heat required to warm the solid, leaving the dryer approximately adiabatic.

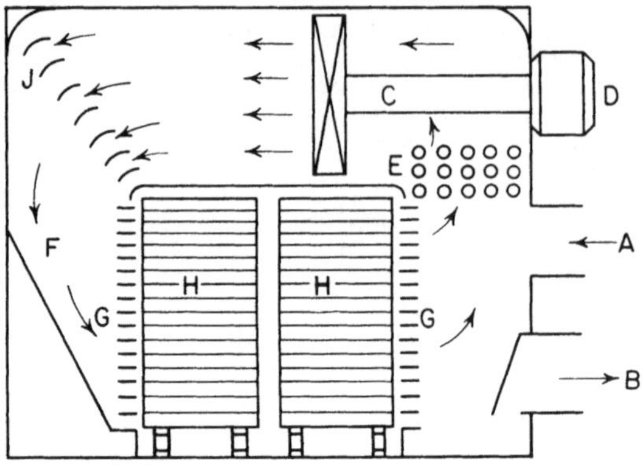

FIG. 12-44 Double-truck tray dryer. (*A*) Air inlet duct. (*B*) Air exhaust duct with damper. (*C*) Adjustable-pitch fan 1 to 15 hp. (*D*) Fan motor. (*E*) Fin heaters. (*F*) Plenum chamber. (*G*) Adjustable air blast nozzles. (*H*) Trucks and trays. (*J*) Turning vanes.

As with many drying calculations, the most reliable design method is to perform experimental tests and to scale up. By measuring performance on a single tray with similar layer depth, air velocity, and temperature, the SDR (specific drying rate) concept can be applied to give the total area and number of trays required for the full-scale dryer.

Performance data for direct heat tray dryers A standard two-truck dryer is illustrated in Fig. 12-44. Adjustable baffles or a perforated distribution plate is normally employed to develop 0.3 to 1.3 cm of water pressure drop at the wall through which air enters the truck enclosure. This will enhance the uniformity of air distribution, from top to bottom, among the trays. In three (or more) truck ovens, air reheat coils may be placed between trucks if the evaporative load is high. Means for reversing airflow direction may also be provided in multiple-truck units.

Performance data on some typical tray and compartment dryers are tabulated in Table 12-14. These indicate that an overall rate of evaporation of 0.0025 to 0.025 kg water/(s·m^2) of tray area may be expected from tray and tray-truck dryers. The thermal efficiency of this type of dryer will vary from 20 to 50 percent, depending on the drying temperature used and the humidity of the exhaust air. In drying to very low moisture contents under temperature restrictions, the thermal efficiency may be on the order of 10 percent. The major operating cost for a tray dryer is the labor involved in loading and unloading the trays. About 2 labor-hours is required to load and unload a standard

TABLE 12-14 Manufacturer's Performance Data for Tray and Tray-Truck Dryers*

Material	Color	Chrome yellow	Toluidine red	Half-finished Titone	Color
Type of dryer	2-truck	16-tray dryer	16-tray	3-truck	2-truck
Capacity, kg product/h	11.2	16.1	1.9	56.7	4.8
Number of trays	80	16	16	180	120
Tray spacing, cm	10	10	10	7.5	9
Tray size, cm	60 × 75 × 4	65 × 100 × 2.2	65 × 100 × 2	60 × 70 × 3.8	60 × 70 × 2.5
Depth of loading, cm	2.5 to 5	3	3.5	3	
Initial moisture, % bone-dry basis	207	46	220	223	116
Final moisture, % bone-dry basis	4.5	0.25	0.1	25	0.5
Air temperature, °C	85–74	100	50	95	99
Loading, kg product/m^2	10.0	33.7	7.8	14.9	9.28
Drying time, h	33	21	41	20	96
Air velocity, m/s	1.0	2.3	2.3	3.0	2.5
Drying, kg water evaporated/(h·m^2)	0.59	65	0.41	1.17	0.11
Steam consumption, kg/kg water evaporated	2.5	3.0	—	2.75	
Total installed power, kW	1.5	0.75	0.75	2.25	1.5

*Courtesy of Wolverine Proctor & Schwartz, Inc.

two-truck tray dryer. In addition, about one-third to one-fifth of a worker's time is required to supervise the dryer during the drying period. Power for tray and compartment dryers will be approximately 1.1 kW per truck in the dryer. Maintenance will run from 3 to 5 percent of the installed cost per year.

Batch Through-Circulation Dryers These may be either of shallow bed or deep bed type. In the first type of batch through-circulation dryer, heated air passes through a stationary permeable bed of the wet material placed on removable screen-bottom trays suitably supported in the dryer. This type is similar to a standard tray dryer except that hot air passes through the wet solid instead of across it. The pressure drop through the bed of material does not usually exceed about 2 cm of water. In the second type, deep perforated-bottom trays are placed on top of plenum chambers in a closed-circuit hot air circulating system. In some food-drying plants, the material is placed in finishing bins with perforated bottoms; heated air passes up through the material and is removed from the top of the bin, reheated, and recirculated. The latter types involve a pressure drop through the bed of material of 1 to 8 cm of water at relatively low air rates. Table 12-15 gives performance data on three applications of batch through-circulation dryers. Batch through circulation dryers are restricted in application to granular materials (particle size typically 1 mm or greater) that permit free flow-through circulation of air. Drying times are usually much shorter than in parallel-flow tray dryers. Design methods are included in the subsection "Continuous Through-Circulation Dryers."

Contact Tray and Vacuum-Shelf Dryers Vacuum-shelf dryers are indirectly heated batch dryers consisting of a vacuum-tight chamber usually constructed of cast iron or steel plate, heated, supporting shelves within the chamber, a vacuum source, and usually a condenser. One or two doors are provided, depending on the size of the chamber. The doors are sealed with resilient gaskets of rubber or similar material. It is also possible, but much less common, to operate at atmospheric pressure without vacuum.

Hollow shelves of flat steel plate are fastened permanently inside the vacuum chamber and are connected in parallel to inlet and outlet headers. The heating medium, entering through one header and passing through the hollow shelves to the exit header, is generally steam, ranging in pressure from 700 kPa gauge to subatmospheric pressure for low-temperature operations. Low temperatures can be provided by circulating hot water, and high temperatures can be obtained by circulating hot oil or Dowtherm. Some small dryers employ electrically heated shelves. The material to be dried is placed in pans or trays on the heated shelves. The trays are generally of metal to ensure good heat transfer between the shelf and the tray.

TABLE 12-15 Performance Data for Batch Through-Circulation Dryers*

Kind of material	Granular polymer	Vegetable	Vegetable seeds
Capacity, kg product/h	122	42.5	27.7
Number of trays	16	24	24
Tray spacing, cm	43	43	43
Tray size, cm	91.4 × 104	91.4 × 104	85 × 98
Depth of loading, cm	7.0	6	4
Physical form of product	Crumbs	0.6-cm diced cubes	Washed seeds
Initial moisture content, % dry basis	11.1	669.0	100.0
Final moisture content, % dry basis	0.1	5.0	9.9
Air temperature, °C	88	77 dry-bulb	36
Air velocity, superficial, m/s	1.0	0.6 to 1.0	1.0
Tray loading, kg product/m²	16.1	5.2	6.7
Drying time, h	2.0	8.5	5.5
Overall drying rate, kg water evaporated/(h·m²)	0.89	11.86	1.14
Steam consumption, kg/kg water evaporated	4.0	2.42	6.8
Installed power, kW	7.5	19	19

*Courtesy of Wolverine Proctor & Schwartz, Inc.

Vacuum-shelf dryers may vary in size from 1 to 24 shelves, the largest chambers having overall dimensions of 6 m wide, 3 m long, and 2.5 m high.

Vacuum is applied to the chamber, and vapor is removed through a large pipe which is connected to the chamber in such a manner that if the vacuum is broken suddenly, the in-rushing air will not greatly disturb the bed of material being dried. This line leads to a condenser where moisture or solvent that has been vaporized is condensed. The noncondensable exhaust gas goes to the vacuum source, which may be a wet or dry vacuum pump or a steam-jet ejector.

Vacuum-shelf dryers are used extensively for drying pharmaceuticals, temperature-sensitive or easily oxidizable materials, and materials so valuable that labor cost is insignificant. They are particularly useful for handling small batches of materials wet with toxic or valuable solvents. Recovery of the solvent is easily accomplished without danger of passing through an explosive range. Dusty materials may be dried with negligible dust loss. Hygroscopic materials may be completely dried at temperatures below that required in atmospheric dryers. The equipment is employed also for freeze-drying processes, for metallizing-furnace operations, and for the manufacture of semiconductor parts in controlled atmospheres. All these latter processes demand much lower operating pressures than do ordinary drying operations.

Design methods for vacuum-shelf dryers Heat is transferred to the wet material by conduction through the shelf and bottom of the tray and by radiation from the shelf above. The critical moisture content will not be necessarily the same as for atmospheric tray drying, as the heat-transfer mechanisms are different.

During the constant-rate period, moisture is rapidly removed. Often 50 percent of the moisture will evaporate in the first hour of a 6- to 8-h cycle. The drying time has been found to be proportional to between the first and second power of the depth of loading. Shelf vacuum dryers operate in the range of 1 to 25 mmHg pressure. For size-estimating purposes, a heat-transfer coefficient of 20 J/(m²·s·K) may be used. The area employed in this case should be the shelf area in direct contact with the trays. Trays should be maintained as flatly as possible to obtain maximum area of contact with the heated shelves. For the same reason, the shelves should be kept free from scale and rust. Air vents should be installed on steam-heated shelves to vent noncondensable gases. The heating medium should not be applied to the shelves until after the air has been evacuated from the chamber, to reduce the possibility of the material's overheating or boiling at the start of drying. Case hardening can sometimes be avoided by retarding the rate of drying in the early part of the cycle.

Performance data for vacuum-shelf dryers The purchase price of a vacuum-shelf dryer depends upon the cabinet size and number of shelves per cabinet. For estimating purposes, typical prices (1985) and auxiliary equipment requirements are given in Table 12-16. Installed cost of the equipment will be roughly 100 percent of the carbon-steel purchase cost.

The thermal efficiency of a vacuum-shelf dryer is usually on the order of 60 to 80 percent. Table 12-17 gives operating data for one organic color and two inorganic compounds. Labor may constitute 50 percent of the operating cost; maintenance, 20 percent. Annual maintenance costs amount to 5 to 10 percent of the total installed cost. Actual labor costs will depend on drying time, facilities for loading and unloading trays, etc. The power required for these dryers is only that for the vacuum system; for vacuums of 680 to 735 mmHg, the power requirements are on the order of 0.06 to 0.12 kW/m² tray surface.

Continuous Tray and Gravity Dryers Continuous tray dryers are equivalent to batch tray dryers, but with the solids moving between trays by a combination of mechanical movement and gravity. Gravity (moving-bed) dryers are normally through-circulation convective dryers with no internal trays where the solids gradually descend by gravity. In all these types, the net movement of solids is vertically downward.

Classification Continuous; nonagitated (except for turnover when falling between trays); layer; convective (cross-circulation or through-circulation) or contact/conduction; vertical solids movement by gravity and mechanical agitation.

Turbo-Tray Dryers The turbo-tray dryer (also known as rotating tray, rotating shelf, or Wyssmont TURBO-DRYER®) is a continuous

TABLE 12-16 Standard Vacuum-Shelf Dryers*

Shelf area, m^2	Floor space, m^2	Weight average, kg	Pump capacity, m^3/s	Pump motor, kW	Condenser area, m^2	Price/m^2 (1995)	
						Carbon steel	304 stainless steel
0.4–1.1	4.5	540	0.024	1.12	1	$110	$170
1.1–2.2	4.5	680	0.024	1.12	1	75	110
2.2–5.0	4.6	1130	0.038	1.49	4	45	65
5.0–6.7	5.0	1630	0.038	1.49	4	36	65
6.7–14.9	6.4	3900	0.071	2.24	9	27	45
16.7–21.1	6.9	5220	0.071	2.24	9	22	36

*Stokes Vacuum, Inc.

dryer consisting of a stack of rotating annular shelves in the center of which turbo-type fans revolve to circulate the air over the shelves. Wet material enters through the roof, falling onto the top shelf as it rotates beneath the feed opening. After completing 1 r, the material is wiped by a stationary wiper through radial slots onto the shelf below, where it is spread into a uniform pile by a stationary leveler. The action is repeated on each shelf, with transfers occurring once in each revolution. From the last shelf, material is discharged through the bottom of the dryer (Fig. 12-45). The steel-frame housing consists of removable insulated panels for access to the interior. All bearings and lubricated parts are exterior to the unit with the drives located under the housing. Parts in contact with the product may be of steel or special alloy. The trays can be of any sheet material.

The rate at which each fan circulates air can be varied by changing the pitch of the fan blades. In final drying stages, in which diffusion controls or the product is light and powdery, the circulation rate is considerably lower than in the initial stage, in which high evaporation rates prevail. In the majority of applications, air flows through the dryer upward in counterflow to the material. In special cases, required drying conditions dictate that airflow be cocurrent or both countercurrent and cocurrent with the exhaust leaving at some level between solids inlet and discharge. A separate cold-air-supply fan is provided if the product is to be cooled before being discharged.

By virtue of its vertical construction, the turbo-type tray dryer has a stack effect, the resulting draft being frequently sufficient to operate the dryer with natural draft. Pressure at all points within the dryer is maintained close to atmospheric. Most of the roof area is used as a breeching, lowering the exhaust velocity to settle dust back into the dryer.

Heaters can be located in the space between the trays and the dryer housing, where they are not in direct contact with the product, and thermal efficiencies up to 3500 kJ/kg (1500 Btu/lb) of water evaporated can be obtained by reheating the air within the dryer. For materials which have a tendency to foul internal heating surfaces, an external heating system is employed.

The turbo-tray dryer can handle materials from thick slurries [1 million N·s/m^2 (100,000 cP) and over] to fine powders. Filterpress cakes are granulated before feeding. Thixotropic materials are fed directly from a rotary filter by scoring the cake as it leaves the drum. Pastes can be extruded onto the top shelf and subjected to a hot blast of air to make them firm and free-flowing after 1 r.

TABLE 12-17 Performance Data of Vacuum-Shelf Dryers

Material	Sulfur black	Calcium carbonate	Calcium phosphate
Loading, kg dry material/m^2	25	17	33
Steam pressure, kPa gauge	410	410	205
Vacuum, mmHg	685–710	685–710	685–710
Initial moisture content, % (wet basis)	50	50.3	30.6
Final moisture content, % (wet basis)	1	1.15	4.3
Drying time, h	8	7	6
Evaporation rate, kg/(s·m^2)	8.9×10^{-4}	7.9×10^{-4}	6.6×10^{-4}

The turbo-tray dryer is manufactured in sizes from package units 2 m in height and 1.5 m in diameter to large outdoor installations 20 m in height and 11 m in diameter. Tray areas range from 1 m^2 up to about 2000 m^2. The number of shelves in a tray rotor varies according to space available and the minimum rate of transfer required, from as few as 12 shelves to as many as 58 in the largest units. Standard construction permits operating temperatures up to 615 K, and high-temperature heaters permit operation at temperatures up to 925 K.

A recent innovation has enabled TURBO-DRYER® to operate with very low inert gas makeup. Wyssmont has designed a tank housing that is welded up around the internal structure rather than the column-and-gasket panel design that has been the Wyssmont standard for many years. In field-erected units, the customer does the welding in the field; in packaged units, the tank-type welding is done in the shop. The tank-type housing finds particular application for operation under positive pressure. On the standard design, doors with explosion latches and gang latch operators are used. In the tank-type design,

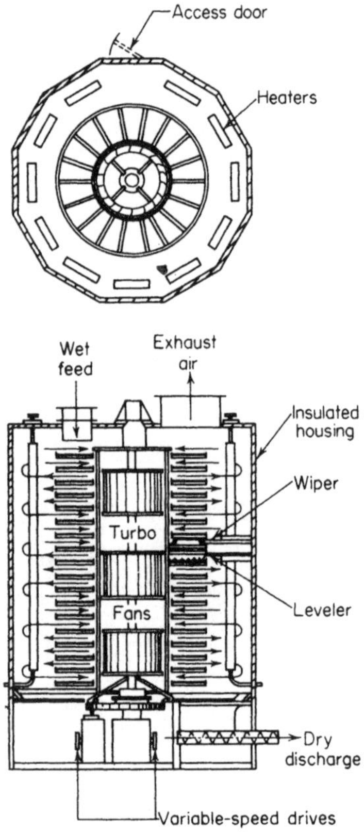

FIG. 12-45 TURBO-DRYER®. (*Wyssmont Company, Inc.*)

tight-sealing manway-type openings permit access to the interior. Tank-type housing designs have been requested when drying solvent wet materials and for applications where the material being dried is highly toxic and certainty is required that no toxic dust get out.

Design methods for turbo-tray dryers The heat- and mass-transfer mechanisms are similar to those in batch tray dryers, except that constant turning over and mixing of the solids significantly improve drying rates. Design must usually be based on previous installations or pilot tests by the manufacturer; apparent heat-transfer coefficients are typically 30 to 60 J/(m²·s·K) for dry solids and 60 to 120 J/(m²·s·K) for wet solids. Turbo-tray dryers have been employed successfully for the drying and cooling of calcium hypochlorite, urea crystals, calcium chloride flakes, and sodium chloride crystals. The Wyssmont "closed-circuit" system, as shown in Fig. 12-46, consists of the turbo-tray dryer with or without internal heaters, recirculation fan, condenser with receiver and mist eliminators, and reheater. Feed and discharge are through a sealed wet feeder and lock, respectively. This method is used for continuous drying without leakage of fumes, vapors, or dust to the atmosphere. A unified approach for scaling up dryers such as turbo-tray, plate, conveyor, or any other dryer type that forms a defined layer of solids next to a heating source is the SDR (specific drying rate) method described by Moyers [*Drying Technol.* **12**(1 & 2): 393–417 (1994)].

Performance and cost data for turbo-tray dryers Performance data for four applications of closed-circuit drying are included in Table 12-18. Operating, labor, and maintenance costs compare favorably with those of direct heat rotating equipment.

Plate Dryers The plate dryer is an indirectly heated, fully continuous dryer available for three modes of operation: atmospheric, gastight, or full vacuum. The dryer is of vertical design, with horizontal, heated plates mounted inside the housing. The plates are heated by hot water, steam, or thermal oil, with operating temperatures up to 320°C possible. The product enters at the top and is conveyed through the dryer by a product transport system consisting of a central-rotating shaft with arms and plows. (See dryer schematic, Fig. 12-47.) The thin product layer [approximately ½-in (12-mm) depth] on the surface of the plates, coupled with frequent product turnover by the conveying system, results in short retention times (approximately 5 to 40 min), true plug flow of the material, and uniform drying. The vapors are removed from the dryer by a small amount of heated purge gas or by vacuum. The material of construction of the plates and housing is normally stainless steel, with special metallurgies also available. The drive unit is located at the bottom of the dryer and supports the central-rotating shaft. Typical speed of the dryer is 1 to 7 rpm. Full-opening doors are located on two adjacent sides of the dryer for easy access to dryer internals.

The plate dryer may vary in size from 5 to 35 vertically stacked plates with a heat-exchange area between 3.8 and 175 m². The largest unit available has overall dimensions of 3 m (w) by 4 m (l) by 10 m (h). Depending upon the loose-bulk density of the material and the overall retention time, the plate dryer can process up to 5000 kg/h of wet product.

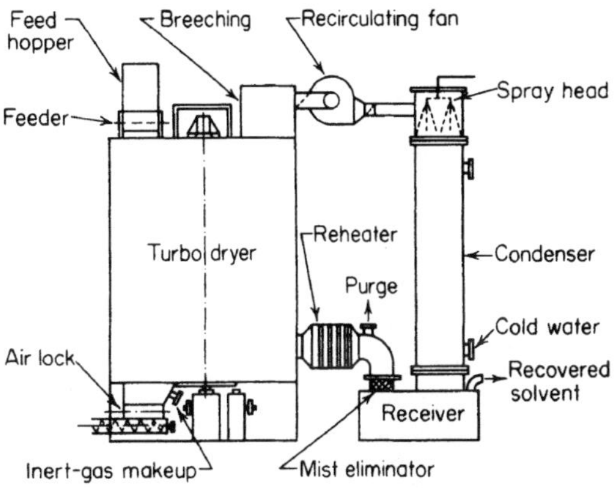

FIG. 12-46 TURBO-DRYER® in closed circuit for continuous drying with solvent recovery. *(Wyssmont Company, Inc.)*

The plate dryer is limited in its scope of applications only in the consistency of the feed material (the products must be friable, free-flowing, and not undergo phase changes) and drying temperatures up to 320°C. Applications include specialty chemicals, pharmaceuticals, foods, polymers, pigments, etc. Initial moisture or volatile level can be as high as 65 percent, and the unit is often used as a final dryer to take materials to a bone-dry state, if necessary. The plate dryer can also be used for heat treatment, removal of waters of hydration (bound moisture), solvent removal, and as a product cooler.

The atmospheric plate dryer is a dust-tight system. The dryer housing is an octagonal, panel construction, with operating pressure in the range of ±0.5 kPa gauge. An exhaust air fan draws the purge air through the housing for removal of the vapors from the drying process. The purge air velocity through the dryer is in the range of 0.1 to 0.15 m/s, resulting in minimal dusting and small dust filters for the exhaust air. The air temperature is normally equal to the plate temperature. The vapor-laden exhaust air is passed through a dust filter or a scrubber (if necessary) and is discharged to the atmosphere. Normally, water is the volatile to be removed in this type of system.

The gastight plate dryer, together with the components of the gas recirculation system, forms a closed system. The dryer housing is semicylindrical and is rated for a nominal pressure of 5 kPa gauge. The flow rate of the recirculating purge gas must be sufficient to absorb the vapors generated from the drying process. The gas temperature must be adjusted according to the specific product characteristics and the

TABLE 12-18 Turbo-Dryer® Performance Data in Wyssmont Closed-Circuit Operations*

Material dried	Antioxidant	Water-soluble polymer	Antibiotic filter cake	Petroleum coke
Dried product, kg/h	500	85	2400	227
Volatiles composition	Methanol and water	Xylene and water	Alcohol and water	Methanol
Feed volatiles, % wet basis	10	20	30	30
Product volatiles, % wet basis	0.5	4.8	3.5	0.2
Evaporation rate, kg/h	53	16	910	302
Type of heating system	External	External	External	External
Heating medium	Steam	Steam	Steam	Steam
Drying medium	Inert gas	Inert gas	Inert gas	Inert gas
Heat consumption, J/kg	0.56×10^6	2.2×10^6	1.42×10^6	1.74×10^6
Power, dryer, kW	1.8	0.75	12.4	6.4
Power, recirculation fan, kW	5.6	5.6	37.5	15
Materials of construction	Stainless-steel interior	Stainless-steel interior	Stainless-steel interior	Carbon steel
Dryer height, m	4.4	3.2	7.6	6.5
Dryer diameter, m	2.9	1.8	6.0	4.5
Recovery system	Shell-and-tube condenser	Shell-and-tube condenser	Direct-contact condenser	Shell-and-tube condenser
Condenser cooling medium	Brine	Chilled water	Tower water	Chilled water
Location	Outdoor	Indoor	Indoor	Indoor
Approximate cost of dryer (2004)	$300,000	$175,000	$600,000	$300,000
Dryer assembly	Packaged unit	Packaged unit	Field-erected unit	Field-erected unit

*Courtesy of Wyssmont Company, Inc.

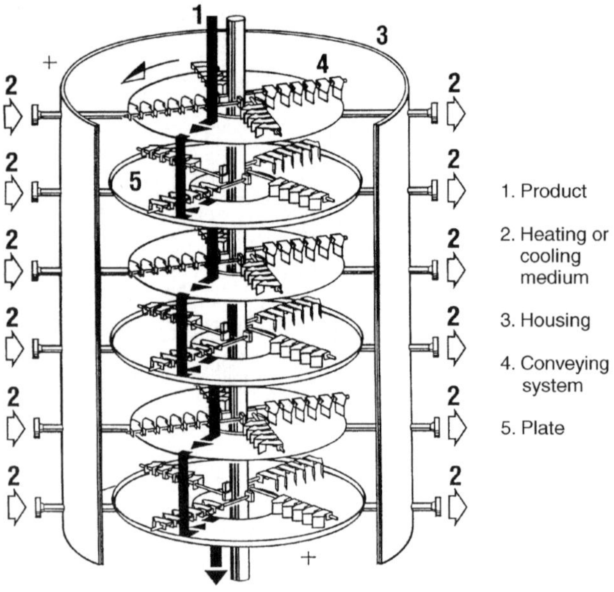

FIG. 12-47 Indirect heat continuous plate dryer for atmospheric, gastight, or full-vacuum operation. *(Krauss Maffei.)*

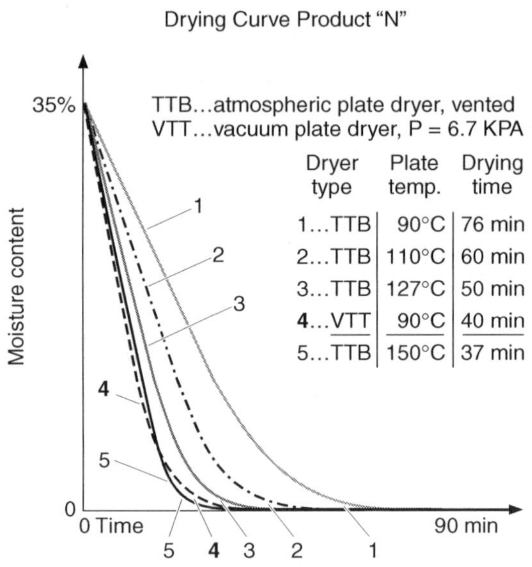

FIG. 12-48 Plate dryer drying curves demonstrating impact of elevated temperature and/or operation under vacuum. *(Krauss Maffei.)*

type of volatile. After condensation of the volatiles, the purge gas (typically nitrogen) is recirculated back to the dryer via a blower and heat exchanger. Solvents such as methanol, toluene, and acetone are normally evaporated and recovered in the gastight system.

The vacuum plate dryer is provided as part of a closed system. The vacuum dryer has a cylindrical housing and is rated for full-vacuum operation (typical pressure range of 3 to 27 kPa absolute). The exhaust vapor is evacuated by a vacuum pump and is passed through a condenser for solvent recovery. There is no purge gas system required for operation under vacuum. Of special note in the vacuum-drying system are the vacuum feed and discharge locks, which allow for continuous operation of the plate dryer under full vacuum.

Comparison data—plate dryers Comparative studies have been done on products under both atmospheric and vacuum drying conditions. See Fig. 12-48. These curves demonstrate (1) the improvement in drying achieved with elevated temperature and (2) the impact to the drying process obtained with vacuum operation. Note that curve 4 at 90°C, pressure at 6.7 kPa absolute, is comparable to the atmospheric curve at 150°C. Also, the comparative atmospheric curve at 90°C requires 90 percent more drying time than the vacuum condition. The dramatic improvement with the use of vacuum is important to note for heat-sensitive materials.

The above drying curves have been generated via testing on a plate dryer simulator. The test unit duplicates the physical setup of the production dryer; therefore linear scale-up from the test data can be made to the full-scale dryer. Because of the thin product layer on each plate, drying in the unit closely follows the normal type of drying curve in which the constant-rate period (steady evolution of moisture or volatiles) is followed by the falling-rate period of the drying process. This results in higher heat-transfer coefficients and specific drying capacities on the upper plates of the dryer as compared to the lower plates. The average specific drying capacity for the plate dryer is in the range of 2 to 20 kg/(m²·h) (based on final dry product). Performance data for typical applications are shown on Table 12-19.

TABLE 12-19 Plate Dryer Performance Data for Three Applications*

Product	Plastic additive	Pigment	Foodstuff
Volatiles	Methanol	Water	Water
Production rate, dry	362 kg/hr	133 kg/hr	2030 kg/hr
Inlet volatiles content	30%	25%	4%
Final volatiles content	0.1%	0.5%	0.7%
Evaporative rate	155 kg/hr	44 kg/hr	70 kg/hr
Heating medium	Hot water	Steam	Hot water
Drying temperature	70°C	150°C	90°C
Dryer pressure	11 kPa abs	Atmospheric	Atmospheric
Air velocity	NA	0.1 m/sec	0.2 m/sec
Drying time, min	24	23	48
Heat consumption, kcal/kg dry product	350	480	100
Power, dryer drive	3 kW	1.5 kW	7.5 kW
Material of construction	SS 316L/316Ti	SS 316L/316Ti	SS 316L/316Ti
Dryer height	5 m	2.6 m	8.2 m
Dryer footprint	2.6 m diameter	2.2 m by 3.0 m	3.5 m by 4.5 m
Location	Outdoors	Indoors	Indoors
Dryer assembly	Fully assembled	Fully assembled	Fully assembled
Power, exhaust fan	NA	2.5 kW	15 kW
Power, vacuum pump	20 kW	NA	NA

*Krauss Maffei

Gravity or Moving-Bed Dryers A body of solids in which the particles, consisting of granules, pellets, beads, or briquettes, flow downward by gravity at substantially their normal settled bulk density through a vessel in contact with gases is defined frequently as a **moving-bed** or **tower dryer.** Moving-bed equipment is frequently used for grain drying and plastic pellet drying, and it also finds application in blast furnaces, shaft furnaces, and petroleum refining. Gravity beds are also employed for the cooling and drying of extruded pellets and briquettes from size enlargement processes.

A gravity dryer consists of a stationary vertical, usually cylindrical housing with openings for the introduction of solids (at the top) and removal of solids (at the bottom), as shown schematically in Fig. 12-49. Gas flow is through the solids bed and may be cocurrent or countercurrent and, in some instances, cross-flow. By definition, the rate of gas flow upward must be less than that required for fluidization.

Fields of application One of the major advantages of the gravity-bed technique is that it lends itself well to true intimate countercurrent contacting of solids and gases. This provides for efficient heat transfer and mass transfer. Gravity-bed contacting also permits the use of the solid as a heat-transfer medium, as in pebble heaters.

Gravity vessels are applicable to coarse granular free-flowing solids which are comparatively dust-free. The solids must possess physical properties in size and surface characteristics so that they will not stick together, bridge, or segregate during passage through the vessel. The presence of significant quantities of fines or dust will close the passages among the larger particles through which the gas must penetrate, increasing pressure drop. Fines may also segregate near the sides of the bed or in other areas where gas velocities are low, ultimately completely sealing off these portions of the vessel. The high efficiency of gas-solids contacting in gravity beds is due to the uniform distribution of gas throughout the solids bed; hence choice of feed and its preparation are important factors to successful operation. Preforming techniques such as pelleting and briquetting are employed frequently for the preparation of suitable feed materials.

Gravity vessels are suitable for low-, medium-, and high-temperature operation; in the last case, the housing will be lined completely with refractory brick. Dust recovery equipment is minimized in this type of operation since the bed actually performs as a dust collector itself, and dust in the bed will not, in a successful application, exist in large quantities.

Other advantages of gravity beds include flexibility in gas and solids flow rates and capacities, variable retention times from minutes to several hours, space economy, ease of start-up and shutdown, the potentially large number of contacting stages, and ease of control by using the inlet and exit gas temperatures.

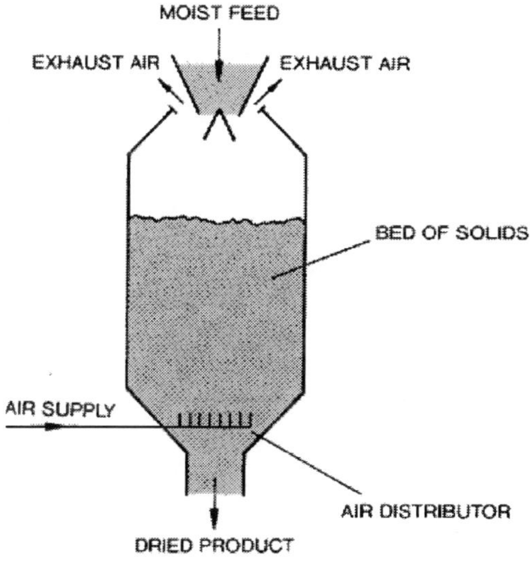

FIG. 12-49 Moving-bed gravity dryer.

Maintenance of a uniform rate of solids movement downward over the entire cross-section of the bed is one of the most critical operating problems encountered. For this reason gravity beds are designed to be as high and narrow as practical. In a vessel of large cross section, discharge through a conical bottom and center outlet will usually result in some degree of "ratholing" through the center of the bed. Flow through the center will be rapid while essentially stagnant pockets are left around the sides. To overcome this problem, multiple outlets may be provided in the center and around the periphery; table unloaders, rotating plows, wide moving grates, and multiple-screw unloaders are employed; insertion of inverted cone baffles in the lower section of the bed, spaced so that flushing at the center is retarded, is also a successful method for improving uniformity of solids movement. Fortunately, the problems are less critical in gravity dryers, which are usually for slow drying of large particles, than in applications such as catalytic reactors, where disengagement of gas from solids at the top of the tower can also present serious difficulties.

Continuous Band and Tunnel Dryers This group of dryers is variously known as band, belt, conveyor, or tunnel dryers.

Classification Continuous; nonagitated; layer; convective (cross-circulation or through-circulation) or contact/conduction; horizontal movement by mechanical means.

Continuous tunnels are **batch truck** or **tray compartments, operated in series.** The solids to be processed are placed in trays or on trucks which move progressively through the tunnel in contact with hot gases. Operation is **semicontinuous;** when the tunnel is filled, one truck is removed from the discharge end as each new truck is fed into the inlet end. In some cases, the trucks move on tracks or monorails, and they are usually conveyed mechanically, employing chain drives connecting to the bottom of each truck.

Belt-conveyor and **screen-conveyor (band) dryers** are **truly continuous** in operation, carrying a layer of solids on an endless conveyor.

Continuous tunnel and conveyor dryers are more suitable than (multiple) batch compartments for large-quantity production, usually giving investment and installation savings. In the case of truck and tray tunnels, labor savings for loading and unloading are not significant compared with those for batch equipment. Belt and screen conveyors which are truly continuous represent major labor savings over batch operations but require additional investment for automatic feeding and unloading devices.

Airflow can be totally **cocurrent, countercurrent,** or a combination of both. In addition, **cross-flow** designs are employed frequently, with the heating air flowing back and forth across the trucks or belt in series. Reheat coils may be installed after each cross-flow pass to maintain constant-temperature operation; large propeller-type circulating fans are installed at each stage, and air may be introduced or exhausted at any desirable points. Tunnel equipment possesses maximum flexibility for any combination of airflow and temperature staging. When handling granular, particulate solids which do not offer high resistance to airflow, perforated or screen-type belt conveyors are employed with **through-circulation** of gas to improve heat- and mass-transfer rates, almost invariably in cross-flow. Contact drying is also possible, usually under vacuum, with the bands resting on heating plates (vacuum band dryer).

Tunnel Dryers In tunnel equipment, the solids are usually heated by direct contact with hot gases. In high-temperature operations, radiation from walls and refractory lining may be significant also. The air in a direct heat unit may be heated directly or indirectly by combustion or, at temperature below 475 K, by finned steam coils.

Applications of tunnel equipment are essentially the same as those for batch tray and compartment units previously described, namely, practically all forms of particulate solids and large solid objects. Continuous tunnel or conveyor ovens are employed also for drying refractory shapes and for drying and baking enameled pieces. In many of these latter, the parts are suspended from overhead chain conveyors.

Auxiliary equipment and the special design considerations discussed for batch trays and compartments apply also to tunnel equipment. For size-estimating purposes, tray and truck tunnels and furnaces can be treated in the same manner as discussed for batch equipment.

Ceramic tunnel kilns handling large irregular-shaped objects must be equipped for precise control of temperature and humidity conditions to prevent cracking and condensation on the product. The internal mechanism causing cracking when drying clay and ceramics have been studied extensively. Information on ceramic tunnel kiln operation and design is reported fully in publications such as *The American Ceramic Society Bulletin, Ceramic Industry,* and *Transactions of the British Ceramic Society*.

Another use of tunnel dryers is for drying leather. Moisture content is initially around 50 percent but must not be reduced below about 15 percent, or else the leather will crack and be useless. To avoid this, a high-humidity atmosphere is maintained by gas recycle, giving a high equilibrium moisture content.

Continuous Through-Circulation Band Dryers Continuous through-circulation dryers operate on the principle of blowing hot air through a permeable bed of wet material passing continuously through the dryer. Drying rates are high because of the large area of contact and short distance of travel for the internal moisture.

The most widely used type is the **horizontal conveyor dryer** (also called perforated band or conveying-screen dryer), in which wet material is conveyed as a layer, 2 to 15 cm deep (sometimes up to 1 m), on a horizontal mesh screen, belt, or perforated apron, while heated air is blown either upward or downward through the bed of material. This dryer consists usually of a number of individual sections, complete with fan and heating coils, arranged in series to form a housing or tunnel through which the conveying screen travels. As shown in the sectional view in Fig. 12-50, the air circulates through the wet material and is reheated before reentering the bed. It is not uncommon to circulate the hot gas upward in the wet end and downward in the dry end, as shown in Fig. 12-51. A portion of the air is exhausted continuously by one or more exhaust fans, not shown in the sketch, which handle air from several sections. Since each section can be operated independently, extremely flexible operation is possible, with high temperatures usually at the wet end, followed by lower temperatures; in some cases a unit with cooled or specially humidified air is employed for final conditioning. The **maximum pressure drop** that can be taken through the bed of solids without developing leaks or air bypassing is roughly 50 mm of water.

Through-circulation drying requires that the wet material be in a state of granular or pelleted subdivision so that hot air may be readily blown through it. Many materials meet this requirement without special preparation. Others require special and often elaborate pretreatment to render them suitable for through-circulation drying. The process of converting a wet solid to a form suitable for through-circulation of air is called **preforming,** and often the success or failure of this contacting method depends on the preforming step. Fibrous, flaky, and coarse granular materials are usually amenable to drying without preforming. They can be loaded directly onto the conveying screen by suitable spreading feeders of the oscillating-belt or vibrating type or by spiked drums or belts feeding from bins.

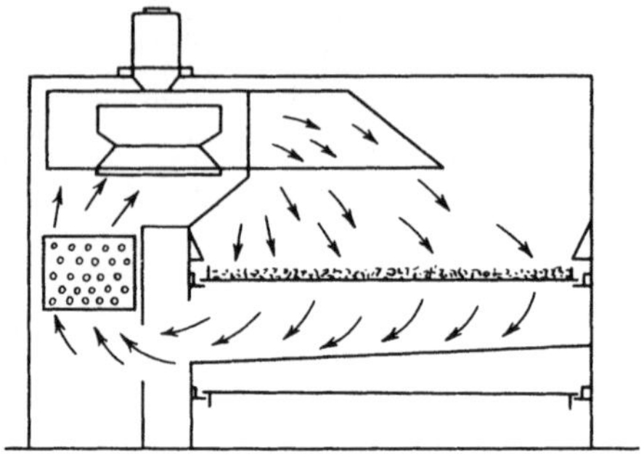

FIG. 12-50 Section view of a continuous through-circulation conveyor dryer. *(Proctor & Schwartz, Inc.)*

When materials must be preformed, several methods are available, depending on the physical state of the wet solid.

1. Relatively dry materials such as centrifuge cakes can sometimes be granulated to give a suitably porous bed on the conveying screen.
2. Pasty materials can often be preformed by extrusion to form spaghettilike pieces, about 6 mm in diameter and several centimeters long.
3. Wet pastes that cannot be granulated or extruded may be predried and preformed on a steam-heated finned drum. Preforming on a finned drum may be desirable also in that some predrying is accomplished.
4. Thixotropic filter cakes from rotary vacuum filters that cannot be preformed by any of the above methods can often be scored by knives on the filter, the scored cake discharging in pieces suitable for through-circulation drying.
5. Material that shrinks markedly during drying is often reloaded during the drying cycle to 2 to 6 times the original loading depth. This is usually done after a degree of shrinkage which, by opening the bed, has destroyed the effectiveness of contact between the air and solids.
6. In a few cases, powders have been pelleted or formed in briquettes to eliminate dustiness and permit drying by through-circulation. Table 12-20 gives a list of materials classified by preforming methods suitable for through-circulation drying.

Steam-heated air is the usual heat-transfer medium employed in these dryers, although combustion gases may be used also. Temperatures above 600 K are not usually feasible because of the problems of lubricating the conveyor, chain, and roller drives. Recirculation of air is in the range of 60 to 90 percent of the flow through the bed. Conveyors may be made of wire-mesh screen or perforated-steel plate. The minimum practical screen opening size is about 30-mesh (0.5 mm). Multiple bands in series may be used.

Vacuum band dryers utilize heating by conduction and are a continuous equivalent of vacuum tray (shelf) dryers, with the moving

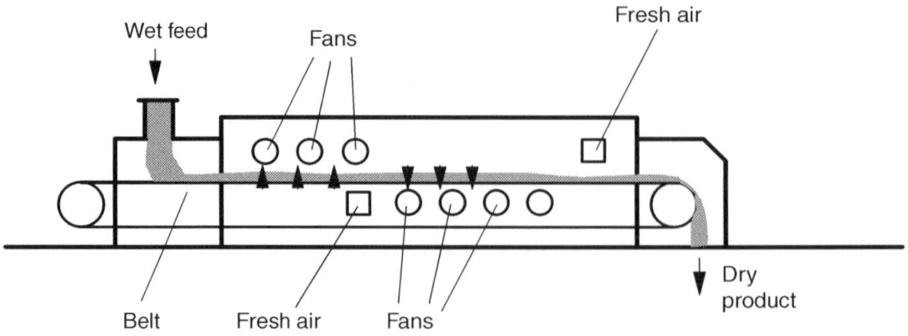

FIG. 12-51 Longitudinal view of a continuous through-circulation conveyor dryer with intermediate airflow reversal.

TABLE 12-20 Methods of Preforming Some Materials for Through-Circulation Drying

No preforming required	Scored on filter	Granulation	Extrusion	Finned drum	Flaking on chilled drum	Briquetting and squeezing
Cellulose acetate	Starch	Kaolin	Calcium carbonate	Lithopone	Soap flakes	Soda ash
Silica gel	Aluminum hydrate	Cryolite	White lead	Zinc yellow		Cornstarch
Scoured wool		Lead arsenate	Lithopone	Calcium carbonate		Synthetic rubber
Sawdust		Cornstarch	Titanium dioxide	Magnesium carbonate		
Rayon waste		Cellulose acetate	Magnesium carbonate			
Fluorspar		Dye intermediates	Aluminum stearate			
Tapioca			Zinc stearate			
Breakfast food						
Asbestos fiber						
Cotton linters						
Rayon staple						

bands resting on heating plates. Drying is usually relatively slow, and it is common to find several bands stacked above one another, with material falling to the next band and flowing in opposite directions on each pass, to reduce dryer length and give some product turnover.

Design Methods for Continuous Band Dryers In actual practice, design of a continuous through-circulation dryer is best based upon data taken in pilot-plant tests. Loading and distribution of solids on the screen are rarely as nearly uniform in commercial installations as in test dryers; 50 to 100 percent may be added to the test drying time for commercial design.

A mathematical method of a through-circulation dryer has been developed by Thygeson [*Am. Inst. Chem. Eng. J.* **16**(5):749 (1970)]. Rigorous modeling is possible with a two-dimensional incremental model, with steps both horizontally along the belt and vertically through the layer; nonuniformity of the layer across the belt could also be allowed for if desired. Heat-transfer coefficients are typically in the range of 100 to 200 W/(m^2·K) and the relationship $h_c = 12(\rho_g U_g/d_p)^{0.5}$ may be used for a first estimate, where ρ_g is gas density (kg/m^3); U_g, local gas velocity (m/s); and d_p, particle diameter (m). For 5-mm particles and air at 1 m/s, 80°C and 1 kg/m^3 [mass flux 1 kg/(m^2·s)] this gives h_c = 170 W/(m^2·K).

Performance and Cost Data for Continuous Band and Tunnel Dryers Experimental performance data are given in Table 12-21 for numerous common materials. Performance data from several commercial through-circulation conveyor dryers are given in Table 12-22. Labor requirements vary depending on the time required for feed adjustments, inspection, etc. These dryers may consume as little as 1.1 kg of steam/kg of water evaporated, but 1.4 to 2 is a more common range. Thermal efficiency is a function of final moisture required and percent air recirculation.

Conveying-screen dryers are fabricated with conveyor widths from 0.3- to 4.4-m sections 1.6 to 2.5 m long. Each section consists of a sheet-metal enclosure, insulated sidewalls and roof, heating coils, a circulating fan, inlet air distributor baffles, a fines catch pan under the conveyor, and a conveyor screen (Fig. 12-51). Table 12-23 gives approximate purchase costs for equipment with type 304 stainless-steel hinged conveyor screens and includes steam-coil heaters, fans, motors, and a variable-speed conveyor drive. Cabinet and auxiliary equipment fabrication is of aluminized steel or stainless-steel materials. Prices do not include temperature controllers, motor starters, preform equipment, or auxiliary feed and discharge conveyors. These may add $75,000 to $160,000 to the dryer purchase cost (2005 costs).

Batch Agitated and Rotating Dryers

Description An *agitated dryer* is defined as one on which the housing enclosing the process is stationary while solids movement is accomplished by an internal mechanical agitator. A rotary dryer is one in which the outer housing rotates. Many forms are in use, including batch and continuous versions. The batch forms are almost invariably heated by conduction with operation under vacuum. Vacuum is used in conjunction with drying or other chemical operations when low solids temperatures must be maintained because heat will cause damage to the product or change its nature; when air combines with the product as it is heated, causing oxidation or an explosive condition; when solvent recovery is required; and when materials must be dried to extremely low moisture levels.

Vertical agitated pan, spherical and conical dryers are mechanically agitated; tumbler or double-cone dryers have a rotating shell. All these types are typically used for the drying of solvent or water-wet, free-flowing powders in small batch sizes of 1000 L or less, as frequently found in the pharmaceutical, specialty chemical, and fine chemicals industries. Corrosion resistance and cleanability are often important, and common materials of construction include SS 304 and 316, and Hastelloy. The batch nature of operation is of value in the pharmaceutical industry to maintain batch identification. In addition to pharmaceutical materials, the conical mixer dryer is used to dry polymers, additives, inorganic salts, and many other specialty chemicals. As the size increases, the ratio of jacket heat-transfer surface area to volume falls, extending drying times. For larger batches, horizontal agitated pan dryers are more common, but there is substantial overlap of operating ranges. Drying times may be reduced for all types by heating the internal agitator, but this increases complexity and cost.

Classification Batch; mechanical or rotary agitation; layer; contact/conduction.

Mechanical versus rotary agitation Agitated dryers are applicable to processing solids which are relatively free-flowing and granular when discharged as product. Materials which are not free-flowing in their feed condition can be treated by recycle methods as described in the subsection "Continuous Rotary Dryers." In general, agitated dryers have applications similar to those of rotating vessels. Their chief advantages compared with the latter are twofold. (1) Large-diameter rotary seals are not required at the solids and gas feed and exit points because the housing is stationary, and for this reason gas leakage problems are minimized. Rotary seals are required only at the points of entrance of the mechanical agitator shaft. (2) Use of a mechanical agitator for solids mixing introduces shear forces which are helpful for breaking up lumps and agglomerates. Balling and pelleting of sticky solids, an occasional occurrence in rotating vessels, can be prevented by special agitator design. The problems concerning dusting of fine particles in direct-heat units are identical to those discussed under "Continuous Rotary Dryers."

Vacuum processing All these types of dryer usually operate under vacuum, especially when drying heat-sensitive materials or when removing flammable organic solvents rather than water. The heating medium is hot water, steam, or thermal oil, with most applications in the temperature range of 50 to 150°C and pressures in the range of 3 to 30 kPa absolute. The vapors generated during the drying process are evacuated by a vacuum pump and passed through a condenser for recovery of the solvent. A dust filter is normally mounted over the vapor discharge line as it leaves the dryer, thus allowing any entrapped dust to be pulsed back into the process area. Standard cloth-type dust filters are available, along with sintered metal filters.

In vacuum processing and drying, a major objective is to create a large temperature-driving force between the jacket and the product. To accomplish this purpose at fairly low jacket temperatures, it is necessary to reduce the internal process pressure so that the liquid being removed will boil at a lower vapor pressure. It is not always economical, however,

TABLE 12-21 Experimental Through-Circulation Drying Data for Miscellaneous Materials

Material	Physical form	Moisture contents, kg/kg dry solid			Inlet-air temperature, K	Depth of bed, cm	Loading, kg product/m^2	Air velocity, m/s × 10^1	Experimental drying time, s × 10^{-2}
		Initial	Critical	Final					
Alumina hydrate	Briquettes	0.105	0.06	0.00	453	6.4	60.0	6.0	18.0
Alumina hydrate	Scored filter cake	9.60	4.50	1.15	333	3.8	1.6	11.0	90.0
Alumina hydrate	Scored filter cake	5.56	2.25	0.42	333	7.0	4.6	11.0	108.0
Aluminum stearate	0.7-cm extrusions	4.20	2.60	0.003	350	7.6	6.5	13.0	36.0
Asbestos fiber	Flakes from squeeze rolls	0.47	0.11	0.008	410	7.6	13.6	9.0	5.6
Asbestos fiber	Flakes from squeeze rolls	0.46	0.10	0.0	410	5.1	6.3	9.0	3.6
Asbestos fiber	Flakes from squeeze rolls	0.46	0.075	0.0	410	3.8	4.5	11.0	2.7
Calcium carbonate	Preformed on finned drum	0.85	0.30	0.003	410	3.8	16.0	11.5	12.0
Calcium carbonate	Preformed on finned drum	0.84	0.35	0.0	410	8.9	25.7	11.7	18.0
Calcium carbonate	Extruded	1.69	0.98	0.255	410	1.3	4.9	14.3	9.0
Calcium carbonate	Extruded	1.41	0.45	0.05	410	1.9	5.8	10.2	12.0
Calcium stearate	Extruded	2.74	0.90	0.0026	350	7.6	8.8	5.6	57.0
Calcium stearate	Extruded	2.76	0.90	0.007	350	5.1	5.9	6.0	42.0
Calcium stearate	Extruded	2.52	1.00	0.0	350	3.8	4.4	10.2	24.0
Cellulose acetate	Granulated	1.14	0.40	0.09	400	1.3	1.4	12.7	1.8
Cellulose acetate	Granulated	1.09	0.35	0.0027	400	1.9	2.7	8.6	7.2
Cellulose acetate	Granulated	1.09	0.30	0.0041	400	2.5	4.1	5.6	10.8
Cellulose acetate	Granulated	1.10	0.45	0.004	400	3.8	6.1	5.1	18.0
Clay	Granulated	0.277	0.175	0.0	375	7.0	46.2	10.2	19.2
Clay	1.5-cm extrusions	0.28	0.18	0.0	375	12.7	100.0	10.7	43.8
Cryolite	Granulated	0.456	0.25	0.0026	380	5.1	34.2	9.1	24.0
Fluorspar	Pellets	0.13	0.066	0.0	425	5.1	51.4	11.6	7.8
Lead arsenate	Granulated	1.23	0.45	0.043	405	5.1	18.1	11.6	18.0
Lead arsenate	Granulated	1.25	0.55	0.054	405	6.4	22.0	10.2	24.0
Lead arsenate	Extruded	1.34	0.64	0.024	405	5.1	18.1	9.4	36.0
Lead arsenate	Extruded	1.31	0.60	0.0006	405	8.4	26.9	9.2	42.0
Kaolin	Formed on finned drum	0.28	0.17	0.0009	375	7.6	44.0	9.2	21.0
Kaolin	Formed on finned drum	0.297	0.20	0.005	375	11.4	56.3	12.2	15.0
Kaolin	Extruded	0.443	0.20	0.008	375	7.0	45.0	10.16	18.0
Kaolin	Extruded	0.36	0.14	0.0033	400	9.6	40.6	15.2	12.0
Kaolin	Extruded	0.36	0.21	0.0037	400	19.0	80.7	10.6	30.0
Lithopone (finished)	Extruded	0.35	0.065	0.0004	408	8.2	63.6	10.2	18.0
Lithopone (crude)	Extruded	0.67	0.26	0.0007	400	7.6	41.1	9.1	51.0
Lithopone	Extruded	0.72	0.28	0.0013	400	5.7	28.9	11.7	18.0
Magnesium carbonate	Extruded	2.57	0.87	0.001	415	7.6	11.0	11.4	17.4
Magnesium carbonate	Formed on finned drum	2.23	1.44	0.0019	418	7.6	13.2	8.6	24.0
Mercuric oxide	Extruded	0.163	0.07	0.004	365	3.8	66.5	11.2	24.0
Silica gel	Granular	4.51	1.85	0.15	400	3.8–0.6	3.2	8.6	15.0
Silica gel	Granular	4.49	1.50	0.215	340	3.8–0.6	3.4	9.1	63.0
Silica gel	Granular	4.50	1.60	0.218	325	3.8–0.6	3.5	9.1	66.0
Soda salt	Extruded	0.36	0.24	0.008	410	3.8	22.8	5.1	51.0
Starch (potato)	Scored filter cake	0.866	0.55	0.069	400	7.0	26.3	10.2	27.0
Starch (potato)	Scored filter cake	0.857	0.42	0.082	400	5.1	17.7	9.4	15.0
Starch (corn)	Scored filter cake	0.776	0.48	0.084	345	7.0	26.4	7.4	54.0
Starch (corn)	Scored filter cake	0.78	0.56	0.098	380	7.0	27.4	7.6	24.0
Starch (corn)	Scored filter cake	0.76	0.30	0.10	345	1.9	7.7	6.7	15.0
Titanium dioxide	Extruded	1.2	0.60	0.10	425	3.0	6.8	13.7	6.3
Titanium dioxide	Extruded	1.07	0.65	0.29	425	8.2	16.0	8.6	6.0
White lead	Formed on finned drum	0.238	0.07	0.001	355	6.4	76.8	11.2	30.0
White lead	Extruded	0.49	0.17	0.0	365	3.8	33.8	10.2	27.0
Zinc stearate	Extruded	4.63	1.50	0.005	360	4.4	4.2	8.6	36.0

to reduce the internal pressure to extremely low levels because of the large vapor volumes thereby created. It is necessary to compromise on operating pressure, considering leakage, condensation problems, and the size of the vapor lines and pumping system. Very few vacuum dryers operate below 5 mmHg pressure on a commercial scale. Air in-leakage through gasket surfaces will be in the range of 0.2 kg/(h·linear m of gasketed surface) under these conditions. To keep vapor partial pressure and solids temperature low without pulling excessively high vacuum, a nitrogen bleed may be introduced, particularly in the later stages of drying. The vapor and solids surface temperatures then fall below the vapor boiling point, toward the wet-bulb temperature.

Vertical Agitated Dryers This classification includes vertical pan dryers, filter dryers, and spherical and conical dryers.

Vertical pan dryer The basic vertical pan dryer consists of a short, squat vertical cylinder (Fig. 12-52 and Table 12-24) with an outer heating jacket and an internal rotating agitator, again with the axis vertical, which mixes the solid and sweeps the base of the pan. Heat is supplied by circulation of hot water, steam, or thermal fluid through the jacket; it may also be used for cooling at the end of the batch cycle, using cooling water or refrigerant. The agitator is usually a plain set of solid blades, but may be a ribbon-type screw or internally heated blades. Product is discharged from a door at the lower side of the wall. Sticky materials may adhere to the agitator or be difficult to discharge.

Filter dryer The basic Nutsche filter dryer is like a vertical pan dryer, but with the bottom heated plate replaced by a filter plate. Hence, a slurry can be fed in and filtered, and the wet cake dried in situ. These units are especially popular in the pharmaceutical industry, as containment is good and a difficult wet solids transfer operation is eliminated by carrying out both filtration and drying in the same vessel. Drying times tend to be longer than for vertical pan dryers as the bottom plate is no longer heated. Some types (e.g., Mitchell Thermovac,

TABLE 12-22 Performance Data for Continuous Through-Circulation Dryers*

	Kind of material						
	Inorganic pigment	Cornstarch	Fiber staple		Charcoal briquettes	Gelatin	Inorganic chemical
Capacity, kg dry product/h	712	4536	1724		5443	295	862
			Stage A,	Stage B,			
Approximate dryer area, m^2	22.11	66.42	57.04	35.12	52.02	104.05	30.19
Depth of loading, cm	3	4			16	5	4
Air temperature, °C	120	115 to 140	130 to 100	100	135 to 120	32 to 52	121 to 82
Loading, kg product/m^2	18.8	27.3	3.5	3.3	182.0	9.1	33
Type of conveyor, mm	1.59 by 6.35 slots	1.19 by 4.76 slots	2.57-diameter holes, perforated plate		8.5 × 8.5 mesh screen	4.23 × 4.23 mesh screen	1.59 × 6.35 slot
Preforming method or feed	Rolling extruder	Filtered and scored	Fiber feed		Pressed	Extrusion	Rolling extruder
Type and size of preformed particle, mm	6.35-diameter extrusions	Scored filter cake	Cut fiber		64 × 51 × 25	2-diameter extrusions	6.35-diameter extrusions
Initial moisture content, % bone-dry basis	120	85.2	110		37.3	300	111.2
Final moisture content, % bone-dry basis	0.5	13.6	9		5.3	11.1	1.0
Drying time, min	35	24	11		105	192	70
Drying rate, kg water evaporated/(h·m^2)	38.39	42.97	17.09		22.95	9.91	31.25
Air velocity (superficial), m/s	1.27	1.12	0.66		1.12	1.27	1.27
Heat source per kg water evaporated, steam kg/kg gas (m^3/kg)	Gas 0.11	Steam 2.0	Steam 1.73		Waste heat	Steam 2.83	Gas 0.13
Installed power, kW	29.8	119.3	194.0		82.06	179.0	41.03

*Courtesy of Wolverine Proctor & Schwartz, Inc.

Krauss-Maffei TNT) invert the unit between the filtration and drying stages to avoid this problem.

Spherical dryer Sometimes called the *turbosphere,* this is another agitated dryer with a vertical axis mixing shaft, but rotation is typically faster than in the vertical pan unit, giving improved mixing and heat transfer. The dryer chamber is spherical, with solids discharge through a door or valve near the bottom.

Conical mixer dryer This is a vertically oriented conical vessel with an internally mounted rotating screw. Figure 12-53 shows a schematic of a typical conical mixer dryer. The screw rotates about its own axis (speeds up to 100 rpm) and around the interior of the vessel (speeds up to 0.4 rpm). Because it rotates around the full circumference of the vessel, the screw provides a self-cleaning effect for the heated vessel walls, as well as effective agitation; it may also be internally heated. Either top-drive (via an internal rotating arm) or bottom-drive (via a universal joint) may be used; the former is more common. The screw is cantilevered in the vessel and requires no additional support (even in vessel sizes up to 20-m^3 operating volume). Cleaning of the dryer is facilitated with CIP systems that can be used for cleaning, and/or the vessel can be completely flooded with water or solvents. The dryer makes maximum use of the product-heated areas—the filling volume of the vessel (up to the knuckle of the dished head) is the usable product loading. In some recent applications, microwaves have been used to provide additional energy input and shorten drying times.

In the bottom-drive system, the vessel cover is free of drive components, allowing space for additional process nozzles, manholes, explosion venting, etc., as well as a temperature lance for direct, continuous product temperature measurement in the vessel. The top cover of the vessel is easily heated by either a half-pipe coil or heat tracing, which ensures that no vapor condensation will occur in the process area.

TABLE 12-23 Conveyor-Screen-Dryer Costs*

Length	2.4-m-wide conveyor	3.0-m-wide conveyor
7.5 m	\$8600/m^2	\$7110/m^2
15 m	\$6700/m^2	\$5600/m^2
22.5 m	\$6200/m^2	\$5150/m^2
30 m	\$5900/m^2	\$4950/m^2

*National Drying Machinery Company, 1996.

Because there are no drive components in the process area, the risk of batch failures due to contamination from gear lubricants is eliminated. However, the bottom joint requires especially careful design, maintenance, and sealing. The disassembly of the unit is simplified, as all work on removing the screw can be done without vessel entry. For disassembly, the screw is simply secured from the top, and the drive components are removed from the bottom of the dryer.

Horizontal Pan Dryer This consists of a stationary cylindrical shell, mounted horizontally, in which a set of agitator blades mounted on a revolving central shaft stirs the solids being treated. They tend to be used for larger batches than vertical agitated or batch rotating dryers. Heat is supplied by circulation of hot water, steam, or Dowtherm through the jacket surrounding the shell and, in larger units, through the hollow central shaft. The agitator can be of many different forms, including simple paddles, ploughshare-type blades, a single discontinuous spiral, or a double continuous spiral. The outer blades are set as closely as possible to the wall without touching, usually leaving a gap of 0.3 to 0.6 cm. Modern units occasionally employ spring-loaded shell scrapers mounted on the blades. The dryer is charged through a port at the top and emptied through one or more discharge nozzles at the bottom. Vacuum is applied and maintained by any of the conventional methods, i.e., steam jets, vacuum pumps, etc.

A similar type, the batch indirect rotary dryer, consists of a rotating horizontal cylindrical shell, suitably jacketed. Vacuum is applied to this unit through hollow trunnions with suitable packing glands. Rotary glands must be used also for admitting and removing the heating medium from the jacket. The inside of the shell may have lifting bars, welded longitudinally, to assist agitation of the solids. Continuous rotation is needed while emptying the solids, and a circular dust hood is frequently necessary to enclose the discharge-nozzle turning circle and prevent serious dust losses to the atmosphere during unloading. A typical vacuum rotary dryer is illustrated in Fig. 12-54. Sealing tends to be more difficult where the entire shell rotates compared to the horizontal pan, where only the central agitator shaft rotates, since the seal diameter is smaller in the latter case. Conversely, a problem with a stationary shell is that it can be difficult to empty the final "heel" of material out of the bottom of the cylinder. If batch integrity is important, this is an advantage for the rotary variant over the horizontal pan.

Heated Agitators For all agitated dryers, in addition to the jacket heated area, heating the agitator with the same medium as the

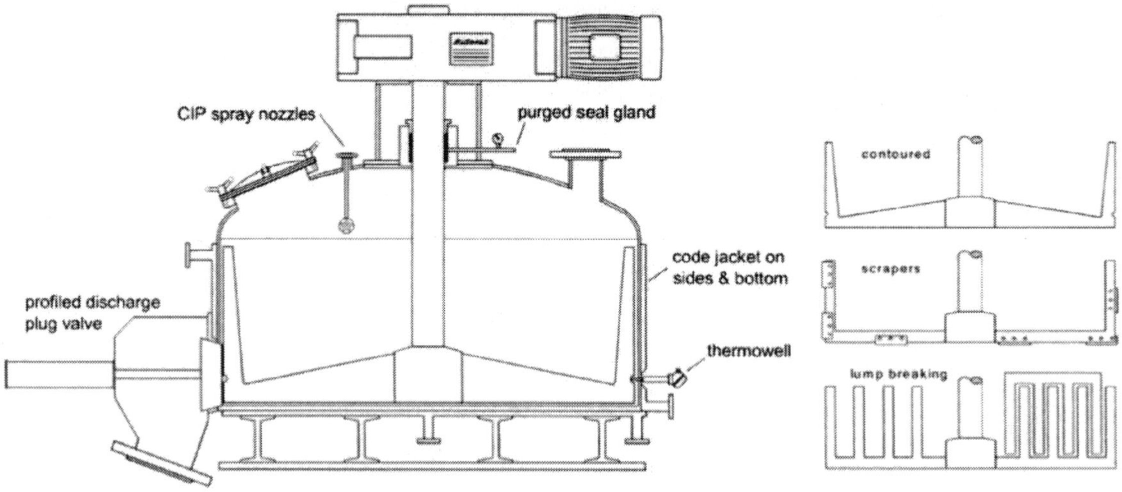

FIG. 12-52 Vertical pan dryer. (*Buflovak Inc.*)

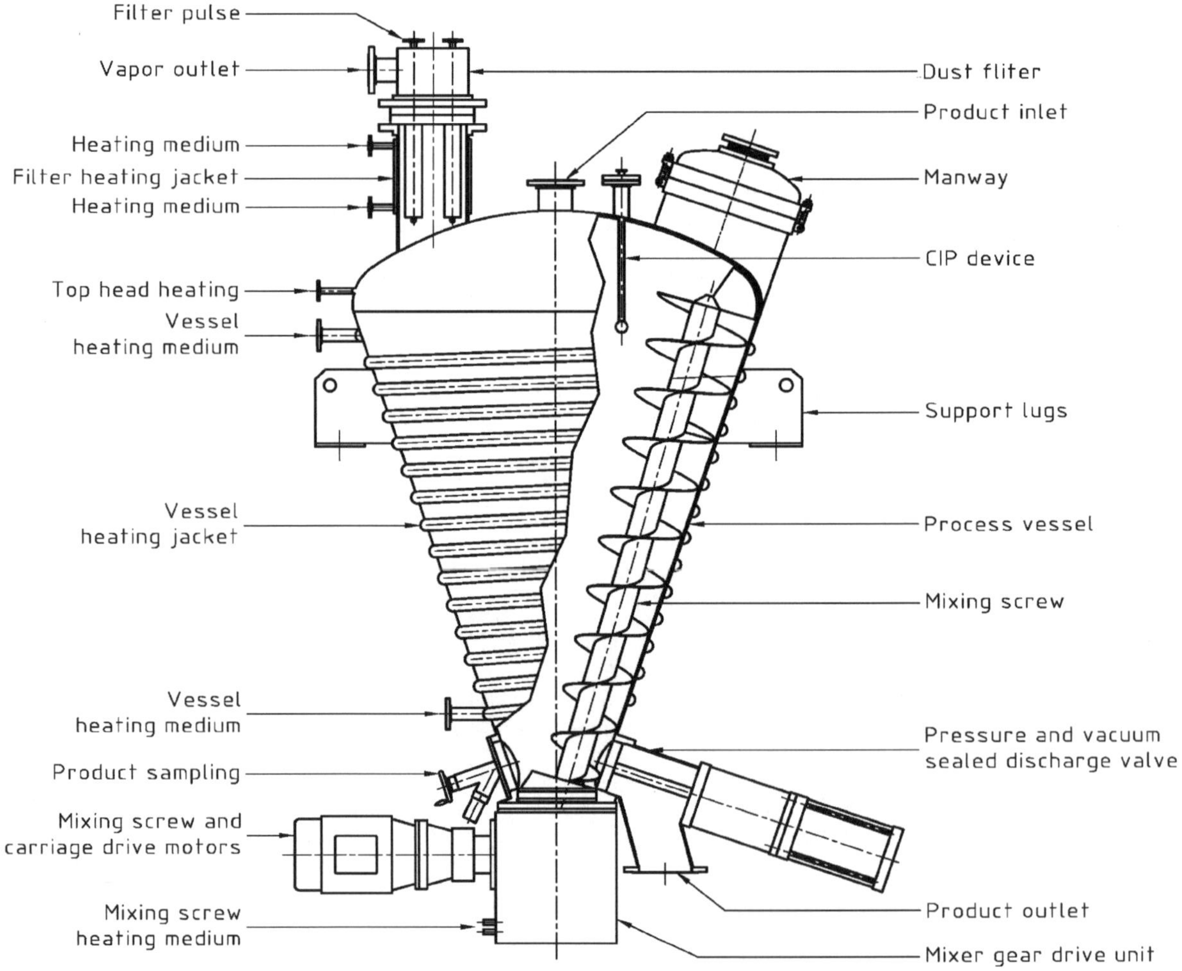

FIG. 12-53 Bottom-drive conical mixer dryer. (*Krauss Maffei.*)

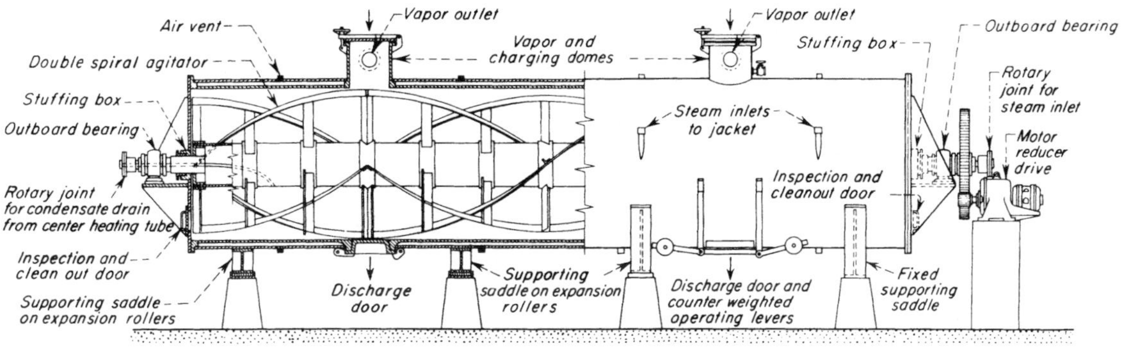

FIG. 12-54 A typical horizontal pan vacuum dryer. *(Blaw-Knox Food & Chemical Equipment, Inc.)*

jacket (hot water, steam, or thermal oil) will increase the heat-exchange area. This is usually accomplished via rotary joints. Obviously, heating the screw or agitator will mean shorter batch drying times, which yields higher productivity and better product quality due to shorter exposure to the drying temperature, but capital and maintenance costs will be increased. In pan and conical dryers the area is increased only modestly, by 15 to 30 percent; but in horizontal pan and paddle dryers, the opportunity is much greater and indeed the majority of the heat may be supplied through the agitator.

Also, the mechanical power input of the agitator can be a significant additional heat source, and microwave assistance has also been used in filter dryers and conical dryers to shorten drying times (and is feasible in other types).

Tumbler or Double-Cone Dryers These are rotating batch vacuum dryers, as shown in Fig. 12-55. Some types are an offset cylinder, but a double-cone shape is more common. They are very common in the pharmaceutical and fine chemicals industries. The gentle rotation can give less attrition than in some mechanically agitated dryers; on the other hand, formation of lumps and balls is more likely. The sloping walls of the cones permit more rapid emptying of solids when the dryer is in a stationary position, compared to a horizontal cylinder, which requires continuous rotation during emptying to convey product to the discharge nozzles. Several new designs of the double-cone type employ internal tubes or plate coils to provide additional heating surface.

On all rotating dryers, the vapor outlet tube is stationary; it enters the shell through a rotating gland and is fitted with an elbow and an upward extension so that the vapor inlet, usually protected by a felt dust filter, will be at all times near the top of the shell.

Design, Scale-up, and Performance Like all batch dryers, agitated and rotating dryers are primarily sized to physically contain the required batch volume. Note that the nominal capacity of most dryers is significantly lower than their total internal volume, because of the headspace needed for mechanical drives, inlet ports, suction lines, dust filters, etc. Care must be taken to determine whether a stated "percentage fill" is based on nominal capacity or geometric volume. Vacuum dryers are usually filled to 50 to 65 percent of their total shell volume.

The standard scoping calculation methods for batch conduction drying apply. The rate of heat transfer from the heating medium through the dryer wall to the solids can be expressed by the usual formula

$$Q = hA\,\Delta T_m \tag{12-87}$$

where Q = heat flux, J/s [Btu/h]; h = overall heat-transfer coefficient, J/(m^2·s·K) [Btu/(h·ft^2 jacket area·°F)]; A = total jacket area, m^2 (ft^2); and ΔT_m = log-mean-temperature driving force from heating medium to the solids, K (°F).

The overall heat-transfer rate is almost entirely dependent upon the film coefficient between the inner jacket wall and the solids, which depends on the dryer type and agitation rate, and to a large extent on the solids characteristics. Overall coefficients may range from 30 to 200 J/(m^2·s·K), based upon total area if the dryer walls are kept reasonably clean. Coefficients as low as 5 or 10 may be encountered if caking on the walls occurs.

For estimating purposes without tests, a reasonable coefficient for ordinary drying, and without taking the product to absolute dryness, may be assumed at h = 50 J/(m^2·s·K) for mechanically agitated dryers (although higher figures have been quoted for conical and spherical dryers) and 35 J/(m^2·s·K) for rotating units. The true heat-transfer coefficient is usually higher, but this conservative assumption makes some allowance for the slowing down of drying during the falling-rate period. However, if at all possible, it is always preferable to do pilot-plant tests to establish the drying time of the actual material. Drying trials are conducted in small pilot dryers (50- to 100-L batch units) to determine material handling and drying retention times. Variables such as drying temperature, vacuum level, and screw speed are analyzed during the

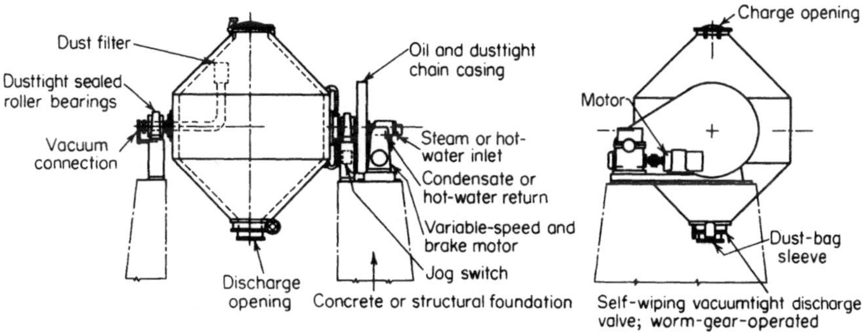

FIG. 12-55 Rotating (double-cone) vacuum dryer. (*Stokes Vacuum, Inc.*)

TABLE 12-24 Dimensions of Vertical Pan Dryers *(Buflovak Inc.)*

					Jacketed area, ft^2				
I.D, ft	Product depth, ft	Working volume, ft^3	USG	Jacketed height, ft	Cylinder wall	Bottom	Total	Discharge door, in	
3	0.75	5.3	40	1.0	9	7	16	5	8
4	1	12.6	94	2.0	25	13	38	6	8
5	1	19.6	147	2.0	31	20	51	8	9
6	1	28.3	212	2.0	38	28	66	8	9
8	1	50.3	377	2.0	50	50	101	8	9
10	1.5	117.8	884	3.0	94	79	173	12	12

test trials. Scale-up to larger units is done based upon the area/volume ratio of the pilot unit versus the production dryer. In most applications, the overall drying time in the production models is in the range of 2 to 24 h.

Agitator or rotation speeds range from 3 to 8 rpm. Faster speeds yield a slight improvement in heat transfer but consume more power and in some cases, particularly in rotating units, can cause more "balling up" and other stickiness-related problems.

In all these dryers, the surface area tends to be proportional to the square of the diameter D^2, and the volume to diameter cubed D^3. Hence the area/volume ratio falls as diameter increases, and drying times increase. It can be shown that the ratio of drying times in the production and pilot-plant dryers is proportional to the cube root of the ratio of batch volumes. However, if the agitator of the production unit is heated, the drying time increase can be reduced or reversed. Table 12-25 gives basic geometric relationships for agitated and rotating batch dryers, which can be used for approximate size estimation or (with great caution) for extrapolating drying times obtained from one dryer type to another. Note that these do not allow for nominal capacity or partial solids fill. For the paddle (horizontal pan) dryer with heated agitator, R is the ratio of the heat transferred through the agitator to that through the walls, which is proportional to the factor hA for each case.

Example 22: Calculations for Batch Dryer For a 10-m^3 batch of material containing 5000 kg dry solids and 30 percent moisture (dry basis), estimate the size of vacuum dryers required to contain the batch at 50 percent volumetric fill. Jacket temperature is 200°C, applied pressure is 100 mbar (0.1 bar), and the solvent is water (take latent heat as 2400 kJ/kg). Assuming the heat-transfer coefficient based on the total surface area to be 50 W/(m^2·K) for all types, calculate the time to dry to 5 percent for (*a*) unhindered (constant-rate) drying throughout, (*b*) first-order falling-rate (hindered) drying throughout, (*c*) if experiment shows the actual drying time for a conical dryer to be 12.5 h and other cases are scaled accordingly. Take $R = 5$ with the heated agitator. Assume material is nonhygroscopic (equilibrium moisture content $X_E = 0$).

Solution: The dryer volume V must be 20 m^3, and the diameter is calculated from column 4 of Table 12-25, assuming the default L/D ratios. Table 12-26 gives the results. Water at 100 mbar boils at 46°C so take ΔT as 200 − 46 = 154°C. Then Q is found from Eq. (12-87). The methods used are given in the section "Equipment—General, Scoping Design." For constant-rate drying throughout, drying time t_{CR} = evaporation rate/heat input rate and was given by Eq. (12-62):

$$t_{CR} = \frac{m_S(X_O - X_I)\lambda_{ev}}{h_{WS}\,\Delta T_{WS}\,A_S} = \frac{5000(0.3 - 0.05)\,2400}{0.05(154A_S)} \qquad (12\text{-}62)$$

This gives t_{CR} as 389,610/A_S s or 108.23/A_S h. Values for A_S and calculated times for the various dryer types are given in Table 12-26.

For falling-rate drying throughout, time t_{FR} is given by Eq. (12-63); the multiplying factor for drying time is 1.2 ln 6 = 2.15 for all dryer types.

$$\frac{t_{FR}}{t_{CR}} = \frac{X_1 - X_E}{X_1 - X_2}\ln\left(\frac{X_1 - X_E}{X_2 - X_E}\right) = \frac{0.3}{0.25}\ln\frac{0.3}{0.05} \qquad (12\text{-}63)$$

If the material showed a critical moisture content, the calculation could be split into two sections for constant-rate and falling-rate drying. Likewise, the experimental drying time t_{expt} for the conical dryer is 12.5 h which is a factor of 3.94 greater than the constant-rate drying time. A very rough estimate of drying times for the other dryer types has then been made by applying the same scaling factor (3.94) to their constant-rate drying times. Two major sources of error are possible: (1) The drying kinetics could differ between dryers; and (2) if the

TABLE 12-25 Calculation of Key Dimensions for Various Batch Contact Dryers (Fig. 12-55a Shows the Geometries)

Dryer type	Volume as $f(D)$	Typical L/D	Diameter as $f(V)$	Surface area as $f(D)$	Ratio A/V
Tumbler/double-cone	$V = \frac{\pi D^3}{12}\left(\frac{L}{D}\right)$	1.5	$D = \left[\frac{12V}{\pi(L/D)}\right]^{1/3}$	$A = \frac{\pi D^2}{2}\left[\left(\frac{L}{D}\right)^2 + 1\right]^{1/2}$	$\frac{A}{V} = \frac{6}{D}\left[1 + \left(\frac{D}{L}\right)^2\right]^{1/2}$
Vertical pan	$V = \frac{\pi D^3}{4}\left(\frac{L}{D}\right)$	0.5	$D = \left[\frac{4V}{\pi(L/D)}\right]^{1/3}$	$A = \pi D^2\left(\frac{L}{D} + \frac{1}{4}\right)$	$\frac{A}{V} = \frac{4}{D}\left(1 + \frac{D}{4L}\right)$
Spherical	$V = \frac{\pi D^3}{6}\left(\frac{L}{D}\right)$	1	$D = \left[\frac{6V}{\pi(L/D)}\right]^{1/3}$	$A = \pi D^2\left(\frac{L}{D}\right)$	$\frac{A}{V} = \frac{6}{D}$
Filter dryer	$V = \frac{\pi D^3}{4}\left(\frac{L}{D}\right)$	0.5	$D = \left[\frac{4V}{\pi(L/D)}\right]^{1/3}$	$A = \pi D^2\left(\frac{L}{D}\right)$	$\frac{A}{V} = \frac{4}{D}$
Conical agitated	$V = \frac{\pi D^3}{12}\left(\frac{L}{D}\right)$	1.5	$D = \left[\frac{12V}{\pi(L/D)}\right]^{1/3}$	$A = \frac{\pi D^2}{2}\left[\left(\frac{L}{D}\right)^2 + \frac{1}{4}\right]^{1/2}$	$\frac{A}{V} = \frac{6}{D}\left[1 + \frac{1}{4}\left(\frac{D}{L}\right)^2\right]^{1/2}$
Paddle (horizontal agitated)	$V = \frac{\pi D^3}{4}\left(\frac{L}{D}\right)$	5	$D = \left[\frac{4V}{\pi(L/D)}\right]^{1/3}$	$A = \pi D^2\left(\frac{L}{D}\right)$	$\frac{A}{V} = \frac{4}{D}$
Paddle, heated agitator	$V = \frac{\pi D^3}{4}\left(\frac{L}{D}\right)$	5	$D = \left[\frac{4V}{\pi(L/D)}\right]^{1/3}$	$A = \pi D^2\left(\frac{L}{D}\right)(1 + R)$	$\frac{A}{V} = \frac{4}{D}(1 + R)$

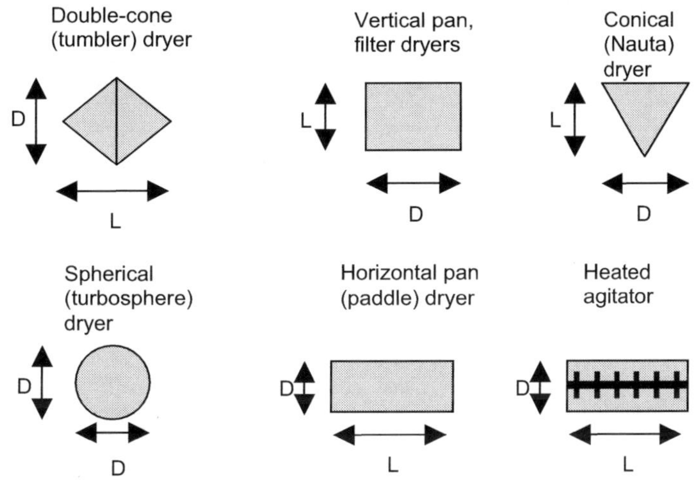

FIG. 12-55a Basic geometries for batch dryer calculations.

estimated heat-transfer coefficient for either the base case or the new dryer type is in error, the scaling factor will be wrong. All drying times have been shown in hours, as this is more convenient than seconds.

The paddle with heated agitator has the shortest drying time, and the filter dryer the longest (because the bottom plate is unheated). Other types are fairly comparable. The spherical dryer would usually have a higher heat-transfer coefficient and shorter drying time than shown.

Performance and Cost Data for Batch Vacuum Rotary Dryers Typical performance data for horizontal pan vacuum dryers are given in Table 12-27. Size and cost data for rotary agitator units are given in Table 12-28. Data for double-cone rotating units are in Table 12-29.

Continuous Agitated Dryers

Description These dryers, often known as paddle or horizontal agitated dryers, consist of one or more horizontally mounted shells with internal mechanical agitators, which may take many different forms. They are a continuous equivalent of the horizontal pan dryer and are similar in construction, but usually of larger dimensions. They have many similarities to continuous indirect rotary dryers and are sometimes classed as rotary dryers, but this is a misnomer because the outer shell does not rotate, although in some types there is an inner shell which does. Frequently, the internal agitator is heated, and a wide variety of designs exist. Often, two intermeshing agitators are used. There are important variants with high-speed agitator rotation and supplementary convective heating by hot air.

Classification Continuous; mechanical agitation and transport; layer; contact/conduction or convective (through-circulation).

The basic differences are in type of agitator, the two key factors being heat-transfer area and solids handling/stickiness characteristics. Unfortunately, the types giving the highest specific surface area (multiple tubes and coils) are often also the ones most liable to fouling and blockage and most difficult to clean. Figure 12-56 illustrates a number of different agitator types.

The most common problem with paddle dryers (and with their closely related cousins, steam-tube and indirect rotary dryers) is the buildup of sticky deposits on the surface of the agitator or outer jacket. This leads, first, to reduced heat-transfer coefficients and slower drying and, second, to blockages and stalling of the rotor. Also, thermal decomposition and loss of product quality can result. The problem is usually most acute at the feed end of the dryer, where the material is wettest and stickiest. A wide variety of different agitator designs have been devised to try to reduce stickiness problems and enhance cleanability while providing a high heat-transfer area. Many designs incorporate a high torque drive combined with rugged shaft construction to prevent rotor stall during processing, and stationary mixing elements are installed in the process housing which continually clean the heat-exchange surfaces of the rotor to minimize any crust buildup and ensure an optimum heat-transfer coefficient at all times. Another alternative is to use two parallel intermeshing shafts, as in the Nara paddle dryer (Fig. 12-57). Suitably designed continuous paddle and batch horizontal pan dryers can handle a wide range of product consistencies (dilute slurries, pastes, friable powders) and can be used for processes such as reactions, mixing, drying, cooling, melting, sublimation, distilling, and vaporizing. Bearing supports are usually provided at both ends of the unit for shaft support.

Design Methods for Paddle Dryers Product trials are conducted in small pilot dryers (8- to 60-L batch or continuous units) to determine material handling and process retention times. Variables such as drying temperature, pressure level, and shaft speed are analyzed during the test trials. For initial design purposes, the heat-transfer coefficient for paddle dryers is typically in the range of 10 W/(m^2·K) (light, free-flowing powders) up to 150 W/(m^2·K) (dilute slurries). However, it is preferable to scale up from the test results, finding the heat-transfer coefficient by backcalculation and scaling up on the basis of total area of heat-transfer surfaces, including heated agitators. Typical length/diameter ratios are between 5 and 8, similar to rotary dryers and greater than some batch horizontal pan dryers.

Continuous Rotary Dryers A rotary dryer consists of a cylinder that rotates on suitable bearings and that is usually slightly inclined to the horizontal. The cylinder length may range from 4 to more than 10 times

TABLE 12-26 Comparative Dimensions and Drying Times for Various Batch Contact Dryers

Dryer type	h, kW/(m^2·K)	L/D	D, m	L, m	A, m^2	t_{CR}, h	t_{FR}, h	t_{expt}, h
Tumbler/double-cone	0.05	1.5	3.71	5.56	38.91	2.78	5.98	11.0
Vertical pan	0.05	0.5	3.71	1.85	32.37	3.34	7.19	13.2
Spherical	0.05	1	3.37	3.37	35.63	3.04	6.54	12.0
Filter dryer	0.05	0.5	3.71	1.85	21.58	5.01	10.77	19.8
Conical agitated	0.05	1.5	3.71	5.56	34.12	3.17	6.82	12.5
Paddle (horizontal agitated)	0.05	5	1.72	8.60	46.50	2.33	5.01	9.2
Paddle, heated agitator	0.05	5	1.72	8.60	278.99	0.39	0.83	1.52

TABLE 12-27 Performance Data of Vacuum Rotary Dryers*

Material	Diameter × length, m	Initial moisture, % dry basis	Steam pressure, Pa × 10^3	Agitator speed, r/min	Batch dry weight, kg	Final moisture, % dry basis	Pa × 10^3	Time, h	Evaporation, kg/(h·m²)
Cellulose acetate	1.5 × 9.1	87.5	97	5.25	610	6	90–91	7	1.5
Starch	1.5 × 9.1	45–48	103	4	3630	12	88–91	4.75	7.3
Sulfur black	1.5 × 9.1	50	207	4	3180	1	91	6	4.4
Fuller's earth/mineral spirit	0.9 × 3.0	50	345	6	450	2	95	8	5.4

*Stokes Vacuum, Inc.

the diameter, which may vary from less than 0.3 to more than 3 m. Solids fed into one end of the drum are carried through it by gravity, with rolling, bouncing and sliding, and drag caused by the airflow either retarding or enhancing the movement, depending on whether the dryer is cocurrent or countercurrent. It is possible to classify rotary dryers into direct-fired, where heat is transferred to the solids by direct exchange between the gas and the solids, and indirect, where the heating medium is separated from physical contact with the solids by a metal wall or tube. Many rotary dryers contain flights or lifters, which are attached to the inside of the drum and which cascade the solids through the gas as the drum rotates.

For handling large quantities of granular solids, a cascading rotary dryer is often the equipment of choice. If the material is not naturally free-flowing, recycling of a portion of the final dry product may be used to precondition the feed, either in an external mixer or directly inside the drum. Hanging link chains and/or scrapper chains are also used for sticky feed materials.

Their operating characteristics when performing heat- and mass-transfer operations make them suitable for the accomplishment of drying, chemical reactions, solvent recovery, thermal decompositions, mixing, sintering, and agglomeration of solids. The specific types included are the following:

Direct cascading rotary dryer (cooler). This is usually a bare metal cylinder but with internal flights (shelves) which lift the material and drop it through the airflow. It is suitable for low- and medium-temperature operations, the operating temperature being limited primarily by the strength characteristics of the metal employed in fabrication.

Direct rotary dryer (cooler). As above but without internal flights.

Direct rotary kiln. This is a metal cylinder lined on the interior with insulating block and/or refractory brick. It is suitable for high-temperature operations.

Indirect steam-tube dryer. This is a bare metal cylinder provided with one or more rows of metal tubes installed longitudinally in the shell. It is suitable for operation up to available steam temperatures or in processes requiring water cooling of the tubes.

Indirect rotary calciner. This is a bare metal cylinder surrounded on the outside by a fired or electrically heated furnace. It is suitable for operation at medium temperatures up to the maximum that can be tolerated by the metal wall of the cylinder, usually 650 to 700 K for carbon steel and 800 to 1025 K for stainless steel.

Direct Roto-Louvre dryer. This is one of the more important special types, differing from the direct rotary unit in that true *through-circulation* of gas through the solids bed is provided. Like the direct rotary, it is suitable for low- and medium-temperature operation.

Direct heat rotary dryer. The direct heat units are generally the simplest and most economical in operation and construction, when the solids and gas can be permitted to be in contact. In design mode, the required gas flow rate can be obtained from a heat and mass balance. Bed cross-sectional area is found from a scoping design calculation (a typical gas velocity is 3 m/s for cocurrent and 2 m/s for countercurrent units). Length is normally between 5 and 10 times drum diameter (an L/D value of 8 can be used for initial estimation) or can be calculated by using an incremental model (see Examples 21 and 23).

A typical schematic diagram of a rotary dryer is shown in Fig. 12-58, while Fig. 12-59 shows typical lifting flight designs.

Classification Continuous; agitation and transport by rotation/gravity; layer (dispersion for cascading rotary dryers); convective (through-circulation) or contact/conduction.

Residence Time, Standard Configuration The residence time in a rotary dryer τ represents the average time that particles are present in the equipment, so it must match the required drying time.

Traditional approaches For rotary kilns, without lifting flights, Sullivan et al. (U.S. Bureau of Mines Tech. Paper 384) gave an early formula:

$$\tau = \frac{106.2L\sqrt{\gamma}}{N_m D \tan\alpha} \tag{12-88}$$

Here, the natural angle of repose of the material is γ, which increases as the material becomes more cohesive and less free-flowing, and the residence time τ is in seconds, but the rotation rate N_m is in revolutions per minute (rpm), not per second. The Friedman and Marshall equation [*Chem. Eng. Progr.* **45**(8): 482 (1949)] is derived from this, with an additional term to account for air drag on the solids:

$$\tau = \frac{13.8L}{N_m^{0.9} D\alpha} \pm \frac{590.6LG}{d_p^{0.5}G} \tag{12-89}$$

Here d_p is the particle size, in micrometers, while F and G are the mass flow rates of solids and gas, respectively. This formula has been frequently reported and includes a correction factor to the initial constant term to reflect actual experimental results. Friedman and Marshall took the angle of repose for the solids to be 40° and introduced a 0.9 power for the rotational speed, which had questionable justification within the accuracy of the data. The second term represents the airflow drag term and is negative for cocurrent flow and positive for countercurrent flow.

TABLE 12-28 Standard Rotary Vacuum Dryers*

Diameter, m	Length, m	Heating surface, m²	Working capacity, m³†	Agitator speed, r/min	Drive, kW	Weight, kg	Purchase price (1995): Carbon steel	Purchase price (1995): Stainless steel (304)
0.46	0.49	0.836	0.028	7½	1.12	540	$ 43,000	$ 53,000
0.61	1.8	3.72	0.283	7½	1.12	1,680	105,000	130,000
0.91	3.0	10.2	0.991	6	3.73	3,860	145,000	180,000
0.91	4.6	15.3	1.42	6	3.73	5,530	180,000	205,000
1.2	6.1	29.2	3.57	6	7.46	11,340	270,000	380,000
1.5	7.6	48.1	6.94	6	18.7	15,880	305,000	440,000
1.5	9.1	57.7	8.33	6	22.4	19,050	330,000	465,000

*Stokes Vacuum, Inc. Prices include shell, 50-lb/in²-gauge jacket, agitator, drive, and motor; auxiliary dust collectors, condensers.
†Loading with product level on or around the agitator shaft.

TABLE 12-29 Standard (Double-Cone) Rotating Vacuum Dryers*

Working capacity, m^3	Total volume, m^3	Heating surface, m^2	Drive, kW	Floor space, m^2	Weight, kg	Purchase cost (1995)	
						Carbon steel	Stainless steel
0.085	0.130	1.11	.373	2.60	730	$ 32,400	$ 38,000
0.283	0.436	2.79	.560	2.97	910	37,800	43,000
0.708	1.09	5.30	1.49	5.57	1810	50,400	57,000
1.42	2.18	8.45	3.73	7.15	2040	97,200	106,000
2.83	4.36	13.9	7.46	13.9	3860	198,000	216,000
4.25	6.51	17.5	11.2	14.9	5440	225,000	243,000
7.08	10.5	*38.7	11.2	15.8	9070	324,000	351,000
9.20	13.9	*46.7	11.2	20.4	9980	358,000	387,000
11.3	16.0	*56.0	11.2	26.0	10,890	378,000	441,000

*Stokes Vacuum, Inc. Price includes dryer, 15-lb/in^2 jacket, drive with motor, internal filter, and trunnion supports for concrete or steel foundations. Horsepower is established on 65 percent volume loading of material with a bulk density of 50 lb/ft^3. Models of 250 ft^3, 325 ft^3, and 400 ft^3 have extended surface area.

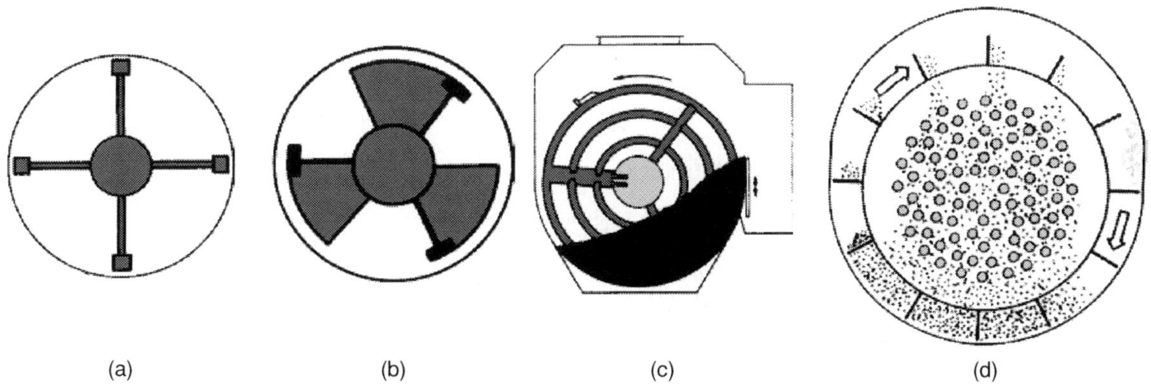

FIG. 12-56 Typical agitator designs for paddle (horizontal agitated) dryers. (*a*) Simple unheated agitator. (*b*) Heated cut-flight agitator. (*c*) Multicoil unit. (*d*) Tube bundle.

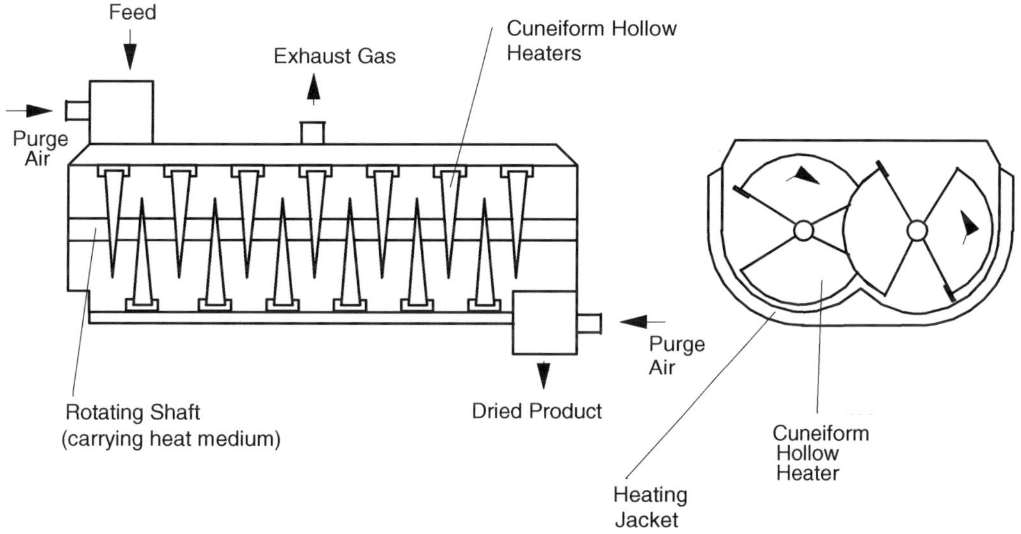

FIG. 12-57 Nara twin-shaft paddle dryer.

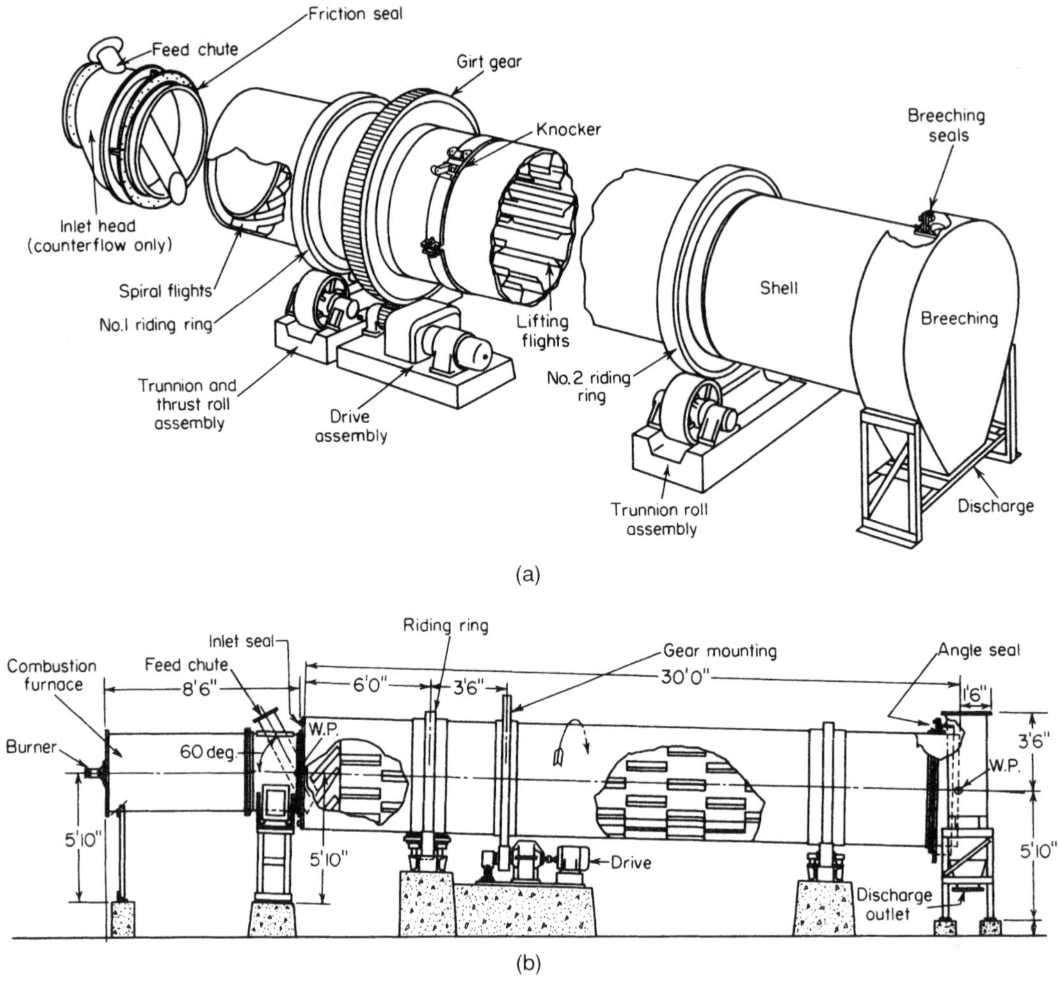

FIG. 12-58 Component arrangement (*a*) and elevation (*b*) of countercurrent direct-heat rotary dryer. (*Air Preheater Company, Raymond® & Bartlett Snow™ Products.*)

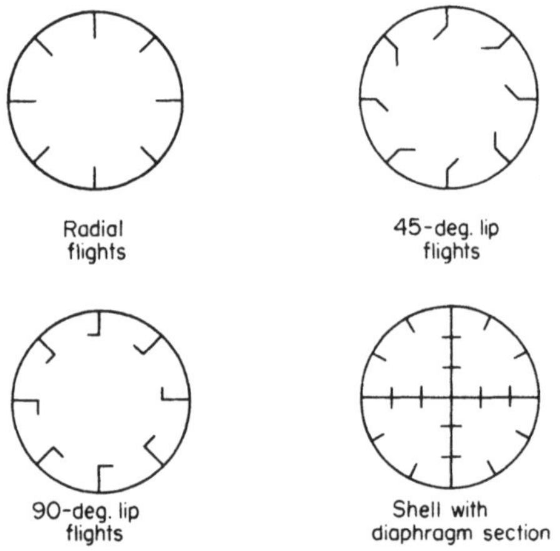

FIG. 12-59 Typical lifting flight designs.

Saeman and Mitchell [*Chem. Eng. Progr.* **50**(9):467 (1954)] proposed the following expression:

$$\tau = \frac{L}{f_H ND(\tan \alpha \pm k_m U_G)} \qquad (12\text{-}90)$$

Here f_H is a *cascade factor*, with values typically between 2 and π, increasing as solids holdup increases, and k_m is an empirical constant (dimensional) for a given material. The superficial gas velocity through the empty drum is U_G. It was assumed that the airborne particle velocity was proportional to the air velocity. Two empirical constants f_H and k_m are also required to use the equation, and these are not generally available.

Schofield and Glikin [*Trans. IChemE* **40:**183 (1962)] analyzed particle motion from flights and airborne drag, obtaining

$$\tau = \frac{L}{\bar{y}\,\bar{\theta}\,N[\sin \alpha - (KU_G^2/g)]} \qquad (12\text{-}91)$$

Here $\bar{y}$ is the mean distance of fall of the particles, $\bar{\theta}$ is the mean angle moved by particles in flights, and K is a dimensional drag constant. At small angles α, sin α ≈ tan α, and they noted that $\bar{y}\,\bar{\theta} \approx 2D$, so their final Eq. (12-91) is similar to that of Saeman and Mitchell (1954).

The above equations mainly differ in whether the drag term is additive or subtractive (as with Friedman and Marshall) or in the denominator (as with Saeman and Mitchell, and Schofield and Glikin). Some workers, including Sullivan et al. (1927), have neglected the effect of air drag completely. However, the general experience with rotary dryers is that the effect of air velocity and hence of air drag is very substantial, suggesting that neglecting air drag in any equation or analysis is unlikely to be sufficient unless the air velocity is very low. The formulas link L and τ, which is reasonably convenient for dryer performance assessment, but inconvenient for dryer design, where neither L nor τ is initially known.

Modern analysis Matchett and Baker [*J. Sep. Proc. Technol.* **8**:11 (1987)] provided a complete analysis of particle motion in rotary dryers. They considered both the airborne phase (particles falling through air) and the dense phase (particles in the flights or the rolling bed at the bottom). Typically, particles spend 90 to 95 percent of the time in the dense phase, but the majority of the drying takes place in the airborne phase. In the direction parallel to the dryer axis, most particle movement occurs through four mechanisms: by gravity and air drag in the airborne phase, and by bouncing, and sliding and rolling, in the dense phase. The combined particle velocity in the airborne phase is U_{P_1}, which is the sum of the gravitational and air drag components for cocurrent dryers and the difference between them for countercurrent dryers. The dense-phase velocity, arising from bouncing, sliding, and rolling, is denoted U_{P_1}.

Papadakis et al. [*Dry. Tech.* **12**(1&2):259 (1994)] rearranged the Matchett and Baker model from its original "parallel" form into a more computationally convenient "series" form. The sum of the calculated residence times in the airborne and dense phases, τ_G and τ_S, respectively, is the total solids residence time. The dryer length is simply the sum of the distances travelled in the two phases.

$$\tau = \tau_G + \tau_S \tag{12-92}$$

$$L = \tau_G U_{P_1} + \tau_S U_{P_2} \tag{12-93}$$

For airborne phase motion, the velocity U^o_{P1} due to the gravitational component is given approximately by

$$U^o_{P_1} = \sqrt{\frac{gD_e}{2\cos\alpha}}\, K_{\text{fall}} \tan\alpha \tag{12-94}$$

where D_e is the effective diameter, which is the distance actually fallen by the particles. When one is designing a dryer, this parameter will not be known until the flight width is decided. And K_{fall} is a parameter that allows for particles falling from a number of positions, with different times of flight and lifting times, and is generally between 0.7 and 1. The velocity $U^o_{P_1}$ due to the gravitational component is most conveniently expressed as

$$U^o_{P_1} = \sqrt{\frac{gD}{2}}\tan\alpha\, K_{\text{fall}}\sqrt{\frac{D_e}{D\cos\alpha}} = \sqrt{\frac{gD}{2}}\tan\alpha K_K \tag{12-95}$$

The drag force gives a velocity component $U^d_{P_1}$ that must be obtained from experimental correlations, and combining these components gives U_{P_1}.

Bouncing, rolling, and sliding are not so easily analyzed theoretically. Matchett and Baker suggested that the dense-phase velocity could be characterized in terms of a dimensionless dense-phase velocity number a, through the equation

$$a = \frac{U_{P_2}}{N \cdot D \cdot \tan\alpha} \tag{12-96}$$

Other workers suggested that, in underloaded and design-loaded dryers, bouncing was a significant transport mechanism, whereas for overloaded dryers, rolling (kilning) was important. Bouncing mechanisms can depend on the airborne phase velocity U_{P_1}, since this affects the angle at which the particles hit the bottom of the kiln and the distance they move forward. Rolling mechanisms would be expected to depend on the depth of the bottom bed, and hence on the difference between the actual holdup H and the design-loaded holdup $H°$.

As an example of the typical numbers involved, Matchett and Baker [*J. Sep. Proc. Technol.*, **9**:5 (1988)] used their correlations to assess the data of Saeman and Mitchell for an industrial rotary dryer with $D = 1.83$ m and $L = 10.67$ m, with a slope of 4°, 0.067 m/m. For a typical run with $U_G = 0.98$ m/s and $N = 0.08$ r/s, they calculated that $U^o_{P_1} = 0.140$ m/s, $U^d_{P_1} = -0.023$ m/s, $U_{P_1} = 0.117$ m/s, and $U_{P_2} = -0.02$ m/s. The dryer modeled was countercurrent and therefore had a greater slope and lower gas velocity than those of a cocurrent unit; for the latter, $U^o_{P_1}$ would be lower and $U^d_{P_1}$ positive and larger. The ratio τ_S/τ_G is approximately 12 in this case, so that the distance traveled in dense-phase motion would be about twice that in the airborne phase.

Kemp and Oakley [*Dry. Tech.*, **20**(9):1699 (2002)] showed that the ratio τ_G/τ_S can be found by comparing the average time of flight from the top of the dryer to the bottom t_f to the average time required for the particles to be lifted by the flights t_d. They derived the following equation:

$$\frac{\tau_S}{\tau_G} = \frac{t_d}{t_f} = \frac{K_{fl}}{N}\sqrt{\frac{g}{D}} \tag{12-97}$$

Here all the unknowns have been rolled into a single dimensionless parameter K_{fl}, given by

$$K_{fl} = \frac{\theta}{\pi\sqrt{2}\sin\theta}\sqrt{\frac{D}{D_e}} \tag{12-98}$$

Here D_e is the effective diameter (internal diameter between lips of flights), and the solids are carried in the flights for an angle 2θ, on average, before falling. Kemp and Oakley concluded that K_{fl} can be taken to be 0.4 to a first (and good) approximation. For overloaded dryers with a large rolling bed, K_{fl} will increase. The form of Eq. (12-97) is very convenient for design purposes since it does not require D_e, which is unknown until a decision has been made on the type and geometry of the flights.

The model of Matchett and Baker has been shown by Kemp (Proc. IDS 2004, B, 790) to be similar in form to that proposed by Saeman and Mitchell:

$$\tau = \frac{1.1L}{ND\left[\tan\alpha \cdot (K_K/K_{fl}\sqrt{2} + a) + (1/K_{fl})\sqrt{\dfrac{1}{gD}}\cdot U^d_{P_1}\right]} \tag{12-99}$$

In Eq. (12-99), $K_K/(K_{fl}\sqrt{2})$ will typically be on the order of unity, and reported values of a are in the range of 1 to 4. The airborne gravity component is usually smaller than the dense-phase motion but is not negligible. The sum of these two terms is essentially equivalent to the factor f_H in Saeman and Mitchell's equation.

Heat- and Mass-Transfer Estimates Many rotary dryer studies have correlated heat- and mass-transfer data in terms of an overall volumetric heat-transfer coefficient $U_v a$ [W/(m³·K)], defined by

$$Q = U_v a \cdot V_{\text{dryer}} \cdot \Delta T_m \tag{12-100}$$

Here Q is the overall rate of heat transfer between the gas and the solids (W), V_{dryer} is the dryer volume (m³), and ΔT_m is an average temperature driving force (K). When one is calculating the average temperature driving force, it is important to distinguish between the case of heat-transfer with dry particles, where the change in the particle temperature is proportional to the change in the gas temperature, and the case of drying particles, where the particle temperature does not change so significantly. Where the particles are dry, the average temperature difference is the logarithmic mean of the temperature differences between the gas and the solids at the inlet and outlet of the dryer, although Miller et al. (1942) took the logarithmic temperature difference as the average temperature difference even when the particles were drying. The volumetric heat-transfer coefficient itself consists of a heat-transfer coefficient U_v based on the effective area of contact between the gas and the solids, and the ratio a of this area to

TABLE 12-30 Values of the Index *n* in Correlations for the Volumetric Heat-Transfer Coefficient (after Baker, 1983)

Author(s)	Exponent n
Saeman and Mitchell (1954)	0
Friedman and Marshall (1949)	0.16
Aiken and Polsak (1982)	0.37
Miller et al. (1942)	0.46–0.60
McCormick (1962)	0.67
Myklestad (1963)	0.80

the dryer volume. Thus, this procedure eliminates the need to specify where most of the heat-transfer occurs (e.g., to material in the air, on the flights, or in the rolling bed). Empirical correlations are of the form

$$U_v a = \frac{K' U^n_{G\text{super}}}{D} \qquad (12\text{-}101)$$

where K' depends on the solids properties, the flight geometry, the rotational speed, and the dryer holdup. Table 12-30 gives the values of n chosen by various authors.

McCormick (1962) reworked the data of Miller et al. (1942), Friedman and Marshall (1949), and Saeman and Mitchell (1954) with a view to obtaining a single correlation of the form of Eq. (12-101) for the volumetric heat-transfer coefficient. He demonstrated that all the data could be correlated with values of the exponent n from 0.46 to 0.67. Although the evidence was far from conclusive, he believed that a value of 0.67 for the exponent n was most reliable. Individual values of the constant K' were obtained from the results of each of the workers cited above. He found that it was a function of the solids properties, the flight geometry, the rotational speed, and the dryer holdup, but that there was insufficient evidence available to relate K' to these parameters.

A comparison between the correlations of various workers was made by Baker (1983), and this is given in Table 12-31. A 2-m-diameter dryer containing 16 flights was chosen as the basis for the comparisons. With the exception of the results of Myklestad (1963), the values of $U_v a$ were calculated by using the values of K' and a value of n of 0.67, as obtained by McCormick (1963). A 17-fold variation in the predicted values of $U_v a$ can be observed at both 1 and 3 m/s. The reason for this is not readily apparent. With the exception of the commercial data correlation of Miller et al. (1942), the results were all obtained in pilot-scale rigs having diameters ranging from 0.2 to 0.3 m. Differences in equipment size are therefore not likely to be the cause of the variation. Hence the variation must be attributed to a combination of experimental errors and differences in the experimental conditions which are unaccounted for in the correlations.

An alternative procedure is the use of a conventional film heat-transfer coefficient h_f[W/(m^2·K)]

$$Q = h_f \cdot A_s \cdot \Delta T \qquad (12\text{-}102)$$

Here Q is the local heat-transfer rate (W), A_s is the total surface area of all the particles (m^2), and ΔT is the temperature difference between the gas and the solids (K). The method has the advantages that h_f can be determined by relatively simple tests (or calculated from appropriate correlations in the literature), variations in operating conditions can be allowed for, and analogies between heat and mass transfer allow the film coefficients for these processes to be related. However, the area for heat transfer must be estimated under the complex conditions of gas-solids interaction present in particle cascades. Schofield and Glikin (1962) estimated this area to be the surface area of particles per unit mass $6/(\rho_P d_P)$, multiplied by the fraction of solids in the drum that are cascading through the gas at any moment, which was estimated as the fraction of time spent by particles cascading through the gas:

$$A_s = \frac{6}{\rho_P\, d_P} \frac{t_f}{t_f + t_d} \qquad (12\text{-}103)$$

Schofield and Glikin estimated the heat-transfer coefficient by using the correlation given by McAdams (1954), which correlates data for gas-to-particle heat transfer in air to about 20 percent over a range of Reynolds numbers (Re_P, defined in the previous section) between 17 and 70,000:

$$\text{Nu}_P = 0.33 \cdot \text{Re}_P^{1/2} \qquad (12\text{-}104)$$

Here the particle Nusselt number is Nu_P, where $\text{Nu}_P = h_f d_P / k_G$, and k_G is the thermal conductivity of the gas [W/(m·K)]. They stated that the heat-transfer rates predicted by this procedure were much larger than those measured on an industrial cooler, which is probably due to the particles on the inside of the cascades not experiencing the full gas velocity. Kamke and Wilson (1986) used a similar approach to model the drying of wood chips, but used the Ranz-Marshall (1952) equation to predict the heat-transfer coefficient:

$$\text{Nu}_P = 2 + 0.6 \cdot \text{Re}_P^{1/2} \cdot \text{Pr}_G \qquad (12\text{-}105)$$

where Pr_G is the Prandtl number of the gas.

Drying Time Estimates Sometimes, virtually all the drying takes place in the airborne phase. Under such circumstances, the airborne-phase residence time τ_G and the drying time are virtually the same, and the required drying time can be estimated from equivalent times in drying kinetics experiments, e.g., using a thin-layer test (Langrish, D.Phil. thesis, 1988).

An example of how to incorporate the concept of the characteristic drying curve into a design calculation is given in Example 23.

Example 23: Sizing of a Cascading Rotary Dryer The average gas velocity passing through a cocurrent, adiabatic, cascading rotary dryer is 4 m/s. The particles moving through the dryer have an average diameter of 5 mm, a solids density of 600 kg/m^{-3}, and a shape factor of 0.75. The particles enter with a moisture content of 0.50 kg/kg (dry basis) and leave with a moisture content of 0.15 kg/kg (dry basis). The drying kinetics may be assumed to be linear, with no unhindered (constant-rate) drying period. In addition, let us assume that the solids are nonhygroscopic (so that the equilibrium moisture content is zero; hygroscopic means that the equilibrium moisture content is nonzero).

The inlet humidity is 0.10 kg/kg (dry basis) due to the use of a direct-fired burner, and the ratio of the flow rates of dry solids to dry gas is unity. The gas temperature at the inlet to the dryer is 800°C, and the gas may be assumed to behave as a pure water vapor/air mixture.

The gas-phase residence time that is required was calculated in the fundamentals section to be 38.0 s.

How does this gas-phase residence time relate to the total residence time that is required and to the dryer dimensions?

Application of residence time calculations (practice): Suppose that this dryer has a slope α of 4° and a diameter D of 1.5 m, operating at a rotational speed N of 0.04 r/s. We already know that the gas velocity through the drum $U_{G\text{super}}$ is 4 m/s, and that the particles have a mean diameter d_P of 5 mm and a particle density ρ_P of 600 kg/m^3. As a first estimate, suppose that the gas density ρ_G is 1 kg/m^3 and the gas viscosity μ_G is 1.8×10^{-5} kg/(m·s).

Now $K_K/(K_{fl}\sqrt{2}) \approx 1$, $K_{fl} \approx 0.4$, $K_{\text{fall}} \approx 1$, and a is within the range of 1 to 4, say, 2.5, and $U^d_{P_1}$ is estimated by the following calculation, for Reynolds numbers up to 220.

$$U^d_{P1} = 7.45 \times 10^{-4}\ \text{Re}^{2.2}\ \frac{\mu U_{G\text{super}} t^*_a}{\rho_P d_P^2} \qquad (12\text{-}106)$$

TABLE 12-31 Summary of the Predictions Using the Correlations for the Volumetric Heat-Transfer Coefficients of Various Authors (after Baker, 1983)

	$U_v a$, W/(m^3·K)	
Author(s)	$U_{G\text{super}_1} = 1$ m/s	$U_{G\text{super}_1} = 3$ m/s
Miller et al. (1942)		
Commercial data	248	516
Pilot-scale data	82	184
Friedman and Marshall (1949)	67	138
Saeman and Mitchell (1954)	495–1155	1032–2410
Myklestad (1963)	423	1019

Above this Reynolds number, the following equation was recommended by Matchett and Baker (1987):

$$U_{P1}^{d} = 125\,\frac{\mu U_{G\text{super}} t_a^{*}}{\rho_P d_P^2} \tag{12-107}$$

Here Re is the Reynolds number ($U_{G\text{super}} d_P \rho/\mu$) and t_f is the average time of flight of a particle in the airborne phase.

$$t_f = \left(\frac{2D}{g\cos\alpha}\right)^{1/2} K_{\text{fall}} \tag{12-108}$$

Substituting in the numbers gives

$$t_f = \left(\frac{2\cdot 1.5\text{ m}}{9.81\text{ m/s}^2\cdot\cos 4^\circ}\right)^{1/2} 1.0 = 0.554\text{ s}$$

$$\text{Re} = \frac{4\text{ m/s}\cdot 0.005\text{ m}\cdot 1\text{ kg/m}^3}{1.8\times 10^5\,(\text{kg/m}\cdot\text{s})} = 1100$$

$$U_{P1}^{d} = 125\;\frac{1.8\times 10^5\text{ kg/(m}\cdot\text{s)}\cdot 4\text{ m/s}\cdot 0.554\text{ s}}{(600\text{ kg/m}^3)(0.005\text{ m})^2}$$

$$= 0.332\text{ m/s}$$

$$\frac{\tau}{L} = \frac{1.1}{0.04\text{ s}^{-1}\cdot 1.5\text{ m}}$$

$$\times\left[\begin{array}{l}\tan 4^\circ\cdot(1+2.5)+\\ \dfrac{1}{0.4}\sqrt{\dfrac{1}{9.81\text{ m/s}^2\cdot 1.5\text{ m}}}\cdot 0.332\text{ m/s}\end{array}\right]$$

$$= 30\text{ s/m}$$

$$\frac{\tau_S}{\tau_G} = \frac{t_d}{t_f} = \frac{K_{fl}}{N}\sqrt{\frac{g}{D}} = \frac{0.4}{0.04\text{ s}^{-1}}\sqrt{\frac{9.81\text{ m/s}^2}{1.5\text{ m}}} = 25.6$$

Now, the required gas-phase residence time τ_G is 38.0 s. The ratio of solids to gas-phase residence times now gives us the required solids-phase residence time τ_S of 25.6 × 38.0 s = 972 s, and a total residence time of 972 + 38 = 1010 s. If the total residence time per unit length is 30 s/m, then the required dryer length is 1010 s/(30 s/m) = 34.2 m. The dryer length/diameter ratio is therefore 34.2 m/1.5 m = 22.8, which is significantly larger than the recommended ratio of between 5:1 and 10:1. The remedy would then be to use a larger dryer diameter and repeat these calculations. The larger dryer diameter would decrease the gas velocity, slowing the particle velocity along the drum, increasing the residence time per unit length, and hence decreasing the required drum length, to give a more normal length/diameter ratio.

Performance and Cost Data for Direct Heat Rotary Dryers Table 12-32 gives estimating-price data for direct rotary dryers employing steam-heated air. Higher-temperature operations requiring combustion chambers and fuel burners will cost more. The total installed cost of rotary dryers including instrumentation, auxiliaries, allocated building space, etc., will run from 150 to 300 percent of the purchase cost. Simple erection costs average 10 to 20 percent of the purchase cost.

Operating costs will include 5 to 10 percent of one worker's time, plus power and fuel required. Yearly maintenance costs will range from 5 to 10 percent of total installed costs. Total power for fans, dryer drive, and feed and product conveyors will be in the range of $0.5D^2$ to $1.0D^2$. Thermal efficiency of a high-temperature direct heat rotary dryer will range from 55 to 75 percent and, with steam-heated air, from 30 to 55 percent.

A representative list of materials dried in direct heat rotary dryers is given in Table 12-33.

Indirect Heat Rotary Steam-Tube Dryers Probably the most common type of indirect heat rotary dryer is the steam-tube dryer (Fig. 12-60). Steam-heated tubes running the full length of the cylinder are fastened symmetrically in one, two, or three concentric rows inside the cylinder and rotate with it. Tubes may be simple pipe with condensate draining by gravity into the discharge manifold or bayonet type. Bayonet-type tubes are also employed when units are used as water-tube coolers. When handling sticky materials, one row of tubes is preferred. These are occasionally shielded at the feed end of the dryer to prevent buildup of solids behind them. Lifting flights are usually inserted behind the tubes to promote solids agitation.

Wet feed enters the dryer through a chute or screw feeder. The product discharges through peripheral openings in the shell in ordinary dryers. These openings also serve to admit purge air to sweep moisture or other evolved gases from the shell. In practically all cases, gas flow is countercurrent to solids flow. To retain a deep bed of material within the dryer, normally 10 to 20 percent fillage, the discharge openings are supplied with removable chutes extending radially into the dryer. These, on removal, permit complete emptying of the dryer.

Steam is admitted to the tubes through a revolving steam joint into the steam side of the manifold. Condensate is removed continuously, by gravity through the steam joint to a condensate receiver and by means of lifters in the condensate side of the manifold. By employing simple tubes, noncondensables are continuously vented at the other ends of the tubes through Sarco-type vent valves mounted on an auxiliary manifold ring, also revolving with the cylinder.

Vapors (from drying) are removed at the feed end of the dryer to the atmosphere through a natural-draft stack and settling chamber or wet scrubber. When employed in simple drying operations with 3.5×10^5 to

TABLE 12-32 Warm-Air Direct-Heat Cocurrent Rotary Dryers: Typical Performance Data*

Dryer size, m × m	1.219 × 7.62	1.372 × 7.621	1.524 × 9.144	1.839 × 10.668	2.134 × 12.192	2.438 × 13.716	3.048 × 16.767
Evaporation, kg/h	136.1	181.4	226.8	317.5	408.2	544.3	861.8
Work, 10^8 J/h	3.61	4.60	5.70	8.23	1.12	1.46	2.28
Steam, kg/h at kg/m² gauge	317.5	408.2	521.6	725.7	997.9	131.5	2041
Discharge, kg/h	408	522	685	953	1270	1633	2586
Exhaust velocity, m/min	70	70	70	70	70	70	70
Exhaust volume, m³/min	63.7	80.7	100.5	144.4	196.8	257.7	399.3
Exhaust fan, kW	3.7	3.7	5.6	7.5	11.2	18.6	22.4
Dryer drive, kW	2.2	5.6	5.6	7.5	14.9	18.6	37.3
Shipping weight, kg	7700	10,900	14,500	19,100	35,800	39,900	59,900
Price, FOB Chicago	\$158,000	\$168,466	\$173,066	\$204,400	\$241,066	\$298,933	\$393,333

*Courtesy of Swenson Process Equipment Inc.

NOTE:

Material: heat-sensitive solid

Maximum solids temperature: 65°C

Feed conditions: 25 percent moisture, 27°C

Product conditions: 0.5 percent moisture, 65°C

Inlet-air temperature: 165°C

Exit-air temperature: 71°C

Assumed pressure drop in system: 200 mm

System includes finned air heaters, transition piece, dryer, drive, product collector, duct, and fan.

Prices are for carbon steel construction and include entire dryer system (November, 1994).

For 304 stainless-steel fabrication, multiply the prices given by 1.5.

TABLE 12-33 Representative Materials Dried in Direct-Heat Rotary Dryers*

Material dried	Moisture content, % (wet basis)		Heat efficiency, %
	Initial	Final	
High-temperature:			
Sand	10	0.5	61
Stone	6	0.5	65
Fluorspar	6	0.5	59
Sodium chloride (vacuum salt)	3	0.04	70–80
Sodium sulfate	6	0.1	60
Ilmenite ore	6	0.2	60–65
Medium-temperature:			
Copperas	7	1 (moles)	55
Ammonium sulfate	3	0.10	50–60
Cellulose acetate	60	0.5	51
Sodium chloride (grainer salt)	25	0.06	35
Cast-iron borings	6	0.5	50–60
Styrene	5	0.1	45
Low-temperature:			
Oxalic acid	5	0.2	29
Vinyl resins	30	1	50–55
Ammonium nitrate prills	4	0.25	30–35
Urea prills	2	0.2	20–30
Urea crystals	3	0.1	50–55

*Taken from *Chem. Eng.*, June 19, 1967, p. 190, Table III.

10×10^5 Pa steam, draft is controlled by a damper to admit only sufficient outside air to sweep moisture from the cylinder, discharging the air at 340 to 365 K and 80 to 90 percent saturation. In this way, shell gas velocities and dusting are minimized. When used for solvent recovery or other processes requiring a sealed system, sweep gas is recirculated through a scrubber-gas cooler and blower.

Steam manifolds for pressures up to 10^6 Pa are of cast iron. For higher pressures, the manifold is fabricated from plate steel, staybolted, and welded. The tubes are fastened rigidly to the manifold faceplate and are supported in a close-fitting annular plate at the other end to permit expansion. Packing on the steam neck is normally graphite asbestos. Ordinary rotating seals are similar in design with allowance for the admission of small quantities of outside air when the dryer is operated under a slight negative internal pressure.

Steam-tube dryers are used for the continuous drying, heating, or cooling of granular or powdery solids which cannot be exposed to ordinary atmospheric or combustion gases. They are especially suitable for fine dusty particles because of the low gas velocities required for purging of the cylinder. Tube sticking is avoided or reduced by employing recycle, shell knockers, etc., as previously described; tube scaling by sticky solids is one of the major hazards to efficient operation. The dryers are suitable for drying, solvent recovery, and chemical reactions. Steam-tube units have found effective employment in soda ash production, replacing more expensive indirect-heat rotary calciners.

Special types of steam-tube dryers employ packed and purged seals on all rotating joints, with a central solids-discharge manifold through the steam neck to reduce the seal diameter. This manifold contains the product discharge conveyor and a passage for the admission of sweep gas. Solids are removed from the shell by special volute lifters and dropped into the discharge conveyor. Units have been fabricated for operation at 76 mm of water, internal shell pressure, with no detectable air leakage.

Design methods for indirect heat rotary steam-tube dryers Heat-transfer coefficients in steam-tube dryers range from 30 to 85 W/(m^2·K). Coefficients will increase with increasing steam temperature because of increased heat transfer by radiation. In units carrying saturated steam at 420 to 450 K, the heat flux UT will range from 6300 W/m^2 for difficult-to-dry and organic solids to 1890 to 3790 W/m^2 for finely divided inorganic materials. The effect of steam pressure on heat-transfer rates up to 8.6×10^5 Pa is illustrated in Fig. 12-61.

Performance and cost data for indirect heat rotary steam-tube dryers Table 12-34 contains data for a number of standard sizes of steam-tube dryers. Prices tabulated are for ordinary carbon steel construction. Installed costs will run from 150 to 300 percent of purchase cost.

The thermal efficiency of steam-tube units will range from 70 to 90 percent, if a well-insulated cylinder is assumed. This does not allow for boiler efficiency, however, and is therefore not directly comparable with direct heat units such as the direct heat rotary dryer or indirect heat calciner.

Operating costs for these dryers include 5 to 10 percent of one person's time. Maintenance will average 5 to 10 percent of total installed cost per year.

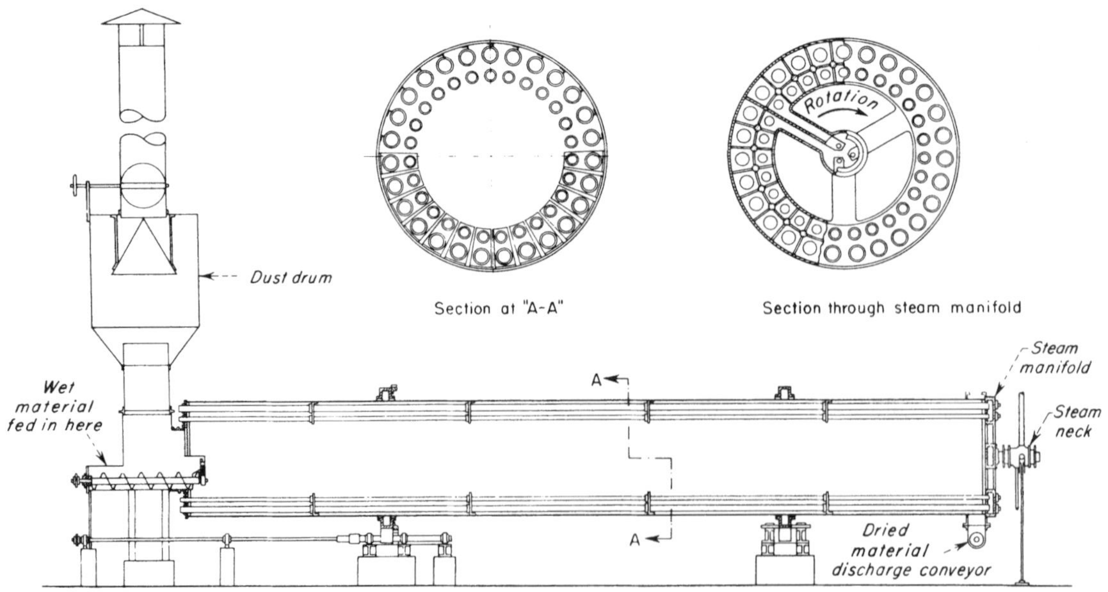

FIG. 12-60 Steam-tube rotary dryer.

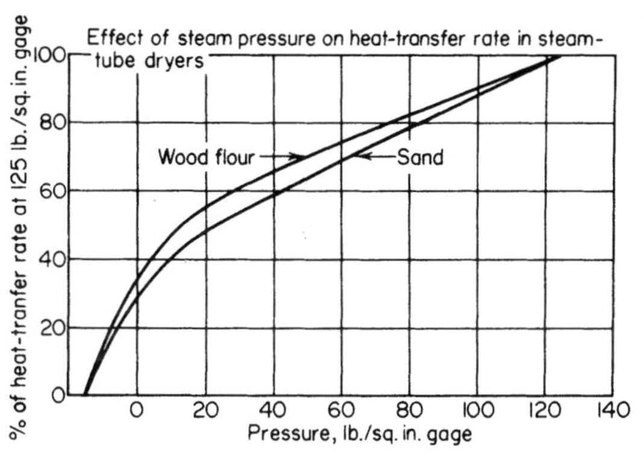

FIG. 12-61 Effect of steam pressure on the heat-transfer rate in steam-tube dryers.

Table 12-35 outlines typical performance data from three drying applications in steam-tube dryers.

Indirect Rotary Calciners and Kilns These large-scale rotary processors are used for very high temperature operations. Operation is similar to that of rotary dryers. For additional information, refer to Perry's 7th Edition, pages 12-56 to 12-58.

Indirect Heat Calciners Indirect heat rotary calciners, either batch or continuous, are employed for heat treating and drying at higher temperatures than can be obtained in steam-heated rotating equipment. They generally require a minimum flow of gas to purge the cylinder, to reduce dusting, and are suitable for gas-sealed operation with oxidizing, inert, or reducing atmospheres. Indirect calciners are widely utilized, and some examples of specific applications are as follows:

1. Activating charcoal
2. Reducing mineral high oxides to low oxides
3. Drying and devolatilizing contaminated soils and sludges
4. Calcination of alumina oxide-based catalysts
5. Drying and removal of sulfur from cobalt, copper, and nickel
6. Reduction of metal oxides in a hydrogen atmosphere
7. Oxidizing and "burning off" of organic impurities
8. Calcination of ferrites

This unit consists essentially of a cylindrical retort, rotating within a stationary insulation-lined furnace. The latter is arranged so that fuel combustion occurs within the annular ring between the retort and the furnace. The retort cylinder extends beyond both ends of the furnace. These end extensions carry the riding rings and drive gear. Material may be fed continuously at one end and discharged continuously at the other. Feeding and solids discharging are usually accomplished with screw feeders or other positive feeders to prevent leakage of gases into or out of the calciner.

In some cases in which it is desirable to cool the product before removal to the outside atmosphere, the discharge end of the cylinder is provided with an additional extension, the exterior of which is water-spray-cooled. In cocurrent flow calciners, hot gases from the interior of the heated portion of the cylinder are withdrawn through a special extraction tube. This tube extends centrally through the cooled section to prevent flow of gas near the cooled-shell surfaces and possible condensation. Frequently a separate cooler is used, isolated from the calciner by an air lock.

To prevent sliding of solids over the smooth interior of the shell, agitating flights running longitudinally along the inside wall are frequently provided. These normally do not shower the solids as in a direct heat vessel but merely prevent sliding so that the bed will turn over and constantly expose new surface for heat and mass transfer. To prevent scaling of the shell interior by sticky solids, cylinder scraper and knocker arrangements are occasionally employed. For example, a scraper chain is fairly common practice in soda ash calciners, while knockers are frequently utilized on metallic-oxide calciners.

Because indirect heat calciners frequently require close-fitting gas seals, it is customary to support all parts on a self-contained frame, for sizes up to approximately 2 m in diameter. The furnace can employ electric heating elements or oil and/or gas burners as the heat source for the process. The hardware would be zoned down the length of the furnace to match the heat requirements of the process. Process control is normally by shell temperature, measured by thermocouples or radiation pyrometers. When a special gas atmosphere must be maintained inside the cylinder, positive rotary gas seals, with one or more pressurized and purged annular chambers, are employed. The diaphragm-type seal is suitable for pressures up to 5 cm of water, with no detectable leakage.

In general, the temperature range of operation for indirect heat calciners can vary over a wide range, from 475 K at the low end to approximately 1475 K at the high end. All types of carbon steel, stainless, and

TABLE 12-34 Standard Steam-Tube Dryers*

Size, diameter × length, m	Tubes		m^2 of free area	Dryer speed, r/min	Motor size, hp	Shipping weight, kg	Estimated price
	No. OD (mm)	No. OD (mm)					
0.965 × 4.572	14 (114)		21.4	6	2.2	5,500	$152,400
0.965 × 6.096	14 (114)		29.3	6	2.2	5,900	165,100
0.965 × 7.620	14 (114)		36.7	6	3.7	6,500	175,260
0.965 × 9.144	14 (114)		44.6	6	3.7	6,900	184,150
0.965 × 10.668	14 (114)		52.0	6	3.7	7,500	196,850
1.372 × 6.096	18 (114)	18 (63.5)	58.1	4.4	3.7	10,200	203,200
1.372 × 7.620	18 (114)	18 (63.5)	73.4	4.4	3.7	11,100	215,900
1.372 × 9.144	18 (114)	18 (63.5)	88.7	5	5.6	12,100	228,600
1.372 × 10.668	18 (114)	18 (63.5)	104	5	5.6	13,100	243,840
1.372 × 12.192	18 (114)	18 (63.5)	119	5	5.6	14,200	260,350
1.372 × 13.716	18 (114)	18 (63.5)	135	5.5	7.5	15,000	273,050
1.829 × 7.62	27 (114)	27 (76.2)	118	4	5.6	19,300	241,300
1.829 × 9.144	27 (114)	27 (76.2)	143	4	5.6	20,600	254,000
1.829 × 10.668	27 (114)	27 (76.2)	167	4	7.5	22,100	266,700
1.829 × 12.192	27 (114)	27 (76.2)	192	4	7.5	23,800	278,400
1.829 × 13.716	27 (114)	27 (76.2)	217	4	11.2	25,700	292,100
1.829 × 15.240	27 (114)	27 (76.2)	242	4	11.2	27,500	304,800
1.829 × 16.764	27 (114)	27 (76.2)	266	4	14.9	29,300	317,500
1.829 × 18.288	27 (114)	27 (76.2)	291	4	14.9	30,700	330,200
2.438 × 12.192	90 (114)		394	3	11.2	49,900	546,100
2.438 × 15.240	90 (114)		492	3	14.9	56,300	647,700
2.438 × 18.288	90 (114)		590	3	14.9	63,500	736,600
2.438 × 21.336	90 (114)		689	3	22.4	69,900	838,200
2.438 × 24.387	90 (114)		786	3	29.8	75,300	927,100

*Courtesy of Swenson Process Equipment Inc. (prices from November, 1994). Carbon steel fabrication; multiply by 1.75 for 304 stainless steel.

TABLE 12-35 Steam-Tube Dryer Performance Data

	Class 1	Class 2	Class 3
Class of materials handled	High-moisture organic, distillers' grains, brewers' grains, citrus pulp	Pigment filter cakes, blanc fixe, barium carbonate, precipitated chalk	Finely divided inorganic solids, water-ground mica, water-ground silica, flotation concentrates
Description of class	Wet feed is granular and damp but not sticky or muddy and dries to granular meal	Wet feed is pasty, muddy, or sloppy; product is mostly hard pellets	Wet feed is crumbly and friable; product is powder with very few lumps
Normal moisture content of wet feed, % dry basis	233	100	54
Normal moisture content of product, % dry basis	11	0.15	0.5
Normal temperature of wet feed, K	310–320	280–290	280–290
Normal temperature of product, K	350–355	380–410	365–375
Evaporation per product, kg	2	1	0.53
Heat load per lb product, kJ	2250	1190	625
Steam pressure normally used, kPa gauge	860	860	860
Heating surface required per kg product, m^2	0.34	0.4	0.072
Steam consumption per kg product, kg	3.33	1.72	0.85

alloy construction are used, depending upon temperature, process, and corrosion requirements. Fabricated-alloy cylinders can be used over the greater part of the temperature range; however, the greater creep-stress abilities of cast alloys makes their use desirable for the highest calciner cylinder temperature applications.

Design methods for calciners In indirect heat calciners, heat transfer is primarily by radiation from the cylinder wall to the solids bed. The thermal efficiency ranges from 30 to 65 percent. By utilization of the furnace exhaust gases for preheated combustion air, steam production, or heat for other process steps, the thermal efficiency can be increased considerably. The limiting factors in heat transmission lie in the conductivity and radiation constants of the shell metal and solids bed. If the characteristics of these are known, equipment may be accurately sized by employing the Stefan-Boltzmann radiation equation. Apparent heat-transfer coefficients will range from 17 $W/(m^2 \cdot K)$ in low-temperature operations to 85 $W/(m^2 \cdot K)$ in high-temperature processes.

Cost data for calciners Power, operating, and maintenance costs are similar to those previously outlined for direct and indirect heat rotary dryers. Estimating purchase costs for preassembled and frame-mounted rotary calciners with carbon steel and type 316 stainless-steel cylinders are given in Table 12-36 together with size, weight, and motor requirements. Sale price includes the cylinder, ordinary angle seals, furnace, drive, feed conveyor, burners, and controls. Installed cost may be estimated, not including building or foundation costs, at up to 50 percent of the purchase cost. A layout of a typical continuous calciner with an extended cooler section is illustrated in Fig. 12-62.

Small batch retorts, heated electrically or by combustion, are widely used as carburizing furnaces and are applicable also to chemical processes involving the heat treating of particulate solids. These are mounted on a structural-steel base, complete with cylinder, furnace, drive motor, burner, etc. Units are commercially available in diameters from 0.24 to 1.25 m and lengths of 1 to 2 m. Continuous retorts with helical internal spirals are employed for metal heat-treating purposes. Precise retention control is maintained in these operations. Standard diameters are 0.33, 0.5, and 0.67 m with effective lengths up to 3 m. These vessels are employed in many small-scale chemical process operations which require accurate control of retention. Their operating characteristics and applications are identical to those of the larger indirect heat calciners.

Direct Heat Roto-Louvre Dryer One of the more important special types of rotating equipment is the Roto-Louvre dryer. As illustrated in Fig. 12-63, hot air (or cooling air) is blown through louvres in a double-wall rotating cylinder and up through the bed of solids. The latter moves continuously through the cylinder as it rotates. Constant turnover of the bed ensures uniform gas contacting for heat and mass transfer. The annular gas passage behind the louvres is partitioned so that contacting air enters the cylinder only beneath the solids bed. The number of louvres covered at any one time is roughly 30 percent. Because air circulates through the bed, fillages of 13 to 15 percent or greater are employed.

Roto-Louvre dryers range in size from 0.8 to 3.6 m in diameter and from 2.5 to 11 m long. The largest unit is reported capable of evaporating 5500 kg/h of water. Hot gases from 400 to 865 K may be employed. Because gas flow is through the bed of solids, high pressure drop, from 7 to 50 cm of water, may be encountered within the shell. For this reason, both a pressure inlet fan and an exhaust fan are provided in most applications to maintain the static pressure within the equipment as closely as possible to atmospheric. This prevents excessive in-leakage or blowing of hot gas and dust to the outside. For pressure control, one fan is usually operated under fixed conditions, with an automatic damper control on the other, regulated by a pressure detector-controller.

In heating or drying applications, when cooling of the product is desired before discharge to the atmosphere, cool air is blown through

TABLE 12-36 Indirect-Heat Rotary Calciners: Sizes and Purchase Costs*

Diameter, ft	Overall cylinder length	Heated cylinder length	Cylinder drive motor hp	Approximate Shipping weight, lb	Approximate sale price in carbon steel construction†	Approximate sale price in No. 316 stainless construction
4	40 ft	30 ft	7.5	50,000	$275,000	$325,000
5	45 ft	35 ft	10	60,000	375,000	425,000
6	50 ft	40 ft	20	75,000	475,000	550,000
7	60 ft	50 ft	30	90,000	550,000	675,000

*ABB Raymond (Bartlett-Snow™).
†Prices for November, 1994.

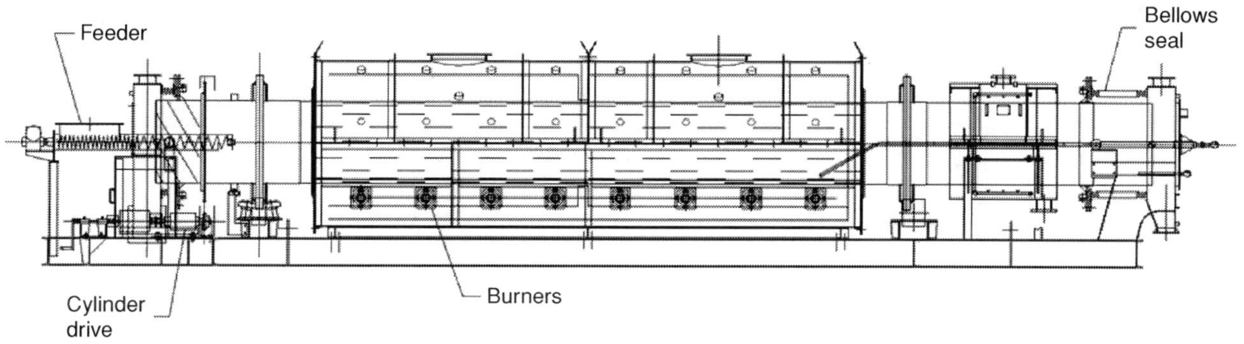

FIG. 12-62 Gas-fired rotary calciner with integral cooler. (*Air Preheater Company, Raymond® & Bartlett Snow™ Products.*)

a second annular space, outside the inlet hot-air annulus, and released through the louvres at the solids discharge end of the shell.

Roto-Louvre dryers are suitable for processing coarse granular solids which do not offer high resistance to airflow, do not require intimate gas contacting, and do not contain significant quantities of dust.

Heat transfer and mass transfer from the gas to the surface of the solids are extremely efficient; hence the equipment size required for a given duty is frequently less than that required when an ordinary direct heat rotary vessel with lifting flights is used. Purchase price savings are partially balanced, however, by the more complex construction of the Roto-Louvre unit. A Roto-Louvre dryer will have a capacity roughly 1.5 times that of a single-shell rotary dryer of the same size under equivalent operating conditions. Because of the cross-flow method of heat exchange, the average t is not a simple function of inlet and outlet t's. There are currently no published data which permit the sizing of equipment without pilot tests as recommended by the manufacturer. Three applications of Roto-Louvre dryers are outlined in Table 12-37. Installation, operating, power, and maintenance costs will be similar to those experienced with ordinary direct heat rotary dryers. *Thermal efficiency* will range from 30 to 70 percent.

Additional Reading

Aiken and Polsak, "A Model for Rotary Dryer Computation," in Mujumdar (ed.), *Drying '82*, Hemisphere, New York, 1982, pp. 32–35.

Baker, "Cascading Rotary Dryers," Chap. 1 in Mujumdar (ed.), *Advances in Drying*, vol. 2, pp. 1–51, Hemisphere, New York, 1983.

Friedman and Mahall, "Studies in Rotary Drying. Part 1. Holdup and Dusting. Part 2. Heat and Mass Transfer," *Chem. Eng. Progr.*, **45**:482–493, 573–588 (1949).

Hirosue and Shinohara, "Volumetric Heat Transfer Coefficient and Pressure Drop in Rotary Dryers and Coolers," 1st Int. Symp. on Drying, **8** (1978).

Kamke and Wilson, "Computer Simulation of a Rotary Dryer. Part 1. Retention Time. Part 2. Heat and Mass Transfer," *AIChE J.* **32**:263–275 (1986).

Kemp, "Comparison of Particle Motion Correlations for Cascading Rotary Dryers," *Drying 2004—Proceedings of the 14th International Drying Symposium* (IDS 2004), São Paulo, Brazil, Aug. 22–25, 2004, vol. B., pp. 790–797.

Kemp and Oakley, "Modeling of Particulate Drying in Theory and Practice," *Drying Technol.* **20**(9):1699–1750 (2002).

Langrish, "The Mathematical Modeling of Cascading Rotary Dryers," DPhil Thesis, University of Oxford, 1988.

Matchett and Baker, "Particle Residence Times in Cascading Rotary Dryers. Part 1—Derivation of the Two-Stream Model," *J. Separ. Proc. Technol.* **8:** 11–17 (1987).

Matchett and Baker, "Particle Residence Times in Cascading Rotary Dryers. Part 2—Application of the Two-Stream Model to Experimental and Industrial Data," *J. Separ. Proc. Technol.* **9:**5 (1988).

McCormick, "Gas Velocity Effects on Heat Transfer in Direct Heat Rotary Dryers," *Chem. Eng. Progr.* **58:**57–61 (1962).

Miller, Smith, and Schuette, "Factors Influencing the Operation of Rotary Dryers. Part 2. The Rotary Dryer as a Heat Exchanger," *Trans. AIChE* **38:** 841–864 (1942).

Myklestad, "Heat and Mass Transfer in Rotary Dryers," *Chem. Eng. Progr. Symp. Series* **59:**129–137 (1963).

Papadakis et al., "Scale-up of Rotary Dryers," *Drying Technol.* **12**(1&2): 259–278 (1994).

Ranz and Marshall, "Evaporation from Drops, Part 1," *Chem. Eng. Progr.* **48:** 123–142, 251–257 (1952).

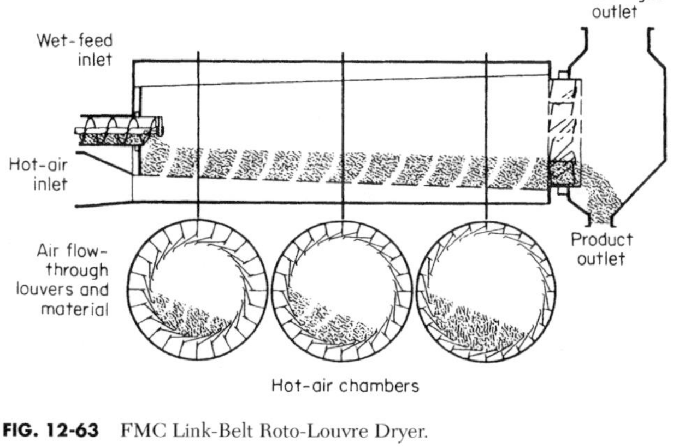

FIG. 12-63 FMC Link-Belt Roto-Louvre Dryer.

TABLE 12-37 Manufacturer's Performance Data for FMC Link-Belt Roto-Louvre Dryers*

Material dried	Ammonium sulfate	Foundry sand	Metallurgical coke
Dryer diameter	2 ft 7 in	6 ft 4 in	10 ft 3 in
Dryer length	10 ft	24 ft	30 ft
Moisture in feed, % wet basis	2.0	6.0	18.0
Moisture in product, % wet basis	0.1	0.5	0.5
Production rate, lb/h	2500	32,000	38,000
Evaporation rate, lb/h	50	2130	8110
Type of fuel	Steam	Gas	Oil
Fuel consumption	255 lb/h	4630 ft³/h	115 gal/h
Calorific value of fuel	837 Btu/lb	1000 Btu/ft³	150,000 Btu/gal
Efficiency, Btu, supplied per lb evaporation	4370	2170	2135
Total power required, hp	4	41	78

*Material Handling Systems Division, FMC Corp. To convert British thermal units to kilojoules, multiply by 1.06; to convert horsepower to kilowatts, multiply by 0.746.

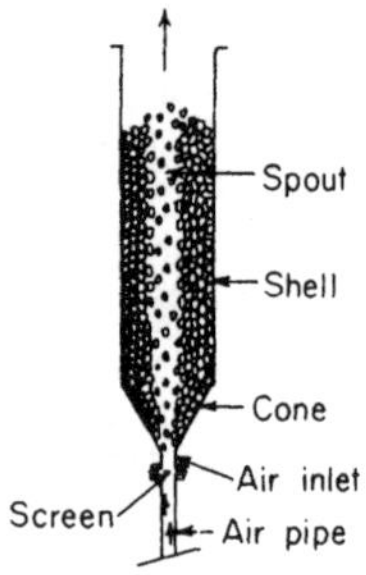

FIG. 12-64 Schematic diagram of sported bed. [*Mathur and Gishler,* Am. Inst. Chem. Eng. J., ***1**, 2, 15 (1955).*]

Saeman and Mitchell, "Analysis of Rotary Dryer Performance," *Chem. Eng. Progr.* **50**(9):467–475 (1954).
Schofield and Glikin, "Rotary Dryers and Coolers for Granular Fertilisers," *Trans. IChemE* **40:**183–190 (1962).
Sullivan, Maier, and Ralston, "Passage of Solid Particles through Rotary Cylindrical Kilns," U.S. Bureau of Mines Tech. Paper, 384, 44 (1927).

Fluidized and Spouted Bed Dryers

Spouted Beds The spouted bed technique was developed primarily for solids which are too coarse to be handled in fluidized beds.

Although their applications overlap, the methods of gas-solids mixing are completely different. A schematic view of a spouted bed is given in Fig. 12-64. Mixing and gas-solids contacting are achieved first in a fluid "spout," flowing upward through the center of a loosely packed bed of solids. Particles are entrained by the fluid and conveyed to the top of the bed. They then flow downward in the surrounding annulus as in an ordinary gravity bed, countercurrently to gas flow. The mechanisms of gas flow and solids flow in spouted beds were first described by Mathur and Gishler [*Am. Inst. Chem. Eng. J.* **1**(2): 157–164 (1955)]. Drying studies have been carried out by Cowan [*Eng. J.* **41:**5, 60–64 (1958)], and a theoretical equation for predicting the minimum fluid velocity necessary to initiate spouting was developed by Madonna and Lama [*Am. Inst. Chem. Eng. J.* **4**(4):497 (1958)]. Investigations to determine maximum spoutable depths and to develop theoretical relationships based on vessel geometry and operating variables have been carried out by Lefroy [*Trans. Inst. Chem. Eng.* **47**(5):T120–128 (1969)] and Reddy [*Can. J. Chem. Eng.* **46**(5):329–334 (1968)].

Gas flow in a spouted bed is partially through the spout and partially through the annulus. About 30 percent of the gas entering the system immediately diffuses into the downward-flowing annulus. Near the top of the bed, the quantity in the annulus approaches 66 percent of the total gas flow; the gas flow through the annulus at any point in the bed equals that which would flow through a loosely packed solids bed under the same conditions of pressure drop. Solids flow in the annulus is both downward and slightly inward. As the fluid spout rises in the bed, it entrains more and more particles, losing velocity and gas into the annulus. The volume of solids displaced by the spout is roughly 6 percent of the total bed.

On the basis of experimental studies, Mathur and Gishler derived an empirical correlation to describe the minimum fluid flow necessary for spouting, in 3- to 12-in-diameter columns:

$$u = \frac{D_p}{D_c}\left(\frac{D_o}{D_c}\right)^{0.33}\left[\frac{2gL(\rho_s - \rho_f)}{\rho_f}\right]^{0.5} \tag{12-109}$$

where u = superficial fluid velocity through the bed, ft/s; D_p = particle diameter, ft; D_c = column (or bed) diameter, ft; D_o = fluid inlet orifice diameter, ft; L = bed height, ft; ρ_s absolute solids density, lb/ft^3; ρ_f fluid density, lb/ft^3; and g = 32.2 ft/s^2, gravity acceleration. To convert feet per second to meters per second, multiply by 0.305; to convert pounds per cubic foot to kilograms per cubic meter, multiply by 16. In SI units, g = 9.8 m/s^2. The inlet orifice diameter, air rate, bed diameter, and bed depth were all found to be critical and interdependent:

1. In a given-diameter bed, deeper beds can be spouted as the gas inlet orifice size is decreased. Using air, a 12-in-diameter bed containing 0.125- by 0.250-in wheat can be spouted at a depth of over 100 in with a 0.8-in orifice, but at only 20 in with a 2.4-in orifice.
2. Increasing bed diameter increases spoutable depth. By employing a bed/orifice diameter ratio of 12 for air spouting, a 9-in-diameter bed was spouted at a depth of 65 in while a 12-in-diameter bed was spouted at 95 in.
3. As indicated by Eq. (12-109), the superficial fluid velocity required for spouting increases with bed depth and orifice diameter and decreases as the bed diameter is increased.

Employing wood chips, Cowan's drying studies indicated that the volumetric heat-transfer coefficient obtainable in a spouted bed is at least twice that in a direct heat rotary dryer. By using 20- to 30-mesh Ottawa sand, fluidized and spouted beds were compared. The volumetric coefficients in the fluid bed were 4 times those obtained in a spouted bed. Mathur dried wheat continuously in a 12-in-diameter spouted bed, followed by a 9-in-diameter spouted bed cooler. A drying rate of roughly 100 lb/h of water was obtained by using 450 K inlet air. Six hundred pounds per hour of wheat was reduced from 16 to 26 percent to 4 percent moisture. Evaporation occurred also in the cooler by using sensible heat present in the wheat. The maximum drying bed temperature was 118°F, and the overall thermal efficiency of the system was roughly 65 percent. Some aspects of the spouted bed technique are covered by patent (U.S. Patent 2,786,280).

Cowan reported that significant size reduction of solids occurred when cellulose acetate was dried in a spouted bed, indicating its possible limitations for handling other friable particles.

Direct Heat Vibrating Conveyor Dryers Information on vibrating conveyors and their mechanical construction is given in Sec. 19, "Solid-Solid Operations and Equipment." The vibrating conveyor dryer is a modified form of fluidized-bed equipment, in which fluidization is maintained by a combination of pneumatic and mechanical forces. The heating gas is introduced into a plenum beneath the conveying deck through ducts and flexible hose connections and passes up through a screen, perforated, or slotted conveying deck, through the fluidized bed of solids, and into an exhaust hood (Fig. 12-65). If ambient air is employed for cooling, the sides of the plenum may be open and a simple exhaust system used; however, because the gas distribution plate may be designed for several inches of water pressure drop to ensure a uniform velocity distribution through the bed of solids, a combination pressure-blower exhaust-fan system is desirable to balance the pressure above the

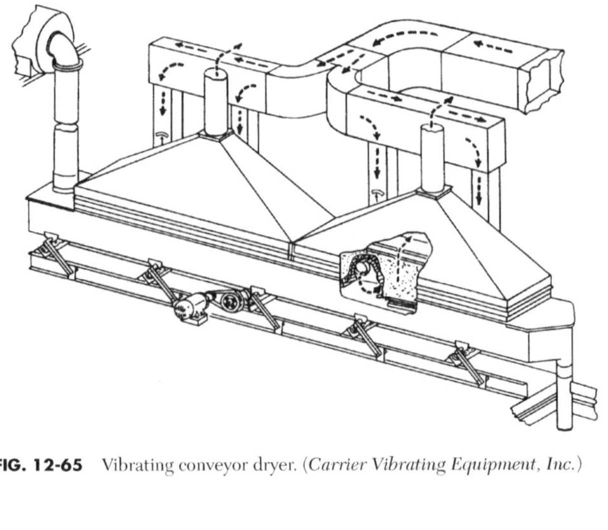

FIG. 12-65 Vibrating conveyor dryer. (*Carrier Vibrating Equipment, Inc.*)

TABLE 12-38 Table for Estimating Maximum Superficial Air Velocities through Vibrating-Conveyor Screens*

Mesh size	Velocity, m/s	
	2.0 specific gravity	1.0 specific gravity
200	0.22	0.13
100	0.69	0.38
50	1.4	0.89
30	2.6	1.8
20	3.2	2.5
10	6.9	4.6
5	11.4	7.9

*Carrier Vibrating Equipment, Inc.

deck with the outside atmosphere and prevent gas in-leakage or blowing at the solids feed and exit points.

Units are fabricated in widths from 0.3 to 1.5 m. Lengths are variable from 3 to 50 m; however, most commercial units will not exceed a length of 10 to 16 m per section. Power required for the vibrating drive will be approximately 0.4 kW/m^2 of deck.

In general, this equipment offers an economical heat-transfer area for first cost as well as operating cost. Capacity is limited primarily by the air velocity which can be used without excessive dust entrainment. Table 12-38 shows limiting air velocities suitable for various solids particles. Usually, the equipment is satisfactory for particles larger than 100 mesh in size. [The use of indirect heat conveyors eliminates the problem of dust entrainment, but capacity is limited by the heat-transfer coefficients obtainable on the deck (see Sec. 11)].

When a stationary vessel is employed for fluidization, all solids being treated must be fluidized; nonfluidizable fractions fall to the bottom of the bed and may eventually block the gas distributor. The addition of mechanical vibration to a fluidized system offers the following advantages:

1. Equipment can handle nonfluidizable solids fractions. Although these fractions may drop through the bed to the screen, directional-throw vibration will cause them to be conveyed to the discharge end of the conveyor. Prescreening or sizing of the feed is less critical than in a stationary fluidized bed.

2. Because of mechanical vibration, incipient channeling is reduced.

3. Fluidization may be accomplished with lower pressures and gas velocities. This has been evidenced on vibratory units by the fact that fluidization stops when the vibrating drive is stopped. Vibrating conveyor dryers are suitable for free-flowing solids containing mainly surface moisture. Retention is limited by conveying speeds which range from 0.02 to 0.12 m/s. Bed depth rarely exceeds 7 cm, although units are fabricated to carry 30- to 46-cm-deep beds; these also employ plate and pipe coils suspended in the bed to provide additional heat-transfer area. Vibrating dryers are not suitable for fibrous materials which mat or for sticky solids which may ball or adhere to the deck.

For estimating purposes for direct heat drying applications, it can be assumed that the average exit gas temperature leaving the solids bed will approach the final solids discharge temperature on an ordinary unit carrying a 5- to 15-cm-deep bed. Calculation of the heat load and selection of an inlet air temperature and superficial velocity (Table 12-38) will then permit approximate sizing, provided an approximation of the minimum required retention time can be made.

Vibrating conveyors employing direct contacting of solids with hot, humid air have also been used for the agglomeration of fine powders, chiefly for the preparation of agglomerated water-dispersible food products. Control of inlet air temperature and dew point permits the uniform addition of small quantities of liquids to solids by condensation on the cool incoming-particle surfaces. The wetting section of the conveyor is followed immediately by a warm-air-drying section and particle screening.

Fluidized-Bed Dryers The basic principles of fluid-bed technology are thoroughly described in Sec. 17, "Gas-Solid Operations and Equipment." Originally conceived as a heterogeneous chemical reactor, the use of this technology in connection with drying processes has increased considerably during the last several decades. The technology offers the following advantages when compared with other drying methods:

- It has no moving parts.
- It provides rapid heat and mass exchange between gas and particles.
- It provides high heat-transfer rates between the gas/particle bed and immersed objects such as heating panels.
- It provides intensive mixing of solids, leading to homogeneous conditions and reliable control of the drying process.

The fluid-bed technology can be applied to continuous as well as batch processes.

As described in Sec. 17, the process parameter of the highest importance is the gas velocity in the fluidized bed, referred to as the fluidizing velocity or the superficial gas velocity. This velocity is of nominal character since the flow field will be disturbed and distorted by the presence of the solid phase and the turbulent fluctuations created by the gas/solid interaction.

The fluid bed consists of a layer of particles suspended partly by a bed plate with perforations or a grid and partly by the fluidizing gas flowing through the bed plate. If the gas velocity is low, the gas will merely percolate through a bed of particles that appear to be fixed. At a higher velocity the particles will start to move under influence of the aerodynamic forces, and at the point where the pressure drop reaches the equivalent of the weight per unit of area, all the particles will tend to be moving in suspension, also called the incipiently fluidized state. The particle layer behaves as a liquid, and the bed volume expands considerably. At even higher velocities the motion will be stronger, and the excess gas flow will tend to appear as bubbles. In this state the particle layer will undergo vigorous mixing, while still appearing as a dense layer of fluidlike material or a boiling liquid. If the gas velocity is further increased, the solid phase will change into a slugging mode where gas bubbles throw lumps of solid material away from the bed surface. Even further increase will result in the solid phase being entrained by the gas flow and will appear as a lean phase undergoing transport or movement by the gas phase. Most fluid-bed drying processes are adjusted to operate safely below slugging conditions. Figure 12-66 shows a view into an operating drying fluid bed.

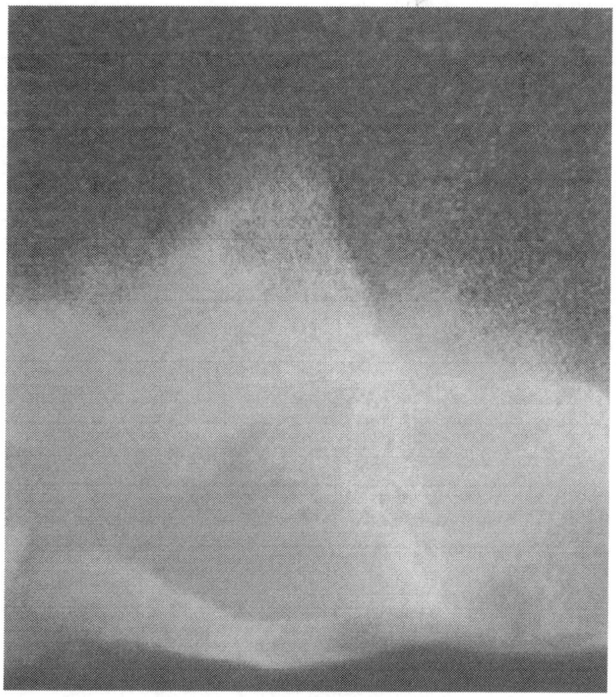

FIG. 12-66 Fluid bed for drying in operation. (*Niro A/S.*)

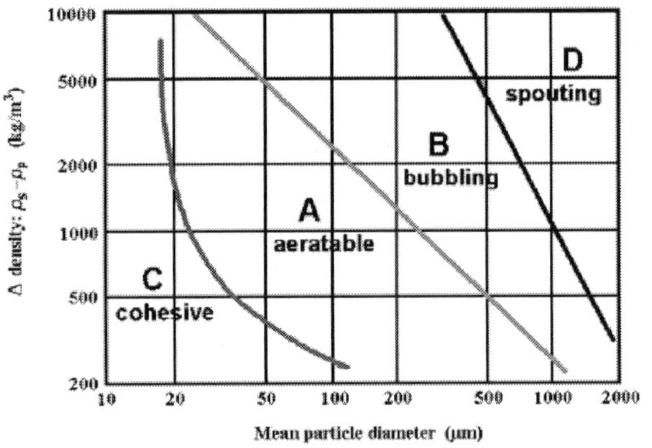

FIG. 12-67 Geldart diagram.

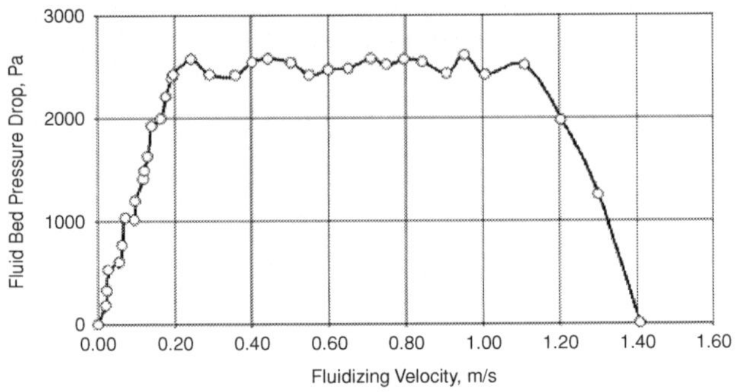

FIG. 12-69 Fluid-bed pressure drop versus fluidizing velocity. (*Niro A/S.*)

Proper design and operation of a fluid bed installation for drying requires consideration of several important topics. Among these are

- Ability of the material to be fluidized
- Drying characteristics of the material
- The fluidization velocity
- The design of the fluid-bed plate
- The operating conditions
- The mode of operation

Ability of the material to be fluidized has been investigated by Geldart, resulting in the well-known Geldart diagram, a version of which is shown in Fig. 12-67. The general knowledge to be derived from the Geldart diagram is that particulate material can be handled successfully in a fluid bed only if it is not too fine or too coarse. It must also have flowability. Fluid beds are best suited for particles that are regular in shape, not too sticky, and with a mean particle size between 20 μm and 10 mm. Particles of needle- or leaflike shape should be considered as nonfluidizable.

Drying characteristics of the material can be difficult to determine, but a test in a small batch fluid bed can reveal the drying curve of the material, as shown in Fig. 12-68. The drying curve clearly shows that the surface moisture is rapidly evaporated while the material is maintained at a low temperature close to the wet-bulb temperature of the drying gas. At a certain time the surface water has disappeared, and the so-called transition point has been reached. From here on the drying rate is controlled by internal diffusion inside the material, and the drying curve becomes characteristic for the individual material. While the moisture content of the material decreases, the bed temperature increases while approaching the inlet temperature of the drying air. The total drying time to reach the final moisture and the heat sensitivity of the material are important parameters for design of an industrial plant.

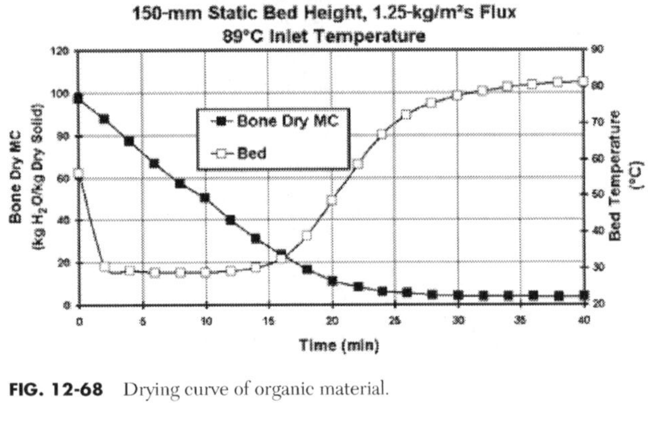

FIG. 12-68 Drying curve of organic material.

The fluidization velocity is of major importance, as indicated in the introduction. Each material will have individual requirements for the gas velocity and pressure drop to provide good fluidization. An investigation of the relationship between fluidization velocity and bed pressure drop for a given material may result in a diagram such as shown in Fig. 12-69. The results are illustrative and intended to give a clear picture of the relationship. The minimum fluidization velocity may be calculated from the Wen and Yu correlation given in Sec. 17.

At a fluidizing velocity below the value required for minimum or incipient fluidization, the pressure drop over the bed will increase proportionally with the velocity. Above a certain critical velocity, the pressure drop corresponds to the weight of the fluidized mass of material and remains roughly at this value even at higher velocities. The critical velocity for a given material may be estimated by methods mentioned in Sec. 17. At a much higher value of the fluidizing velocity, the material in the bed ceases to appear as a moving layer, and it is gradually carried away. Accordingly the pressure drop falls to zero. The fluidizing velocity value that will serve a drying task best cannot be derived exactly from the diagram. However, as a general recommendation, a value between the critical value and the value where the pressure drop falls off will be right. A first choice could be a factor of 2 to 5 times the minimum fluidization velocity. Further clarification must be derived from test work with the actual material.

The **design of the fluid-bed plate** is important for several reasons. First, the plate is responsible for the distribution of the drying or fluidization gas. This requires an even pattern of orifices in the plate and a sufficient pressure drop over the plate. As a general rule, the following guideline may be recommended:

$$\Delta P_{\text{plate}} = \tfrac{1}{2}\,\Delta P_{\text{powder}}$$

with the following limits: ΔP_{plate} minimum 500 Pa, ΔP_{plate} maximum 2500 Pa.

The estimation of the pressure drop in design situations may be difficult except for the case of the traditional perforated sheet with cylindrical holes perpendicular to the plane of the plate, as shown in Fig. 12-70. For this type of plate the formula of McAllister et al. may be useful. A calculation using this formula will show that a plate giving a required pressure drop of 1500 Pa and a typical fluidizing velocity of 0.35 m/s will need an open area of roughly 1 percent. Provided by a plate of 1-mm thickness and 1-mm-diameter holes, this requires approximately 12,500 holes per square meter.

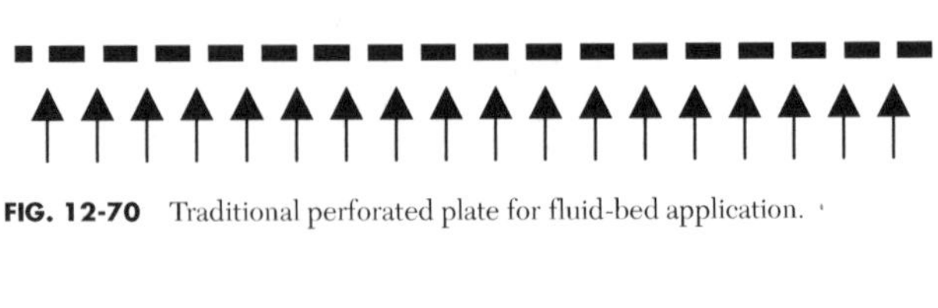

FIG. 12-70 Traditional perforated plate for fluid-bed application.

FIG. 12-71 Conidur® plate for fluid-bed application. (*Hein, Lehmann Trenn- und Fördertechnik GmbH.*)

However, this type of plate is being replaced in most fluid-bed applications due to its inherent disadvantages, which are caused by the difficulties of punching holes of smaller diameter than the thickness of the plate itself. The result is that the plates are weak and are prone to sifting back of the finer particles. The perpendicular flow pattern also means that the plate does not provide a transport capacity for lumps of powder along the plane of the plate.

This transport capacity is provided by plate of so-called gill types of which there are two distinct categories. One category is the type where plates are punched in a very fine regular pattern, not only to provide holes or orifices but also to deform the plate so that each orifice acquires a shape suited for acceleration of the gas flow in magnitude and direction. An example of this type is shown in Fig. 12-71, representing the Conidur® trademark.

The particular feature of Conidur® sheets is the specific hole shape which creates a directional airflow to help in discharging the product and to influence the retention time in the fluid bed. The special method of manufacturing Conidur® sheets enables finishing of fine perforations in sheets with an initial thickness many times over the hole width. Perforations of only 100 μm in an initial sheet thickness of 0.7 mm are possible. With holes this small 1 m^2 of plate may comprise several hundred thousand individual orifices.

The capacity of contributing to the transport of powder in the plane of the plate due to the horizontal component of the gas velocity is also the present for the second category of plates of the gill-type. Figure 12-72 shows an example.

In this type of plate, the holes or orifices are large and the number of gills per square meter is just a few thousand. The gas flow through each of the gills has a strong component parallel to the plate, providing powder transport capacity as well as a cleaning effect. The gills are punched individually or in groups and can be oriented individually to provide a possibility of articulating the horizontal transport effect.

In certain applications in the food and pharmaceutical industries, the nonsifting property of a fluid-bed plate is particularly appreciated. This property of a gill-type plate can be enhanced as illustrated in Fig. 12-73, where the hole after punching is additionally deformed so that the gill overlaps the orifice.

The fifth and final type of fluid-bed plate to be mentioned here is the so-called bubble plate type. Illustrated in Fig. 12-74, it is in principle a gill-type plate. The orifice is cut out of the plate, and the bubble is subsequently pressed so that the orifice is oriented in a predominantly horizontal direction. A fluid-bed plate will typically have only 1600 holes per square meter. By this technology a combination of three key features is established. The plate is nonsifting, it has transport capacity that can be articulated through individual orientation of bubbles, and it is totally free of cracks that may compromise sanitary aspects of the installation.

FIG. 12-72 GILL PLATE™ for fluid-bed application. (*Niro A/S.*)

FIG. 12-73 NON-SIFTING GILL PLATE™. (*Patented by Niro A/S.*)

FIG. 12-74 BUBBLE PLATE™. (*Patented by Niro A/S.*)

The **operating conditions** of a fluid bed are to a high degree dictated by the properties of the material to be dried, as already indicated. One parameter can be chosen regardless of the fluidization process, namely, the fluidization air temperature. For most products, however, the temperature is of primary importance, since the fluidized state results in very high heat-transfer rates so that heat sensitivity may restrict temperature and thereby prolong process time.

To achieve the most favorable combination of conditions to carry out a fluid-bed drying process, it is necessary to consider the different modes of fluid-bed drying available.

Industrial fluid-bed drying The first major distinction between fluid-bed types is the choice of mode: batch or continuous.

Batch fluid beds may appear in several forms. The process chamber has a perforated plate or screen in the bottom and a drying gas outlet at the top, usually fitted with an internal filter. The drying gas enters the fluid bed through a plenum chamber below the perforated plate and leaves through the filter arrangement. The batch of material is enclosed in the process chamber for the duration of the process.

Figure 12-75 shows a sketch of a typical batch fluid-bed dryer as used in the food and pharmaceutical industries. The process chamber is conic in order to create a freeboard velocity in the upper part of the chamber that is lower than the fluidizing velocity just above the plate.

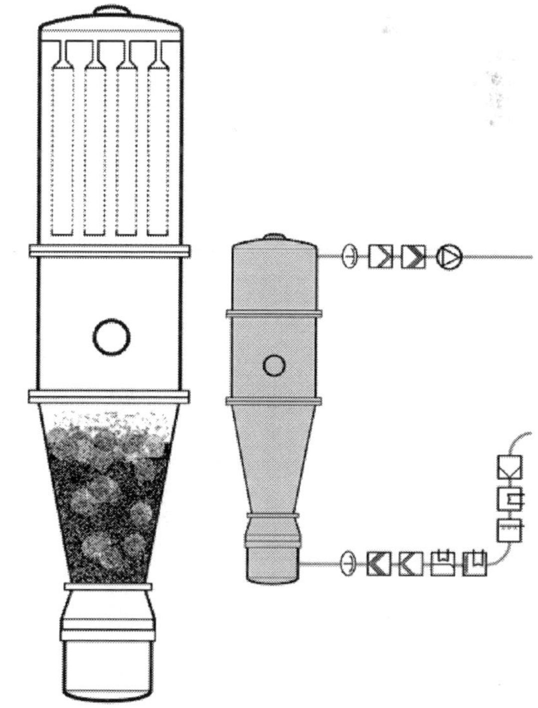

FIG. 12-75 Batch-type fluid-bed. (*Aeromatic-Fielder.*)

The enclosed product batch is prevented from escaping the process chamber and will therefore allow a freer choice of fluidizing velocity than is the case in a continuous fluid bed, as described later.

The right-hand side of Fig. 12-75 illustrates in symbol form the drying gas supply system comprising fan, filters of various grade, preheater, moisturizer, dehumidifier, final heater, and fast-closure valves. This arrangement is necessary for products with extreme quality requirements such as found in pharmaceutical production.

The drying can be carried out very like the process indicated in Fig. 12-68. The versatile drying gas supply system will allow the drying gas temperature and humidity to be controlled throughout the drying process to optimize process time and to minimize overheating of the product.

Continuous fluid beds may be even more varied than batch fluid beds. The main distinction between continuous fluid beds will be according to the solids flow pattern in the dryer. The continuous fluid bed will have an inlet point for moist granular material to be dried and an outlet for the dried material. If the moist material is immediately fluidizable, it can be introduced directly onto the plate and led through the bed in a plug-flow pattern that will enhance control of product residence time and temperature control. If the moist granular material is sticky or cohesive due to surface moisture and therefore needs a certain degree of drying before fluidization, it can be handled by a backmix fluid bed, to be described later.

Continuous plug-flow beds are designed to lead the solids flow along a distinct path through the bed. Baffles will be arranged to prevent or limit solids mixing in the horizontal direction. Thereby the residence time distribution of the solids becomes narrow. The bed may be of cylindrical or rectangular shape.

The temperature and moisture contents of the solids will vary along the path of solids through the bed and thereby enable the solids to come close to equilibrium with the drying gas. A typical plug-flow fluid bed is shown in Fig. 12-76.

Continuous plug-flow beds of stationary as well as vibrating type may benefit strongly from use of the gill-type fluid-bed plates with the capacity for controlling the movement of powder along the plate and around bends and corners created by baffles. Proper use of these means may make it possible to optimize the combination of fluidization velocity, bed layer height, and powder residence time.

Continuous backmix beds are used in particular when the moist granular material needs a certain degree of drying before it can fluidize. By distributing the material over the surface of an operating fluid bed arranged for total solids mixing, also called backmix flow, it will be absorbed by the dryer material in the bed, and lumping as well as sticking to the chamber surfaces will be avoided. The distribution of the feed can be arranged in different ways, among which a rotary thrower is an obvious choice. A typical backmix fluid bed is shown in Fig. 12-77. Backmix fluid beds can be of box-shaped design or cylindrical.

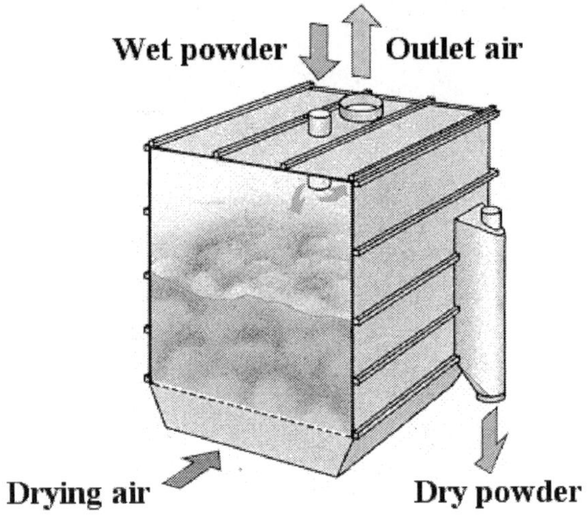

Figure 12-77 Continuous back-mix fluid bed. (*Niro A/S.*)

The whole mass of material in the backmix fluid bed will be totally mixed, and all powder particles in the bed will experience the same air temperature regardless of their position on the drying curve illustrated in Fig. 12-68. The residence time distribution becomes very wide, and part of the material may get a very long residence time while another part may get a very short time.

Continuous contact fluid beds are common in the chemical industry as the solution to the problem arising from materials requiring low fluidizing air temperature due to heat sensitivity and high energy input to complete the drying operation. An illustration of a Niro CONTACT FLUIDIZER™ is shown in Fig. 12-78.

The main feature of the contact fluid bed is the presence of heating panels, which are plate or tube structures submerged in the fluidized-bed layer and heated internally by an energy source such as steam, water, or oil. The fluidized state of the bed provides very high heat-transfer rates between the fluidizing gas, the fluidized material, and any objects submerged in the bed. The result is that a very significant portion of the required energy input can be provided by the heating panels without

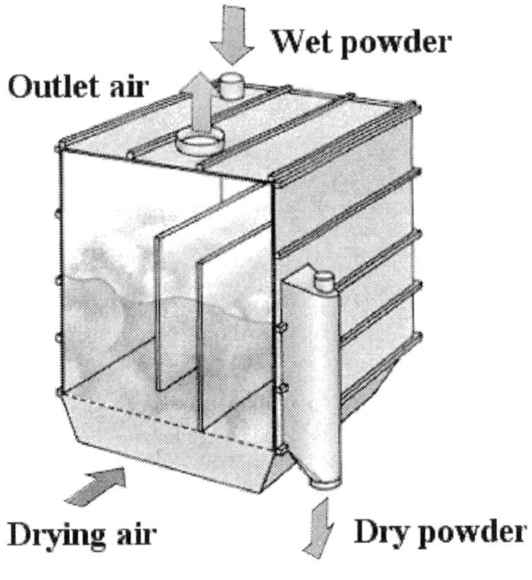

FIG. 12-76 Continuous plug-flow fluid bed. (*Niro A/S.*)

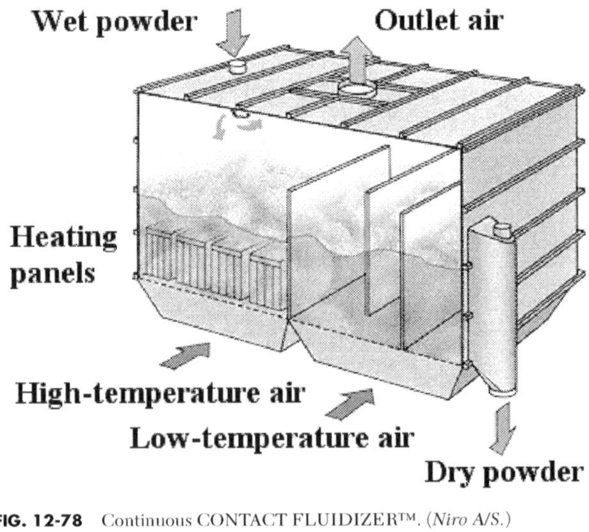

FIG. 12-78 Continuous CONTACT FLUIDIZER™. (*Niro A/S.*)

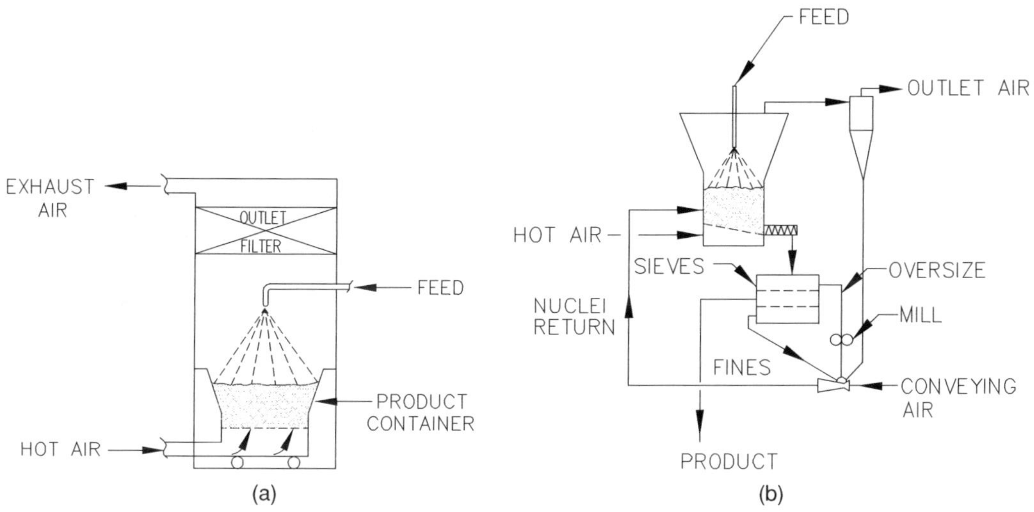

FIG. 12-79 Fluid-bed granulators. (*a*) Batch; (*b*) continuous.

risk of overheating the material. The fluidized state of the bed ensures that the material in the bed will flow with little restriction around the heating panels.

The CONTACT FLUIDIZER™ shown in Fig. 12-78 has a number of other features which in combination lead to compact design, high thermal efficiency, and low gas throughput: The first section of the bed is a backmix bed complete with rotary powder distributor and high-temperature fluidizing air supply. It takes care of the drying of the surface moisture, which is controlled mainly by heat supply. The heating panels are distributed over the whole bed volume of this section. The second section of the bed is a plug-flow bed with a fluidizing gas supply adjusted in both temperature and velocity to fit the requirements for the time-controlled diffusion drying of the powder present in this section.

The CONTACT FLUIDIZER™ is primarily used in the polymer industry for drying of polymer powders in high tonnages. Sizewise the individual units become very large, and units with a total fluid-bed area in excess of 60 m^2 are in operation.

Design methods for fluid beds When fluid-bed technology can be applied to drying of granular products, significant advantages compared to other drying processes can be observed. Design variables such as fluidizing velocity, critical moisture content for fluidization, and residence time required for drying to the specified residual moisture must, however, be established by experimental or pilot test before design steps can be taken. Reliable and highly integrated fluid-bed systems of either batch or continuous type can be designed, but only by using a combination of such pilot test and industrial experience. Scale-up rules are given by Kemp and Oakley (2002).

Additional Reading

Davidson and Harrison, *Fluidized Particles,* Cambridge University Press, 1963.
Geldart, *Powder Technol.* **6:**201–205 (1972).
Geldart, *Powder Technol.* **7:**286–292 (1973).
Grace, "Fluidized-Bed Hydrodynamics," Chap. 8.1 in *Handbook of Multiphase Systems,* McGraw-Hill, New York, 1982.
Gupta and Mujumdar, "Recent Developments in Fluidized Bed Drying," Chap. 5 in Mujumdar (ed.), *Advances in Drying,* vol. 2, Hemisphere, Washington, D.C., 1983, p. 155.
Kemp and Oakley, "Modeling of Particulate Drying in Theory and Practice," *Drying Technol.* **20**(9): 1699–1750 (2002).
McAllister et al., "Perforated-Plate Performance," *Chem. Eng. Sci.* **9:**25–35 (1958).
Poersch, *Aufbereitungs-Technik,* **4:** 205–218 (1983).
Richardson, "Incipient Fluidization and Particulate Systems," Chap. 2 in Davidson and Harrison (eds.), *Fluidization,* Academic Press, London, 1972.
Romankows, "Drying," Chap. 12 in Davidson and Harrison (eds.), *Fluidization,* Academic Press, London, 1972.
Vanacek, Drbohlar, and Markvard, *Fluidized Bed Drying,* Leonard Hill, London, 1965.

Dryers with Liquid Feeds If the feed is a liquid, paste, slurry, or solution, special equipment is required. The available choices are as follows:

Spray Dryers A pumpable feed is atomized into droplets by a rotary or nozzle atomizer, as described under "Entrainment Dryers." An integral fluid bed or belt may be added below the dryer to give longer residence time and some agglomeration. Semibatch and continuous operation is possible.

Fluidized-Bed Granulator A slurry or solution is sprayed onto a fluidized bed of particles, as shown in Fig. 12-79. The difference from the spray fluid bed is that the spray is still liquid when it contacts the particles, so that layered growth or surface agglomeration occurs, producing stronger large particles or agglomerates. Both batch and continuous forms exist (the latter involving continuous solids recycle with classification). In an older variant, the bed may be of inert balls, and the solid forming on the outside is periodically knocked off. Dryer construction and operation are largely as described under "Fluid-Bed Dryers."

Drum (Film-Drum) Dryers A film of liquid or paste is spread onto the outer surface of a rotating, internally heated drum. Drying occurs by conduction, and at the end of the revolution the dry product, which can be in the form of powder, flakes, or chips and typically is 100 to 300 μm thick, is removed by a doctor's knife. Drum dryers cannot handle feedstocks which do not adhere to metal, products which dry to a glazed film, or thermoplastics. The drum is heated normally by condensing steam or in vacuum drum dryers by hot water. Figure 12-80 shows three of the many possible forms. The dip feed system is the simplest and most common arrangement but is not suitable for viscous or pasty materials. The nip feed system is usually employed on double-drum dryers, especially for viscous materials, but it cannot handle lumpy or abrasive solids. The latter are usually applied by roller, and this is also effective for sticky and pasty materials. Spray and splash devices are used for feeding heat-sensitive, low-viscosity materials. Vacuum drum dryers are simply conventional units encased in a vacuum chamber with a suitable air lock for product discharge. Air impingement is also used as a secondary heat source on drum and can dryers, as shown in Fig. 12-81.

Contact Drying (Special thanks to R. B. Keey for the following example of contact drying.) In contact drying, the moist material covers a hot surface which supplies the heat required for the drying process.

Let us consider a moist material lying on a hot flat plate of infinite extent. Figure 12-82 illustrates the temperature profile for the fall in temperature from T_H in the heating fluid to T_G in the surrounding air. It is assumed that the temperatures remain steady, unhindered drying takes place, and there is no air-gap between the material being dried and the heating surface.

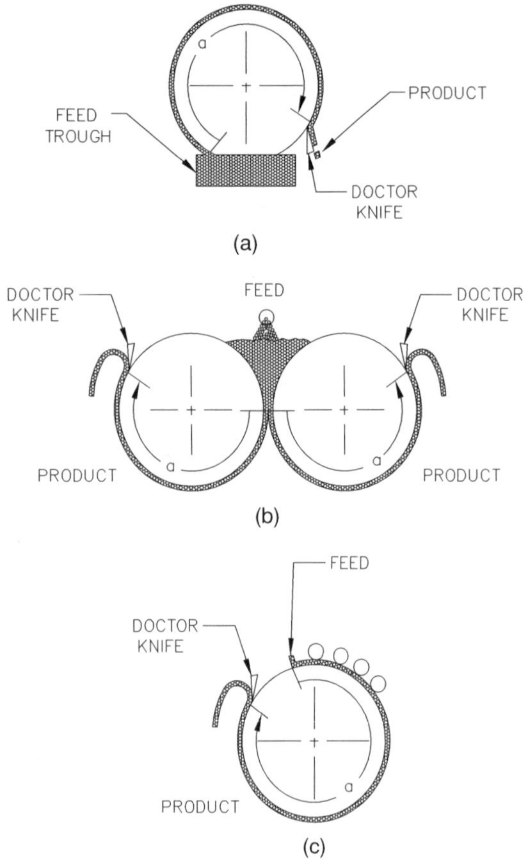

FIG. 12-80 Main types of drum dryers. (*a*) Dip; (*b*) nip; (*c*) roller.

The heat conducted through the wall and material is dissipated by evaporation of moisture and convection from the moist surface to the surrounding air. A heat balance yields

$$U(T_H - T_S) = N_W \Delta H_{VS} + h_C(T_S - T_G) \tag{12-110}$$

where U is the overall heat-transfer coefficient. This coefficient is found from the reciprocal law of summing resistances in series:

$$\frac{1}{U} = \frac{1}{h_H} + \frac{b_B}{\lambda_B} + \frac{b_s}{\lambda_s} \tag{12-111}$$

in which h_H is the heat-transfer coefficient for convection inside the heating fluid. If condensing steam is used, this coefficient is very large normally and the corresponding resistance $1/h_H$ is negligible. Rearrangement of Eq. (12-110) yields an expression for the maximum drying rate

$$N_W = \frac{U(T_H - T_S) - h_C\,(T_S - T_G)}{\Delta H_{VS}} \tag{12-112}$$

Equation (12-112), as it stands, would give an overestimate of the maximum drying rate for the case of contact drying over heated rolls, when there are significant heat losses from the ends of the drum and only part of the drum's surface can be used for drying. In the roller drying arrangements shown in Fig. 12-80, only a fraction a of the drum's periphery is available from the point of pickup to the point where the solids are peeled off.

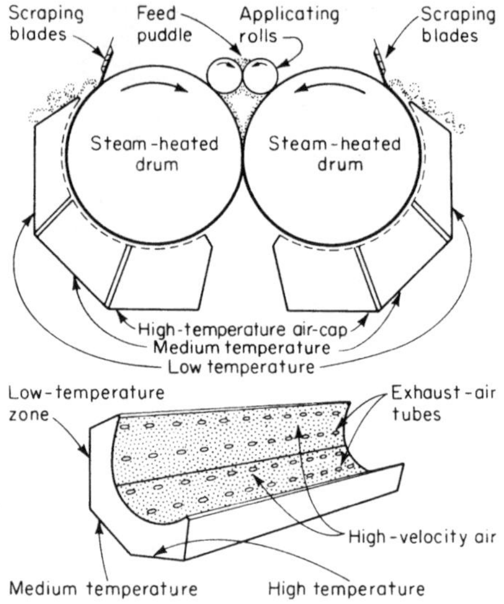

FIG. 12-81 Example of the use of air impingement in drying as a secondary heat source on a double-drum dryer. (Chem. Eng., ***197**, June 19, 1967.*)

Let q_E be the heat loss per unit area from the ends. The ratio of the end areas to cylindrical surface, from a drum of diameter D and length L, is $2(\frac{1}{4}\pi D^2)/\pi DL$ or $D/2L$. Equation (12-112) for the maximum drying rate under roller drying conditions thus becomes

$$N_W = \frac{aU(T_H - T_S) - h_C\,(T_S - T_G) - Dq_E/2L}{\Delta H_{VS}} \tag{12-113}$$

The total evaporation from the drum is $N_w a(\pi DL)$. Equation (12-113) could be refined further, as it neglects the effect caused by the small portion of the drum's surface being covered by the slurry in the feed trough, as well as thermal conduction through the axial shaft to the bearing mounts. The use of Eq. (12-113) to estimate the maximum drying rate is illustrated in Example 24.

Example 24: Heat-Transfer Calculations A single rotating drum of 1.250-m diameter and 3 m wide is internally heated by saturated steam at 0.27 MPa. As the drum rotates, a film of slurry 0.1 mm thick is picked up and dried. The dry product is removed by a knife, as shown in Fig. 12-80*a*. About three-quarters of the drum's surface is available for evaporating moisture. Estimate the maximum drying rate when the outside air temperature T_G is 15°C and the surface temperature 50°C, and compare the effectiveness of the unit with a dryer without end effects and in which all the surface could be used for drying.

Data:

Heat-transfer coefficient h_c 50 W/(m²·K)
Thickness of cylinder wall b_B 10 mm
Thermal conductivity of wall λ_B 40 W/(m·K)
Thermal conductivity of slurry film λ_s 0.10 W/(m·K)
Film transfer coefficient for condensing steam h_H 2.5 kW/(m²·K)

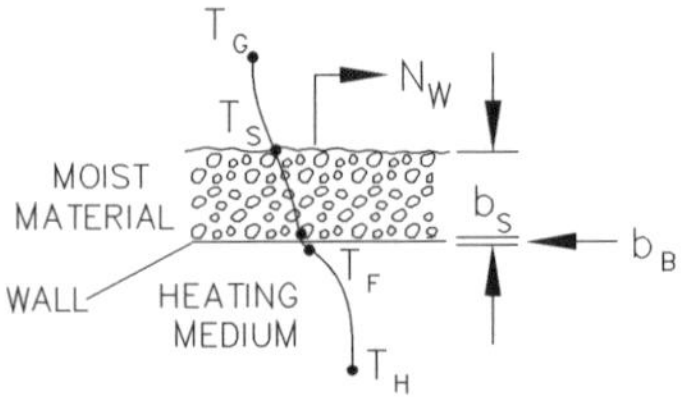

FIG. 12-82 Temperature profile in conductive drying.

Overall heat-transfer coefficient U: The thermal resistances are as follows:

Steamside	$1/2.5 = 0.40$ m²K/kW
Wall	$0.01/0.04 = 0.25$ m²K/kW
Filmside	$0.0001/0.1 \times 10^{-3} = 1.0$ m²K/kW

$\therefore$ Overall resistance $= 0.40 + 0.25 + 1.0 = 1.65$ m²K/kW

$$U = 1/1.65 = 0.606 \text{ kW/(m}^2\cdot\text{K)}$$

Wall temperature T_B: At 0.27 MPa, the steam temperature is 130°C. If it is assumed that the temperature drops between the steam and the film surface are directionally proportional to the respective thermal resistances, it follows that

$$\frac{T_H - T_B}{T_H - T_S} = \frac{0.40 + 0.25}{1.65} = 0.3939$$

$$\therefore T_B = T_H - 0.3939(T_H - T_S)$$
$$= 130 - 0.3939(130 - 50)$$
$$= 98.5°\text{C}$$

Heat losses from ends q_E: For an emissivity ~1 and an air temperature of 15°C with a drum temperature of 98.5°C, one finds [see Eq. (12-119)],

$$q_E = 1184 \text{ W/m}^2$$

Maximum drying rate N_W: From Eq. (12-113),

$$N_W = \frac{aU(T_H - T_S) - h_C(T_S - T_G) - Dq_E/2L}{\Delta H_{VS}}$$

$$= \frac{0.75 \times 0.606(130 - 50) - 0.05(50 - 15) - (1.25 \times 1.184)/6}{2382} \quad (12\text{-}114)$$

$$= 0.0144 \text{ kg/(m}^2\text{·s)}$$

The ideal maximum rate is given by Eq. (12-112) for an endless surface:

$$N_N = \frac{U(T_H - T_S) - h_c(T_S - T_G)}{\Delta H_{VS}}$$

$$= \frac{0.606(130 - 50) - 0.05(50 - 15)}{2382} \quad (12\text{-}115)$$

$$= 0.0196 \text{ kg/(m}^2\cdot\text{s)}$$

Therefore the effectiveness of the dryer is 0.0144/0.0196 = 0.735.

The predicted thermal efficiency η is

$$\eta = 1 - \frac{h_c(T_S - T_G + Dq_E/2L}{aU(T_H - T_S)}$$

$$= 1 - \frac{0.05(50 - 15) + (1.25 \times 1.184)/6}{0.75 \times 0.606(130 - 50)} \quad (12\text{-}116)$$

$$= 0.945$$

These estimates may be compared with the range of values found in practice, as shown in Table 12-39 (Nonhebel and Moss, *Drying of Solids in the Chemical Industry,* Butterworths, London, 1971, p. 168).

The typical performance is somewhat less than the estimated maximum evaporative capacity, although values as high as 25 g/(m²·s) have been reported. As the solids dry out, so the thermal resistance of the film increases and the evaporation falls off accordingly. Heat losses through the bearing of the drum shaft have been neglected, but the effect of radiation is accounted for in the value of h_c taken. In the case of drying organic pastes, the heat losses have been determined to be 2.5 kW/m² over the whole surface, compared with 1.75 kW/m² estimated here for the cylindrical surface. The inside surface of the drum has been assumed to be clean, and scale would reduce the heat transfer markedly.

For constant hygrothermal conditions, the base temperature T_B is directly proportional to the thickness of the material over the hot surface. When the wet-bulb temperature is high and the layer of material is thick enough, the temperature T_B will reach the boiling point of the moisture. Under these conditions, a mixed vapor-air layer interposes between the material and the heating surface. This is known as the Leidenfrost effect, and the phenomenon causes a greatly increased thermal resistance to heat transfer to hinder drying.

TABLE 12-39 Operating Information

	This estimate	Typical range
Specific evaporation, g/(m²·s)	14.4	7–11
Thermal efficiency	0.945	0.4–0.7

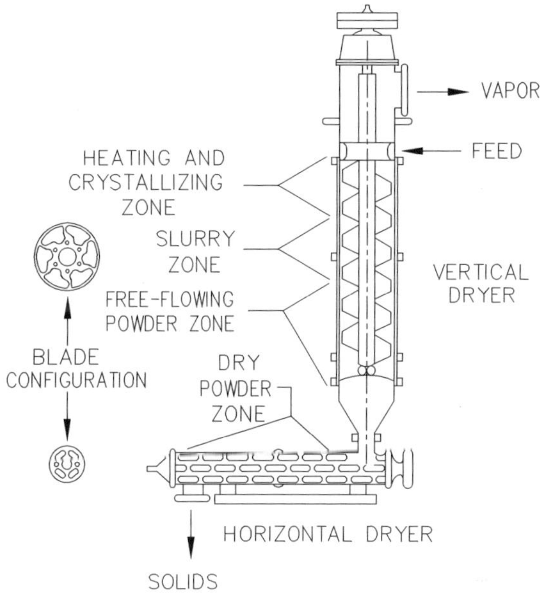

FIG. 12-83 Continuous thin-film dryer.

Thin-Film Dryers Evaporation and drying take place in a single unit, normally a vertical chamber with a vertical rotating agitator which almost touches the internal surface. The feed is distributed in a thin layer over the heated inner wall and may go through liquid, slurry, paste, and wet solid forms before emerging at the bottom as a dry solid. These dryers are based on wiped-film or scraped-surface (Luwa-type) evaporators and can handle viscous materials and deal with the "cohesion peak" experienced by many materials at intermediate moisture contents. They also offer good containment. Disadvantages are complexity, limited throughput, and the need for careful maintenance. Continuous or semibatch operation is possible. A typical unit is illustrated in Fig. 12-83.

Filter Dryers Basically this is a Nutsche filter (Sec. 18, "Liquid-Solid Operations and Equipment") followed by a batch dryer, usually of vertical pan type (see "Batch Agitated and Rotating Dryers" section). They are popular in the pharmaceutical and specialty chemicals industries as two unit operations are performed in the same piece of equipment without intermediate solids transfer, and containment is good.

Centrifuge Dryers Usually they are batch or continuous filtering centrifuges (Sec. 18, "Liquid-Solid Operations and Equipment") with hot air being blown over the solids in the discharge section. Manufacturers include Heinkel and Bird-Humboldt.

Pastelike feeds can be handled by some dryers for particulate materials, if either they do not require free-flowing feeds or some dry product can be backmixed with the wet feed to improve its handling.

Dryers for Films and Sheets The construction of dryers where both the feed and the product are in the form of a sheet, web, or film is markedly different from that for dryers used in handling particulate materials. The main users are the paper and textile industries. Almost invariably the material is formed into a very long sheet (often hundreds or thousands of meters long) which is dried in a continuous process. The sheet is wound onto a bobbin at the exit from the dryer; again, this may be 1 or 2 m in diameter. Alternatively, the sheet may be chopped into shorter sections.

Cylinder Dryers and Paper Machines The most common type of dryer in papermaking is the cylinder dryer (Fig. 12-84), which is a

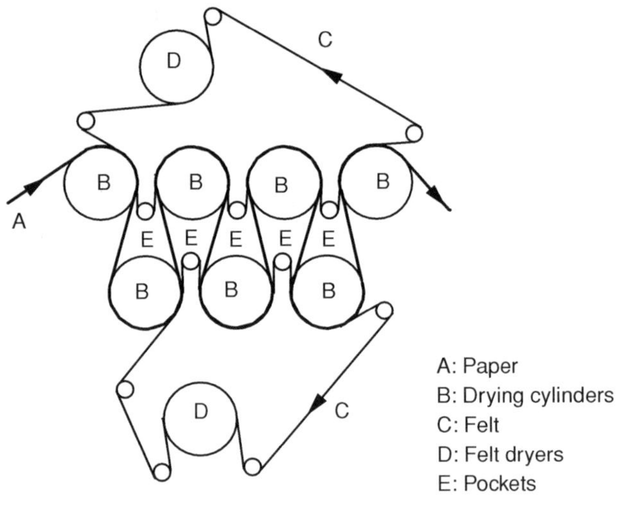

FIG. 12-84 Cylinder dryer (paper machine).

contact dryer. The paper web is taken on a convoluted path during which it wraps around the surface of cylinders which are internally heated by steam or hot water. In papermaking, the sheet must be kept taut, and a large number of cylinders are used, with only short distances between them and additional small unheated rollers to maintain the tension. Normally, a continuous sheet of felt is also used to hold the paper onto the cylinders, and this also becomes damp and is dried on a separate cylinder.

Most of the heating is conductive, through contact with the drums. However, infrared assistance is frequently used in the early stages of modern paper machines. This gets the paper sheet up to the wet-bulb temperature more rapidly, evaporates more surface moisture, and allows the number of cylinders to be reduced for a given throughput. Hot air jets (jet foil dryer) may also be used to supplement heating at the start of the machine. Infrared and dielectric heating may also be used in the later stages to assist the drying of the interior of the sheet.

Although paper is the most common application, multicylinder dryers can also be used for polymer films and other sheet-type feeds.

Convective dryers may be used as well in papermaking. In the Yankee dryer (Fig. 12-85), high-velocity hot airstreams impinging on the

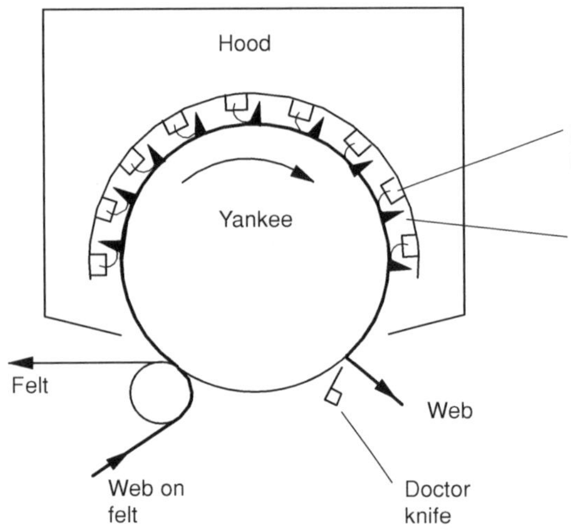

Fig. 12-85 Yankee dryer.

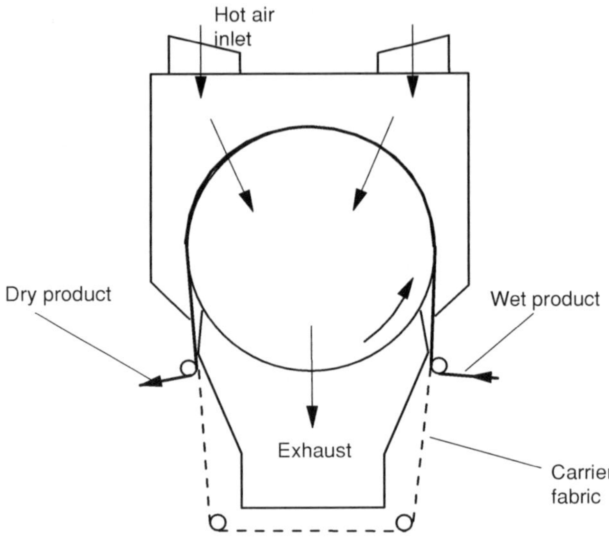

FIG. 12-86 Rotary through-dryer.

web surface give heating by cross-convection. The "Yankees" are barbs holding the web in place. Normally the cylinder is also internally heated, giving additional conduction heating of the lower bed surface. In the rotary through-dryer (Fig. 12-86), the drum surface is perforated and hot air passes from the outside to the center of the drum, so that it is a through-circulation convective dryer.

Another approach to drying of sheets has been to suspend or "float" the web in a stream of hot gas, using the Coanda effect, as illustrated in Fig. 12-87. Air is blown from both sides, and the web passes through as an almost flat sheet (with a slight "ripple"). The drying time is reduced because the heat transfer from the impinging hot air jets is faster than that from stagnant hot air in a conventional oven. It is essential to control the tension of the web very accurately. The technique is particularly useful for drying coated paper, as the expensive surface coating can stick to cylinder dryers.

Stenters (Tenters) and Textile Dryers These are the basic type of dryer used for sheets or webs in the textile industry. The sheet is held by its edges by clips (clip stenter) or pins (pin stenter), which not only suspend the sheet but also keep it taut and regulate its width—a vital consideration in textile drying. Drying is by convection; hot air is introduced from one or both sides, passes over the surface of the sheet, and permeates through it. Infrared panels may also be used to supply additional heat. A schematic diagram of the unit is shown in Fig. 12-88. A typical unit is 1.4 m wide and handles 2 to 4 t/h of material.

Heavy-duty textiles with thick webs may need a long residence time, and the web can be led up and down in "festoons" to reduce dryer length. Substantial improvements in drying rates have been obtained with radio-frequency heating assistance.

Air impingement dryers as in Fig. 12-87 may also be used for textiles.

Spray Dryers Spray drying is a drying process for transformation of a pumpable liquid feed in the form of a solution, dispersion, slurry, or paste into a particulate dried product in one single operation. The process comprises atomization of the feed followed by intense contact with hot air.

Due to the very large surface area created by the atomization of the feed, rapid evaporation occurs from the surface of each particle or droplet in the spray. The magnitude of the surface area can be illustrated by a simple calculation. Atomization of 1 L of water into a uniform spray of 100-μm droplets results in approximately 1.9×10^9 individual particles with a combined surface area of 60 m^2. A realistic spray with variation of the droplet size may have a substantially higher number of droplets and a somewhat higher surface area. The dry

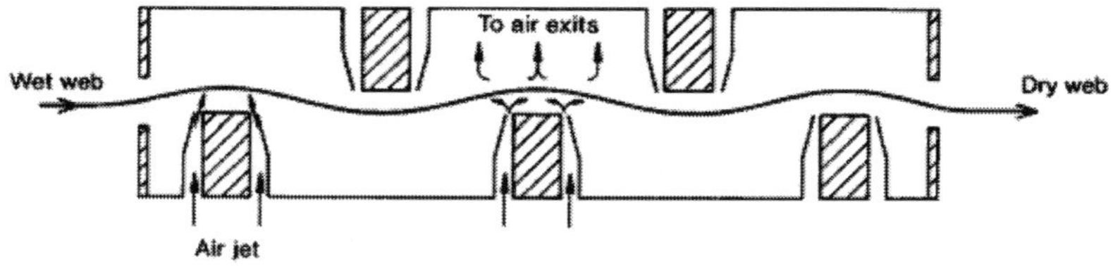

FIG. 12-87 Air flotation (impingement) dryer.

particulate product is formed while the spray droplets are still suspended in the hot drying air. The spray drying process is concluded by product recovery and separation from the drying air.

Spray drying belongs to the family of suspended particle processing (SPP) systems. Other members of this family are fluid-bed drying, flash drying, spray granulation, spray reaction, spray cooling, and spray absorption.

Drying Principles In the spray drying process or operation, the liquid to be removed by drying is predominantly water. Certain special products are produced with use of organic solvents, which are removed in a spray drying process. The drying principles involved for aqueous as well as nonaqueous systems are the same.

The liquid or moisture in a spray droplet is present in two basic forms: bound and unbound moisture. The nature of the solid and the liquid matter determines the drying characteristics of the product.

The category of bound moisture comprises water retained in small capillaries in the solid, water absorbed on solid surfaces, water bound as solutions in cells or fiber walls, and water bound as crystal water in chemical combination with the solid. Bound water exerts an equilibrium vapor pressure lower than that of pure water at the same temperature.

The category of unbound moisture can be described as the moisture in excess of the bound moisture. A hygroscopic material may contain bound as well as unbound moisture. A nonhygroscopic material contains unbound moisture only. The equilibrium vapor pressure of unbound water is equal to that of pure water at the same temperature.

The free moisture in a particle is the moisture in excess of the equilibrium moisture and may consist of unbound and some bound moisture. Only free moisture can be removed by evaporation during spray drying.

The mechanism of moisture flow in a droplet during spray drying is mainly diffusion supplemented by capillary flow. The drying characteristics of the droplet depend on the balance of bound and unbound as each category has distinct features.

The presence of unbound moisture in the droplet means that the drying proceeds at a constant high rate as long as the moisture diffusion within the droplet is able to maintain saturated surface conditions. When the diffusional and capillary flows can no longer maintain these conditions, a critical point is reached and the drying rate will decline until equilibrium moisture content is reached. The evaporation of bound moisture is strongly dependent on the nature of the solid matter in the spray droplet.

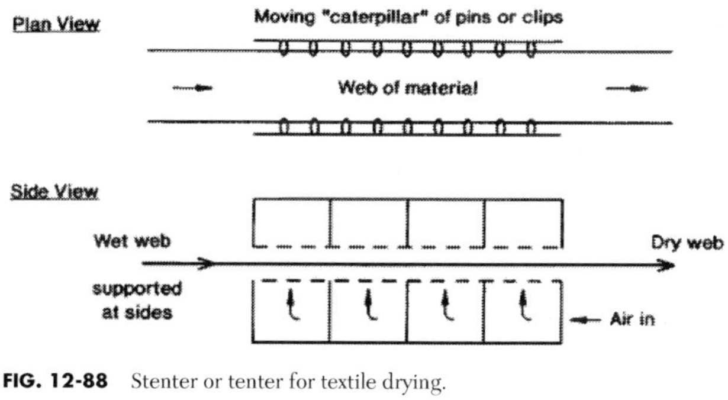

FIG. 12-88 Stenter or tenter for textile drying.

A spray drying plant comprises four process stages, as shown in Table 12-40.

Atomization Stage Spray drying is often used in industrial processes characterized by high production rates. Although the three different methods of atomization indicated in Table 12-40 are the same as those for many other atomization or spray forming processes, the relative weight of the methods is special for spray drying with rotary atomizers and hydraulic pressure nozzles having a very broad application, while two-fluid nozzles are only used to a smaller extent in specialized applications.

Rotary Atomizer Figure 12-89 shows a rotary atomizer in operation. The liquid feed is supplied to the atomizer by gravity or hydraulic pressure. A liquid distributor system leads the feed to the inner part of a rotating wheel. Since the wheel is mounted on a spindle supported by bearings in the atomizer structure, the liquid distributor is usually formed as an annular gap or a ring of holes or orifices concentric with the spindle and wheel. The liquid is forced to follow the wheel either by friction or by contact with internal vanes in the wheel. Due to the high centrifugal forces acting on the liquid, it moves rapidly toward the rim of the wheel, where it is ejected as a film or a series of jets or ligaments. By interaction with the surrounding air the liquid breaks up to form a spray of droplets of varying size. The spray pattern is virtually horizontal with a spray angle said to be 180°. The mean droplet size of the spray depends strongly on the atomizer wheel speed and to a much lesser degree on the feed rate and the feed physical properties such as viscosity. More details about spray characteristics such as droplet size distribution will be given below.

As indicated above, the atomizer wheel speed is the important parameter influencing the spray droplet size and thus the particle size of the final product. The atomizer machine will normally have the capability to operate the wheel at the required speed. More important for the atomization process is the selection of a wheel capable of handling a specific liquid feed with characteristic properties such as abrasiveness, high viscosity, nonnewtonian behavior, or tendency to coagulate.

The most common design of atomizer wheel has radial vanes, as shown in Fig.12-90. This wheel type is widely used in the chemical industry and is virtually blockage-free and simple to operate, even at very high speed. For high-capacity applications, the number and height of the vanes may be increased to maintain limited liquid film thickness conditions on each vane.

Wheels with radial vanes have one important drawback, i.e., their capacity for pumping large amounts of air through the wheel. This so-called air pumping effect causes unwanted product aeration, resulting in powders of low bulk density for some sensitive spray dried products.

TABLE 12-40 Stages of Spray Drying

Process stages of spray drying	Methods
1. Atomization	Rotary atomization Pressure nozzle atomization Two-fluid nozzle atomization
2. Spray/hot air contact	Cocurrent flow Countercurrent flow Mixed flow
3. Evaporation	Drying Particle shape formation
4. Product recovery	Drying chamber Dry collector Wet collectors

FIG. 12-89 Rotary atomizer operation. (*Niro*).

Unwanted air pumping effect and product aeration can be reduced through careful wheel design involving change of the shape of the vanes that may appear as forward-curved. This wheel type is used widely in the dairy industry to produce powders of high bulk density. The powder bulk density may increase as much as 15 percent when a curved vane wheel is replacing a radial vane wheel of standard design.

Another way of reducing the air pumping effect is to reduce the space between the vanes so that the liquid feed takes up a larger fraction of the available cross-sectional area. This feature is used with consequence in the so-called bushing wheels such as shown in Fig.12-91. This wheel combines two important design aspects. The air pumping effect is reduced by reducing the flow area to a number of circular orifices, each 5 to 10 mm in diameter. By placing these orifices or nozzles in replaceable bushings or inserts made of very hard materials such as technical ceramics, i.e., alumina or silicon carbide, a substantially abrasion-resistant atomizer wheel design is achieved. This feature is very important in a number of spray drying applications with abrasive feeds, which would wear down a standard vaned wheel in a matter of hours. With an abrasion-resistant wheel, almost unlimited lifetime can be expected for the atomizer wheel structure and several thousand hours for replaceable bushings.

The rotary atomizer machines are high-speed machines traditionally built with a step-up gear to increase the speed from the 3000 or 3600 rpm of the standard two-pole electric motors to 10,000 to 20,000 rpm normally required to achieve sufficiently fine atomization. Newer designs feature high-speed electric motors with frequency control of the atomizer speed. Table 12-41 gives the main operational parameters for three typical atomizers covering the wide range of capacity and size.

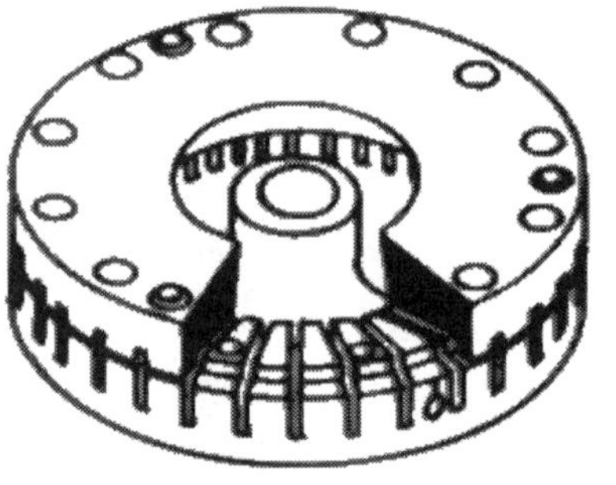

FIG. 12-90 Rotary atomizer wheel with radial vanes. (*Niro.*)

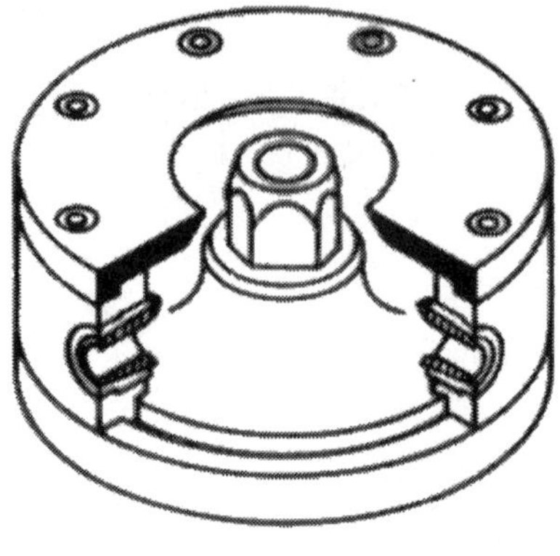

FIG. 12-91 Abrasion-resistant bushing atomizer wheel. (*Niro.*)

The F800 atomizer is the largest rotary atomizer offered to industry today. It has the capability of handling up to 200 t/h in one single atomizer. The capacity limit of an atomizer is normally its maximum power rating. As indicated above, the atomizer wheel speed is the important parameter influencing the spray droplet size. The wheel speed also determines the power consumption of the atomizer. It can be shown that the atomizer power consumption exclusive mechanical losses amount to

$$P_S = \frac{U^2}{3600}$$

where P_s = specific power consumption, kWh/t and U = peripheral velocity, m/s.

Since the atomizer wheel peripheral speed is proportional to the rotational speed, the maximum feed rate that can be handled by a rotary atomizer declines with the square of the rotational speed. The maximum feed rates indicated in Table 12-41 are therefore not available in the higher end of the speed ranges.

The rotary atomizer has one distinct advantage over other means of atomization. The degree or fineness of atomization achieved at a given speed is only slightly affected by changes in the feed rate. In other words, the rotary atomizer has a large turndown capability.

The larger atomizer machines cited in Table 12-41 represent a range of very large rotary atomizers available to industry. They are equipped with epicyclic-type gearboxes complete with a lubrication system. An extensive monitoring system is integrated in each machine.

Many atomization duties involve much lower capacities than foreseen for this range of atomizers. A full range of smaller rotary atomizers are available with nominal capacities down to less than 100 kg/h. Various designs may be seen with either belt drive or worm gears. Designs without gears are available with high-speed electric motor drive. Table 12-41 gives data for a smaller atomizer machine (FS1.5). It belongs to a family of high-speed machines without gears and lubrication systems capable of operating under the strictest requirements

TABLE 12-41 Operational Parameters for Atomizers (Niro)

Rotary atomizer designs				
Atomizer type		FS1.5	F160	F800
Nominal power rating	kW	1.5	160	1000
Maximum feed rate	t/h	0.3	50	200
Atomizer wheel diameter	mm	90	240	350
Typical gear ratio	#	1:1	4.4:1	2.9:1
Minimum speed	rpm	10,000	6,000	8,800
Maximum speed	rpm	30,000	18,200	11,500
Typical peripheral velocity	m/s	141	165	161
Typical specific power	kWh/t	5.5	7.6	7.2

for noncontamination of the product and in explosion-prone environments.

Hydraulic pressure nozzle In hydraulic pressure nozzle atomizers, the liquid feed is fed to the nozzle under pressure. In the nozzle orifice the pressure energy is converted to kinetic energy. The internal parts of the nozzle are normally designed to apply a certain amount of swirl to the feed flow so that it issues from the orifice as a high-speed film in the form of a cone with a desired vertex angle. This film disintegrates readily into droplets due to instability. The vertex or spray angle is normally on the order of 50° to 80°, a much narrower spray pattern than is seen with rotary atomizers. This means that spray drying chamber designs for pressure nozzle atomization differ substantially from designs used with rotary atomizers. The droplet size distribution produced by a pressure nozzle atomizer varies inversely with the pressure and to some degree with feed rate and viscosity. The capacity of a pressure nozzle varies with the square root of the pressure. To obtain a certain droplet size, the pressure nozzle must operate very close to the design pressure and feed rate. This implies that the pressure nozzle has very little turndown capability.

Hydraulic pressure nozzles cannot combine the capability for fine atomization with high feed capacity in one single unit. Many spray dryer applications, where pressure nozzles are applied, therefore require multinozzle systems with the consequence that start-up, operational control, and shutdown procedures become more complicated.

Two-fluid nozzle atomization In two-fluid nozzle atomizers, the liquid feed is fed to the nozzle under marginal or no pressure conditions. An additional flow of gas, normally air, is fed to the nozzle under pressure. Near the nozzle orifice, internally or externally, the two fluids (feed and pressurized gas) are mixed and the pressure energy is converted to kinetic energy. The flow of feed disintegrates into droplets during the interaction with the high-speed gas flow which may have sonic velocity.

The spray angle obtained with two-fluid nozzles is normally on the order of 10° to 20°, a very narrow spray pattern that is related to the spread of a free jet of gas. Spray drying chamber designs for two-fluid nozzle atomization are very specialized according to the application.

The droplet size produced by a two-fluid nozzle atomizer varies inversely with the ratio of gas to liquid and with the pressure of the atomization gas. The capacity of a two-fluid nozzle is not linked to its atomization performance. Therefore two-fluid nozzles can be attributed with some turndown capability.

Two-fluid nozzles share with pressure nozzles the lack of high feed capacity combined with fine atomization in one single unit. Many spray dryer applications with two-fluid nozzle atomization have a very high number of individual nozzles. The main advantage of two-fluid nozzles is the capability to achieve very fine atomization.

Choice of atomizer system The choice of atomizer system for a specific spray drying operation depends upon the particle size distribution required in the final dried product. It also depends upon the physical and chemical properties of the feed liquid.

In cases where the different types of atomizer means produce similar particle size distributions, the rotary atomizer may be preferred due to its greater flexibility and ease of operation. When one is comparing the atomizer types, the rotary atomizer has distinct advantages. (1) It can handle high feed rates in one single unit, (2) it can handle abrasive feeds with minimal wear, and (3) it has negligible blockage tendencies due to the large flow ports in the atomizer wheel. (4) It is a low-pressure system that can be served by a simple feed supply system, and (5) droplet size control is simple through wheel speed adjustment.

Although it lacks the flexibility of the rotary atomizer, the pressure nozzle is nevertheless widely used in spray drying applications. For many products the requirement for nondusty appearance calls for large mean particle size and lack of a fines fraction that cannot be met with a rotary atomizer. In the other end of the particle size range, some products require finer particles than are practically achievable with a rotary atomizer. This is the range where two-fluid nozzles are applied. The following guidelines may be used as an indication of the particle sizes obtainable in spray dryers:

- For spray dryers with rotary atomizer, the mean size of the dried product varies from 40 to 110 μm, although larger product mean sizes can be produced in large-diameter chambers.
- For spray dryers with pressure nozzle atomization, the mean particle size of the dried product varies in the range from 50 to 250 μm.
- For spray dryers with two-fluid nozzle atomization, the mean particle size of the dried product varies in the range from 15 to 50 μm.

The different means of atomization can also be compared in terms of energy power consumption. As indicated in Table 12-41, typical specific power figures for rotary atomizers are in the range of 5 to 11 kWh/t. Similar figures can be calculated for pressure and two-fluid nozzle systems, i.e., the pumping energy of the feed and the compression energy of the atomization gas. Any such calculation will show that similar median particle sizes are obtained for a given atomization energy independent of the means of atomization. None of the three types stand out as being energy-efficient. The hydraulic pressure nozzle is best suited for relatively coarse atomization, because pressures higher than 300 bar are impractical. Rotary atomizers are limited, because the wheel peripheral speeds required for very fine atomization put the wheel material under extreme tensile stress.

Droplet size distributions obtained with any means mentioned here are relatively well represented by a Rosin-Rammler distribution with an exponent of approximately 2. This means that approximately 80 percent of the droplet population mass is in the range of 0.39 to 1.82 times the median droplet size.

Theoretical prediction of mean particle sizes is difficult and of little practical importance, since the selection of spray drying operational parameters is based on experience and pilot-scale test work. The scientific literature, however, contains numerous estimation formulas to help predict the droplet sizes in sprays. Table 12-42 provides nomenclature for these estimation formulas.

For rotary atomizers median droplet sizes can be estimated from the following empirical equation of obscure origin:

$$d_{50} = K_r \times \dot{m}_L^{0.15} \times D^{-0.8} \times N^{-0.05} \times \omega^{-0.75} \times \mu_L^{0.07}$$

For hydraulic pressure nozzles the following formula proposed by Lefebvre may be used:

$$d_{50} = K_p \times \dot{m}_L^{0.25} \times \Delta P_L^{-0.5} \times (\sigma_L \times \mu_L)^{0.25} \times \rho_A^{-0.25}$$

Similarly, a range of equations or formulas are available for prediction of droplet size for sprays from two-fluid nozzles. The most widely cited in the literature is the Nukiyama-Tanasawa equation, which, however, is complicated and of doubtful validity at high flow rates. A much simpler equation has been proposed by Geng Wang et al.:

$$d_{50} = K_t \times \rho_A^{-0.325} \times \left(\frac{\dot{m}_L}{\dot{m}_L \times U_L + \dot{m}_A \times U_A} \right)^{0.55}$$

If any difference between the atomization means mentioned here were to be pointed out, it would be the tendency for two-fluid nozzles to have the wider particle size distribution and narrower pressure nozzles with rotary atomizers in between.

TABLE 12-42 Nomenclature for Atomization Equations

d_{50} = mass median droplet size	m
K_r = empirical factor	0.008
K_p = empirical factor	4.0
K_t = empirical factor	0.1
$\dot{m}_L$ = liquid feed rate	kg/s
D = wheel diameter	m
N = number of vanes	#
ω = atomizer wheel speed	rad/s
μ_L = liquid viscosity	Pa·s
ΔP_L = atomization pressure	Pa
ρ_A = air density	kg/m³
σ_L = liquid surface tension	N/m
$\dot{m}_A$ = atomization gas rate	kg/s
U_L = liquid velocity	m/s
U_A = atomization gas velocity	m/s

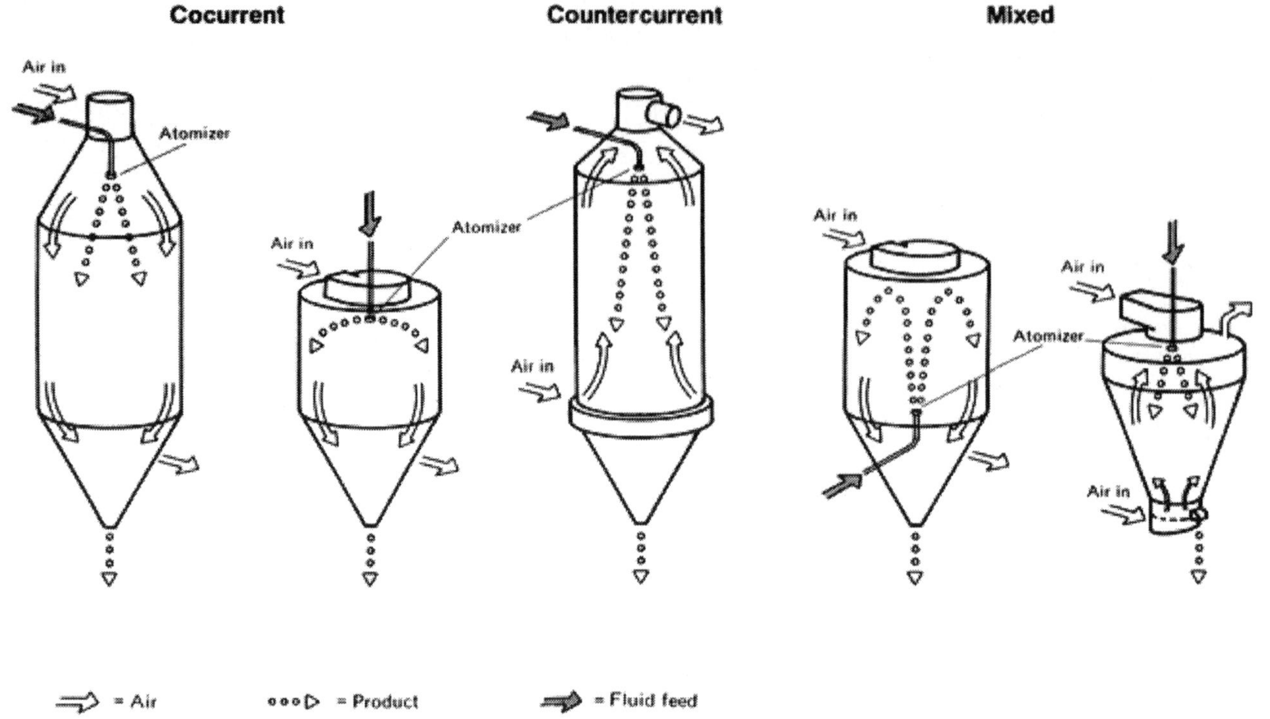

FIG. 12-92 Different forms of spray/hot air contact. (*Niro.*)

Spray/Hot Air Contact Atomization is first and most important process stage in spray drying. The final result of the process does, however, to a very large degree depend on the second stage, the spray/hot air contact. The way the spray of droplets is contacted by the hot air or gas carrying the thermal energy required to evaporate the moisture in the droplets is important for the quality of the product. In general terms three possible forms can be defined. These are as depicted in Fig. 12-92:

- Cocurrent flow
- Countercurrent flow
- Mixed flow

Different drying chamber forms and different methods of hot air introduction accompany the different flow pattern forms and are selected according to

- Required particle size in product specification
- Required particle form
- Temperature or heat sensitivity of the dried particle

In general terms, selection of chamber design and flow pattern form follows these guidelines:

- Use cocurrent spray drying for heat-sensitive products of fine as well as coarse particle size, where the final product temperature must be kept lower than the dryer outlet temperature.
- Use countercurrent spray drying for products which are not heat-sensitive, but may require some degree of heat treatment to obtain a special characteristic, i.e., porosity or bulk density. In this case the final powder temperature may be higher than the dryer outlet temperature.
- Use mixed-flow spray drying when a coarse product is required and the product can withstand short time exposure to heat without adverse effects on dried product quality.

Evaporation Stage Evaporation takes place from a moisture film which establishes on the droplet surface. The droplet surface temperature is kept low and close to the adiabatic saturation temperature of the drying air. As the temperature of the drying air drops off and the solids content of the droplet/particle increases, the evaporation rate is reduced. The drying chamber design must provide a sufficient residence time in suspended condition for the particle to enable completion of the moisture removal.

During the evaporation stage the atomized spray droplet size distribution may undergo changes as the droplets shrink, expand, collapse, fracture, or agglomerate.

Dry Product Recovery Product recovery is the last stage of the spray drying process. Two distinct systems are used:

- In two-point discharge, primary discharge of a coarse powder fraction is achieved by gravity from the base of the drying chamber. The fine fraction is recovered by secondary equipment downstream of the chamber air exit.
- In single-point discharge, total recovery of dry product is accomplished in the dryer separation equipment.

Collection of powder from an airstream is a large subject of its own. In spray drying, dry collection of powder in a nondestructive way is achieved by use of cyclones, filters with textile bags or metallic cartridges, and electrostatic precipitators.

With the current emphasis on environmental protection, many spray dryers are equipped with additional means to collect even the finest fraction. This collection is often destructive to the powder. Equipment in use are wet scrubbers, bag or other kinds of filters, and in a few cases incinerators.

Industrial Designs and Systems Thousands of different products are processed in spray dryers representing a wide range of feed and product properties as well as drying conditions. The flexibility of the spray drying concept, which is the main reason for this wide application, is described by the following systems.

Plant Layouts Figure 12-93*a* shows a standard cocurrent cone-based chamber with roof gas disperser. The chamber can have either single- or two-point discharge and can be equipped with rotary or

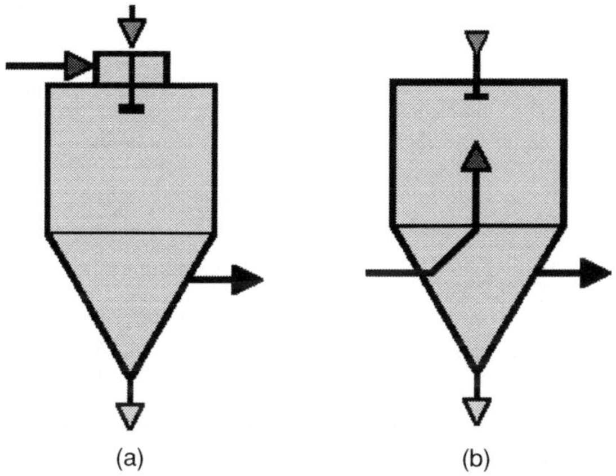

FIG. 12-93 (*a*) Standard cocurrent and (*b*) high-temperature chambers.

nozzle atomization. Fine or moderately coarse powders can be produced. This type of dryer finds application in dairy, food, chemical, pharmaceutical, agrochemical, and polymer industries.

Figure 12-93*b* shows a high-temperature chamber with the hot gas distributor arranged internally on the centerline of the chamber. The atomizer is rotary. Inlet temperature in the range of 600 to 1000°C can be utilized in the drying of non-heat-sensitive products in the chemical and mining industries. Kaolin and mineral flotation concentrates are typical examples.

Figure 12-94*a* shows a cocurrent cone-based tall form chamber with roof gas disperser. This chamber design is used primarily with pressure nozzle atomization to produce powders of large particle sizes with a minimum of agglomeration. The chamber can be equipped with an oversize cone section to maximize powder discharge from the chamber bottom. This type of dryer is used for dyestuffs, baby foods, detergents, and instant coffee powder.

Figure 12-94*b* shows a countercurrent flow chamber with pressure nozzle atomization. This design is in limited use because it cannot produce heat-sensitive products. Detergent powder is the main application.

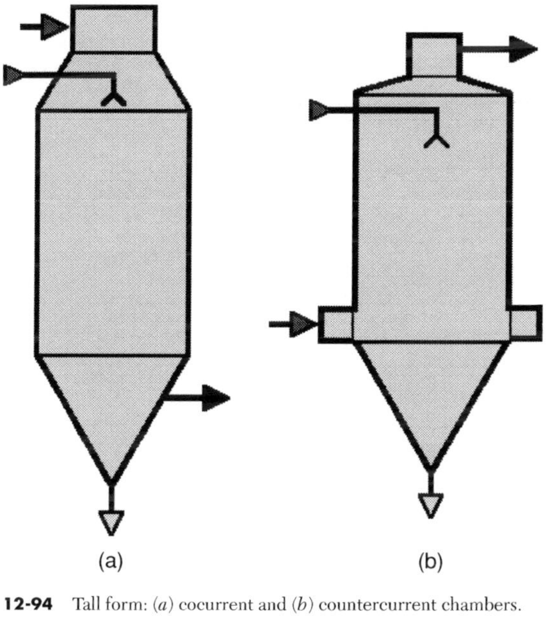

FIG. 12-94 Tall form: (*a*) cocurrent and (*b*) countercurrent chambers.

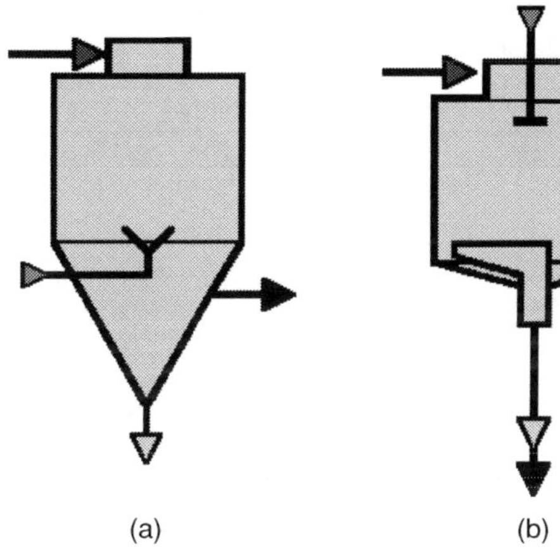

FIG. 12-95 (*a*) Mixed-flow and (*b*) flat chambers.

Figure 12-95*a* shows a mixed-flow chamber with pressure nozzle atomization arranged in so-called fountain mode. This design is ideal for producing a coarse product in a limited-size low-cost drying chamber. This type of dryer is used extensively for ceramic products. Figure 12-95*b* shows a flat-based cocurrent chamber as used with limited building height. Powder removal requires a sweeping suction device. One of few advantages is ease of access for manual cleaning. These are widely used in production of flavoring materials.

Figure 12-96*a* shows an integrated fluid-bed chamber which represents the latest development in spray dryer design. The final stage of the drying process is accomplished in a fluid bed located in the lower cone of the chamber. This type of operation allows lower outlet temperatures to be used, leading to fewer temperature effects on the powder and higher energy efficiency. Figure 12-96*b* shows an integrated belt chamber where product is sprayed onto a moving belt, which also acts as the air exhaust filter. It is highly suitable for slowly crystallizing and high-fat products. Previous operational difficulties derived from hygienic problems on the belt have been overcome, and the integrated belt dryer is now moving the limits of products that can be dried by spray drying.

Atomization/Gas Disperser Arrangement Some of the above-mentioned layouts allow a choice of atomization means while others are restricted to a particular choice. The arrangement of the gas distributor means will be closely related to the choice of atomizer. A

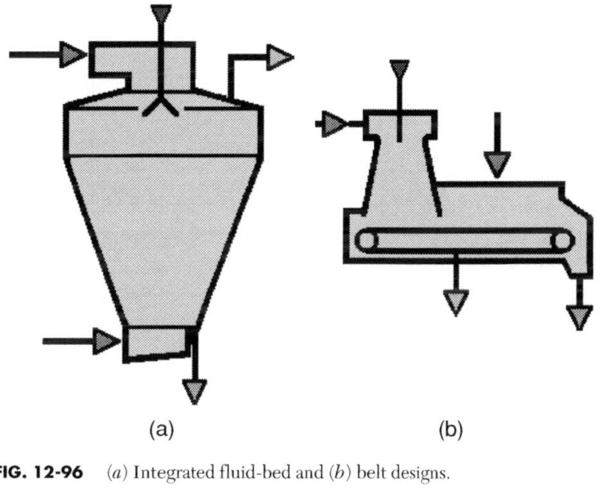

FIG. 12-96 (*a*) Integrated fluid-bed and (*b*) belt designs.

rotary atomizer will generally be arranged in a roof gas disperser as suited for the chambers in Figs. 12-93 and 12-95. The hot gas or air enters through a scroll-shaped housing which distributes the air evenly into an annular gap entry with adjustable guide vanes. The geometry and adjustment of the entry gap may determine the success of the drying process. Figure 12-95*b* shows an alternative arrangement of a rotary atomizer with a central gas disperser such as suited for the high-temperature spray dryer layout.

Hot Air Supply System All the above-mentioned chamber layouts can be used in open-cycle, partial recycle, or closed-cycle layouts. The selection is based on the needs of operation, feed, and powder specification and on environmental considerations.

An open-cycle layout is by far the most common in industrial spray drying. The open layout involves intake of drying air from the atmosphere and discharge of exhaust air to the atmosphere. Drying air can be supplemented by a waste heat source to reduce overall fuel consumption. The heater may be direct, i.e., natural gas burner, or indirect by steam-heated heat exchanger.

A closed-cycle layout is used for drying inflammable or toxic solvent feedstocks. The closed-cycle layout ensures complete solvent recovery and prevents explosion and fire risks. The reason for the use of a solvent system is often to avoid oxidation/degradation of the dried product. Consequently closed-cycle plants are gastight installations operating with an inert drying medium, usually nitrogen. These plants operate at a slight gauge pressure to prevent inward leakage of air.

Partial recycle is used in a plant type applied for products of moderate sensitivity toward oxygen. The atmospheric drying air is heated in a direct fuel-burning heater. Part of the exhaust air, depleted of its oxygen content by the combustion, is condensed in a condenser and recycled to the heater. This type of plant is also designated self-inertizing.

Industrial Applications As mentioned above, thousands of products are spray dried. The most common products may be classified as follows:

- Agrochemicals
- Catalysts
- Ceramics
- Chemicals
- Dyestuffs
- Foodstuffs
- Pharmaceuticals

Table 12-43 shows some of the operational parameters associated with specific and typical products. For each of these product groups and any other product, successful drying depends on the proper selection of a plant concept and proper selection of operational parameters, in particular inlet and outlet temperatures and the atomization method. These parameters are traditionally established through pilot-scale test work, and leading suppliers on the spray drying market often have extensive test stations to support their sales efforts.

Table 12-43 shows the variety of process parameters used in practical applications of spray drying. The air temperatures are traditionally established through experiments and test work. The inlet temperatures reflect the heat sensitivity of the different products, and the outlet temperatures the willingness of the products to release moisture. The percent water in feed parameter is an indication of feed viscosity and other properties that influence the pumpability and behavior under atomization of the individual feeds.

As a consequence, the amount of drying air or gas required for drying one unit of feed or product varies considerably. Table 12-43 shows for the individual products the ratio of drying gas to evaporation as well as the ratio of drying gas to product on a mass basis. The calculation behind the table neglects the variation of thermodynamic properties with temperature and the variation of residual moisture in each product.

A quick scoping estimate of the size of an industrial spray dryer can be made on this basis. The required evaporation rate or product rate can be multiplied by the relevant ratio from the table to give the mass flow rate of the drying gas. The next step would be to calculate the size of a spray drying chamber to allow the drying gas at outlet conditions approximately 25 s of residence time. A cylindrical chamber with diameter D and height H equal to D and a 60° conical bottom has a nominal volume of

$$V_{\text{chamber}} = \frac{\pi}{4} D^2 \times \left(H + \frac{\sqrt{3}}{2} D\right) = 1.47 \times D^3$$

Accordingly a zinc sulfate spray dryer with a drying capacity of 2 t/h would require a drying gas flow rate of approximately 8.45 kg/s. With an outlet gas density of 0.89 kg/m³ and the above-mentioned gas residence time, this results in a required chamber volume of

$$V_{\text{chamber}} = 8.44 \text{ kg/s}/0.89 \text{ kg/m}^3 \times 25 \text{ s} = 237 \text{ m}^3$$

The chamber size now becomes

$$D = \sqrt[3]{\frac{237}{1.47}} = 5.5 \text{ m}$$

A similar calculation for the other products based on a powder capacity of 2 t/h would reveal a variation of gas flow rates from 8.4 to 114 kg/s and chamber diameters from 5.5 to 12.7 m.

The selection of the plant concept involves the drying modes illustrated in Figs. 12-93 through 12-96. For different products a range of plant concepts are available to secure successful drying at the lowest cost. Three different concepts are illustrated in Figs. 12-97, 12-98, and 12-99.

Figure 12-97 shows a traditional spray dryer layout with a cone-based chamber and roof gas disperser. The chamber has two-point discharge and rotary atomization. The powder leaving the chamber bottom as well as the fines collected by the cyclone is conveyed pneumatically to a conveying cyclone from where the product discharges. A bag filter serves as the common air pollution control system.

Figure 12-98 shows closed-cycle spray dryer layout used to dry certain products with a nonaqueous solvent in an inert gas flow. The background for this may be product sensitivity to water and oxygen or severe explosion risk. Typical products can be tungsten carbide or pharmaceuticals.

Figure 12-99 shows an integrated fluid-bed chamber layout of the type used to produce agglomerated product. The drying process is accomplished in several stages, the first being a spray dryer with atomization. The second stage is an integrated static fluid bed located in the lower cone of the chamber. The final stages are completed in external

TABLE 12-43 Some Products That Have Been Successfully Spray Dried

Product	Air temperature, K		Water in feed, %	Air/evap. ratio, kg/kg	Air/prod. ratio, kg/kg	Product	Air temperature, K		Water in feed, %	Air/evap. ratio, kg/kg	Air/prod. ratio, kg/kg
	In	Out					In	Out			
Animal blood	440	345	65	27.6	51.3	Detergent A	505	395	50	25.4	25.4
Yeast	500	335	86	15.7	96.2	Detergent B	510	390	63	22.8	38.8
Zinc sulfate	600	380	55	12.4	15.2	Detergent C	505	395	40	25.8	17.2
Lignin	475	365	63	24.3	41.4	Manganese sulfate	590	415	50	16.3	16.3
Aluminum hydroxide	590	325	93	9.7	128.4	Aluminum sulfate	415	350	70	40.5	94.4
Silica gel	590	350	95	10.9	206.5	Urea resin A	535	355	60	14.8	22.1
Magnesium carbonate	590	320	92	9.5	108.7	Urea resin B	505	360	70	18.3	42.7
Tanning extract	440	340	46	26.4	22.5	Sodium sulfide	500	340	50	16.5	16.5
Coffee extract	420	355	70	40.6	94.8	Pigment	515	335	73	14.4	39.0

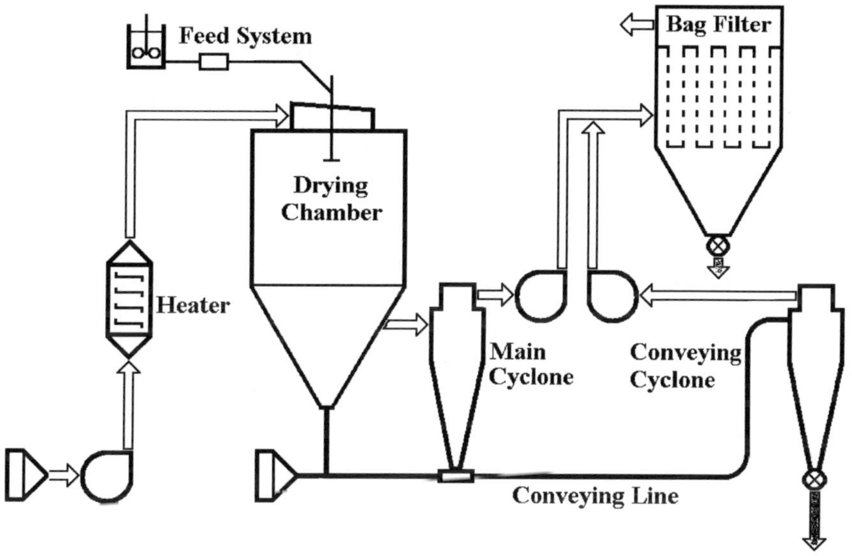

FIG. 12-97 Spray dryer with rotary atomizer and pneumatic powder conveying. (*Niro.*)

fluid beds of the vibrating type. This type of operation allows lower outlet temperatures to be used, leading to fewer temperature effects on the powder and higher energy efficiency. The chamber has a mixed-flow concept with air entering and exiting at the top of the chamber. This chamber is ideal for heat-sensitive, sticky products. It can be used with pressure nozzle as well as rotary atomization. An important feature is the return of fine particles to the chamber to enhance the agglomeration effect. Many products have been made feasible for spray drying by the development of this concept, which was initially aimed at the food and dairy industry. Recent applications have, however, included dyestuffs, agrochemicals, polymers, and detergents.

Additional Reading

Bayvel and Orzechowski, *Liquid Atomization,* Taylor & Francis, New York, 1993.
Geng Wang et al., "An Experimental Investigation of Air-Assist Non-Swirl Atomizer Sprays," *Atomisation and Spray Technol.* **3**:13–36 (1987).
Lefebvre, *Atomization and Sprays,* Hemisphere, New York, 1989.
Marshall, "Atomization and Spray Drying," *Chem. Eng. Prog. Mng. Series* **50**(2) (1954).
Masters, *Spray Drying in Practice,* SprayDryConsult International ApS, Denmark, 2002.
Walzel, "Zerstäuben von Flüssigkeiten," *Chem.-Ing.-Tech.* 62 (1990) Nr. 12, S. 983–994.

Pneumatic Conveying Dryers A gas-solids contacting operation in which the solids phase exists in a dilute condition is termed a *dispersion system.* It is often called a pneumatic system because, in most cases, the quantity and velocity of the gas are sufficient to lift and convey the solids against the forces of gravity and friction. (These systems are sometimes incorrectly called flash dryers when in fact the moisture is not actually "flashed" off. True flash dryers are sometimes used for soap drying to describe moisture removal when pressure is

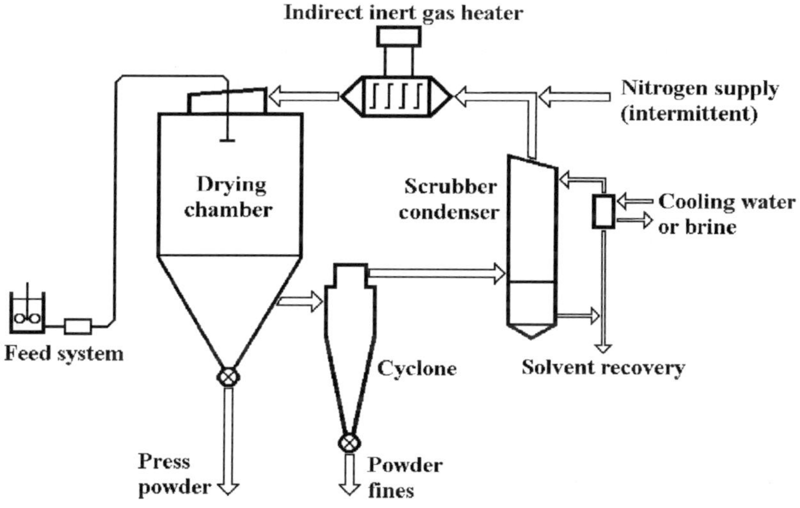

FIG. 12-98 Spray dryer with rotary atomizer and closed-cycle layout. (*Niro.*)

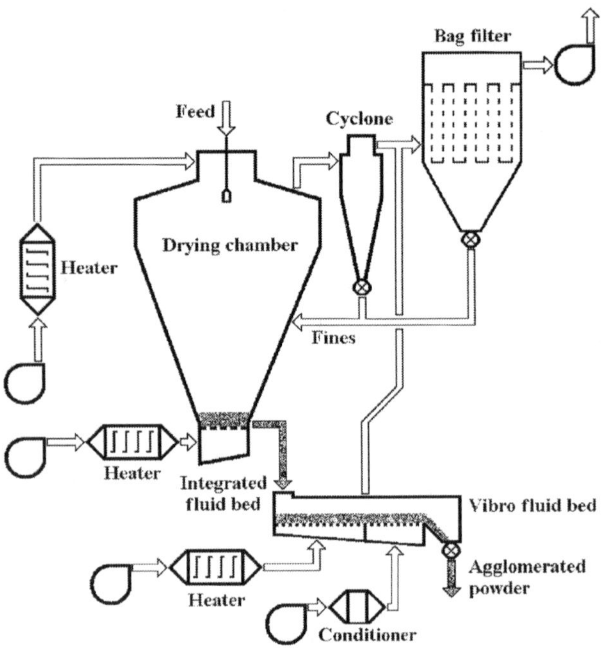

FIG. 12-99 Spray dryer with nozzle atomizer and integrated fluid bed. (*Niro.*)

quickly reduced.) Pneumatic systems may be distinguished by two characteristics:

1. Retention of a given solids particle in the system is on the average very short, usually no more than a few seconds. This means that any process conducted in a pneumatic system cannot be diffusion-controlled. The reaction must be mainly a surface phenomenon, or the solids particles must be so small that heat transfer and mass transfer from the interiors are essentially instantaneous.

2. On an energy-content basis, the system is balanced at all times; i.e., there is sufficient energy in the gas (or solids) present in the system at any time to complete the work on all the solids (or gas) present at the same time. This is significant in that there is no lag in response to control changes or in starting up and shutting down the system; no partially processed residual solids or gas need be retained between runs.

It is for these reasons that pneumatic equipment is especially suitable for processing heat-sensitive, easily oxidized, explosive, or flammable materials which cannot be exposed to process conditions for extended periods.

Gas flow and solids flow are usually cocurrent, one exception being a countercurrent flow spray dryer. The method of gas-solids contacting is best described as through-circulation; however, in the dilute condition, solids particles are so widely dispersed in the gas that they exhibit apparently no effect upon one another, and they offer essentially no resistance to the passage of gas among them.

Pneumatic Conveyor Dryers Pneumatic conveyor dryers, often also referred to as flash dryers, comprise a long tube or duct carrying a gas at high velocity, a fan to propel the gas, a suitable feeder for addition and dispersion of particulate solids in the gas stream, and a cyclone collector or other separation equipment for final recovery of solids from the gas.

The solids feeder may be of any type: Screw feeders, venturi sections, high-speed grinders, and dispersion mills are employed. For pneumatic conveyors, selection of the correct feeder to obtain thorough initial dispersion of solids in the gas is of major importance. For example, by employing an air-swept hammer mill in a drying operation, 65 to 95 percent of the total heat may be transferred within the mill itself if all the drying gas is passed through it. Fans may be of the induced-draft or the forced-draft type. The former is usually preferred because the system can then be operated under a slight negative pressure. Dust and hot gas will not be blown out through leaks in the equipment. Cyclone separators are preferred for low investment. If maximum recovery of dust or noxious fumes is required, the cyclone may be followed by a wet scrubber or bag collector.

In ordinary heating and cooling operations, during which there is no moisture pickup, continuous recirculation of the conveying gas is frequently employed. Also, solvent recovery operations employing continuously recirculated inert gas with intercondensers and gas reheaters are carried out in pneumatic conveyors.

Pneumatic conveyors are suitable for materials which are granular and free-flowing when dispersed in the gas stream, so they do not stick on the conveyor walls or agglomerate. Sticky materials such as filter cakes may be dispersed and partially dried by an air-swept disintegrator in many cases. Otherwise, dry product may be recycled and mixed with fresh feed, and then the two dispersed together in a disintegrator. Coarse material containing internal moisture may be subjected to fine grinding in a hammer mill. The main requirement in all applications is that the operation be instantaneously completed; internal diffusion of moisture must not be limiting in drying operations, and particle sizes must be small enough that the thermal conductivity of the solids does not control during heating and cooling operations. Pneumatic conveyors are rarely suitable for abrasive solids. Pneumatic conveying can result in significant particle size reduction, particularly when crystalline or other friable materials are being handled. This may or may not be desirable but must be recognized if the system is selected. The action is similar to that of a fluid-energy grinder.

Pneumatic conveyors may be single-stage or multistage. The former is employed for evaporation of small quantities of surface moisture. Multistage installations are used for difficult drying processes, e.g., drying heat-sensitive products containing large quantities of moisture and drying materials initially containing internal as well as surface moisture.

Typical single- and two-stage drying systems are illustrated in Figs. 12-100, 12-101, and 12-102. Figure 12-100 illustrates the flow diagram of a single-stage dryer with a paddle mixer, a screw conveyor followed by a rotary disperser for introduction of the feed into the airstream at the throat of a venturi section. The drying takes place in the drying column after which the dry product is collected in a cyclone. A diverter introduces the option of recycling part of the product into the mixer in order to handle somewhat sticky products. The environmental requirements are met with a wet scrubber in the exhaust stream.

Figure 12-101 illustrates a two-stage dryer where the initial feed material is dried in a flash dryer by using the spent drying air from the second stage. This semidried product is then introduced into the second-stage flash dryer for contact with the hottest air. This concept is in use in the pulp and paper industry. Its use is limited to materials that are dry enough on the surface after the first-stage to avoid plugging of the first-stage cyclone. The main advantage of the two-stage concept is the heat economy which is improved considerably over that of the single-stage concept.

Figure 12-102 is an elevation view of an actual single-stage dryer, employing an integral coarse-fraction classifier, used to separate undried particles for recycle.

Several typical products dried in pneumatic conveyors are described in Table 12-44.

Design methods for pneumatic conveyor dryers Depending upon the temperature sensitivity of the product, inlet air temperatures between 125 and 750°C are employed. With a heat-sensitive solid, a high initial moisture content should permit use of a high inlet air temperature. Evaporation of surface moisture takes place at essentially the wet-bulb air temperature. Until this has been completed, by which time the air will have cooled significantly, the surface-moisture film prevents the solids temperature from exceeding the wet-bulb temperature of the air. Pneumatic conveyors are used for solids having initial moisture contents ranging from 3 to 90 percent, wet basis. The air quantity required and solids-to-gas loading are fixed by the moisture load, the inlet air temperature, and, frequently, the exit air humidity. If the last is too great to permit complete drying, i.e., if the

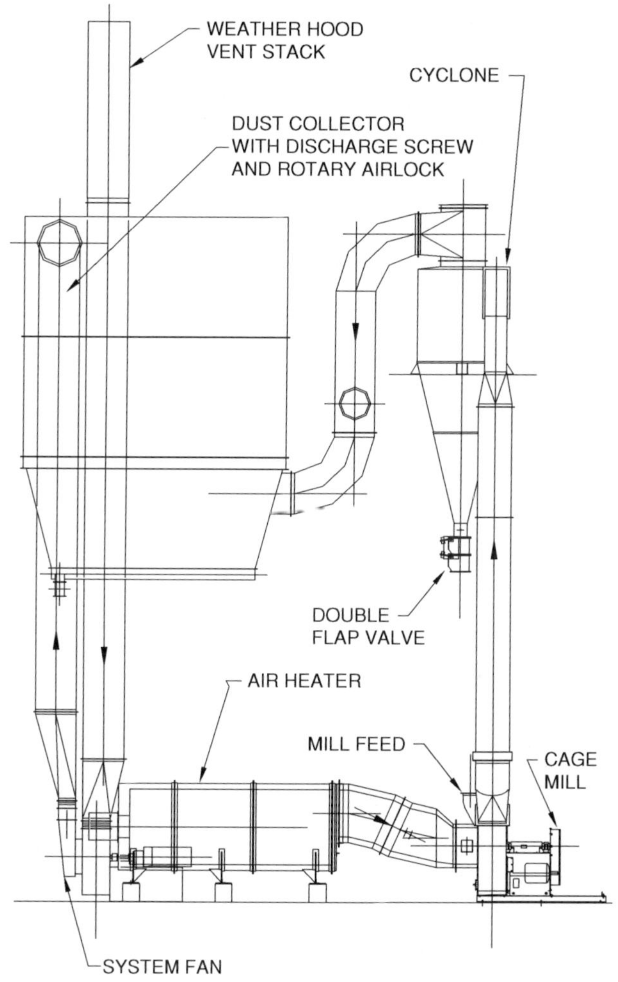

FIG. 12-100 Flow diagram of single-stage flash dryer. (*Air Preheater Company, Raymond® & Bartlett Snow™ Products.*)

exit air humidity is above that in equilibrium with the product at required dryness, then the solids/gas ratio must be reduced together with the inlet air temperature.

The gas velocity in the conveying duct must be sufficient to convey the largest particle. This may be calculated accurately by methods given in Sec. 17, "Gas-Solids Operations and Equipment." For estimating purposes, a velocity of 25 m/s, calculated at the exit air temperature, is frequently employed. If mainly surface moisture is present, the temperature driving force for drying will approach the log mean of the inlet and exit gas wet-bulb depressions. (The exit solids temperature will approach the exit gas dry-bulb temperature.)

Observation of operating conveyors indicates that the solids are rarely uniformly dispersed in the gas phase. With infrequent exceptions, the particles move in a streaklike pattern, following a streamline along the duct wall where the flow velocity is at a minimum. Complete or even partial diffusion in the gas phase is rarely experienced even with low-specific-gravity particles. Air velocities may approach 20 to 30 m/s. It is doubtful, however, that even finer and lighter materials reach more than 80 percent of this speed, while heavier and larger fractions may travel at much slower rates [Fischer, *Mech. Eng.*, **81**(11): 67–69 (1959)]. Very little information and few operating data on pneumatic conveyor dryers which would permit a true theoretical basis for design have been published.

Therefore, firm design always requires pilot tests. It is believed, however, that the significant velocity effect in a pneumatic conveyor is the difference in velocities between gas and solids, which is strongly linked to heat- and mass-transfer coefficients and is the reason why a major part of the total drying actually occurs in the feed input section.

For estimating purposes, the conveyor cross-section is fixed by the assumed air velocity and quantity. The standard scoping design method is used, obtaining the required gas flow rate from a heat and mass balance, and the duct cross-sectional area and diameter from the gas velocity (if unknown, a typical value is 20 m/s). An incremental mode may be used to predict drying conditions along the duct. However, several parameters are hard to obtain, and conditions change rapidly near the feed point. Hence, for reliable estimates of drying time and duct length, pilot-plant tests should always be used. A conveyor length larger than 50 diameters is rarely required. The length of the full-scale dryer should always be somewhat larger than required in pilot-plant tests, because wall effects are higher in small-diameter ducts. This gives greater relative velocity (and thus higher heat transfer) and lower particle velocity in the pilot-plant dryer, both effects giving a shorter length than the full-scale dryer for a given amount of drying. If desired, the length difference on scale-up can be predicted by using the incremental model and using the pilot-plant data to back-calculate the uncertain parameters; see Kemp, *Drying Technol.* **12**(1&2):279 (1994) and Kemp and Oakley (2002).

An alternative method of estimating dryer size very roughly is to estimate a volumetric heat-transfer coefficient [typical values are around 2000 $J/(m^3 \cdot s \cdot K)$] and thus calculate dryer volume.

Pressure drop in the system may be computed by methods described in Sec. 6, "Fluid and Particle Dynamics." To prevent excessive leakage into or out of the system, which may have a total pressure drop of 2000 to 4000 Pa, rotary air locks or screw feeders are employed at the solids inlet and discharge.

The conveyor and collector parts are thoroughly insulated to reduce heat losses in drying and other heating operations. Operating control is maintained usually by control of the exit gas temperature, with the inlet gas temperature varied to compensate for changing feed conditions. A constant solids feed rate must be maintained.

Ring Dryers The ring dryer is a development of flash, or pneumatic conveyor, drying technology, designed to increase the versatility of application of this technology and overcome many of its limitations.

One of the great advantages of flash drying is the very short retention time, typically no more than a few seconds. However, in a conventional flash dryer, residence time is fixed, and this limits its application to materials in which the drying mechanism is not diffusion-controlled and where a range of moisture within the final product is acceptable. The ring dryer offers two advantages over the flash dryer. First, residence time is controlled by the use of an adjustable internal classifier that allows fine particles, which dry quickly, to leave while larger particles, which dry slowly, have an extended residence time within the system. Second, the combination of the classifier with an internal mill can allow simultaneous grinding and drying with control of product particle size and moisture. Available with a range of different feed systems to handle a variety of applications, the ring dryer provides wide versatility.

The essential difference between a conventional flash dryer and the ring dryer is the manifold centrifugal classifier. The manifold provides classification of the product about to leave the dryer by using differential centrifugal force. The manifold, as shown in Fig. 12-103, uses the centrifugal effect of an airstream passing around the curve to concentrate the product into a moving layer, with the dense material on the outside and the light material on the inside.

This enables the adjustable splitter blades within the manifold classifier to segregate the denser, wetter material and return it for a further circuit of drying. Fine, dried material is allowed to leave the dryer with the exhaust air and to pass to the product collection system. This selective extension of residence time ensures a more evenly dried material than is possible from a conventional flash

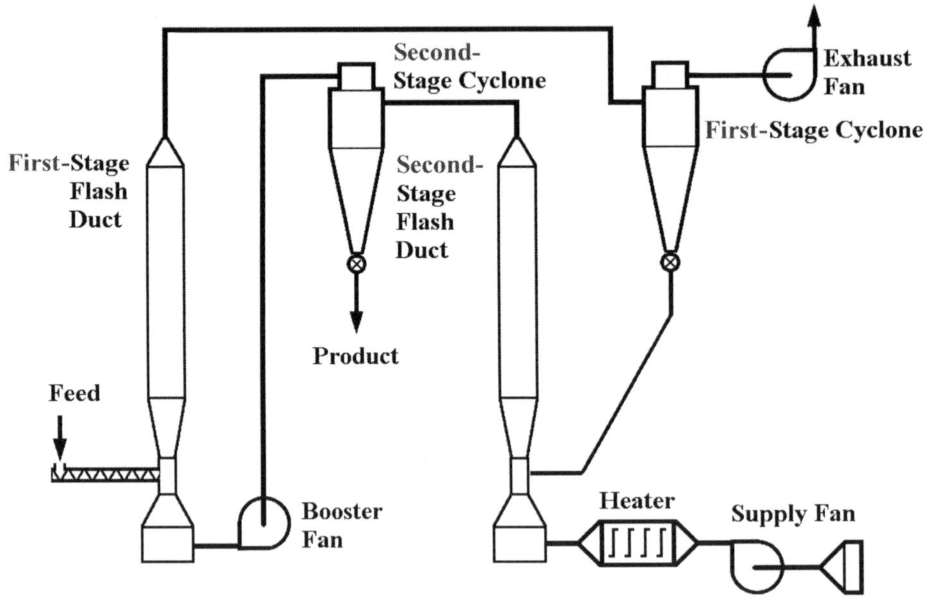

FIG. 12-101 Flow diagram of countercurrent two-stage flash dryer. (*Niro.*)

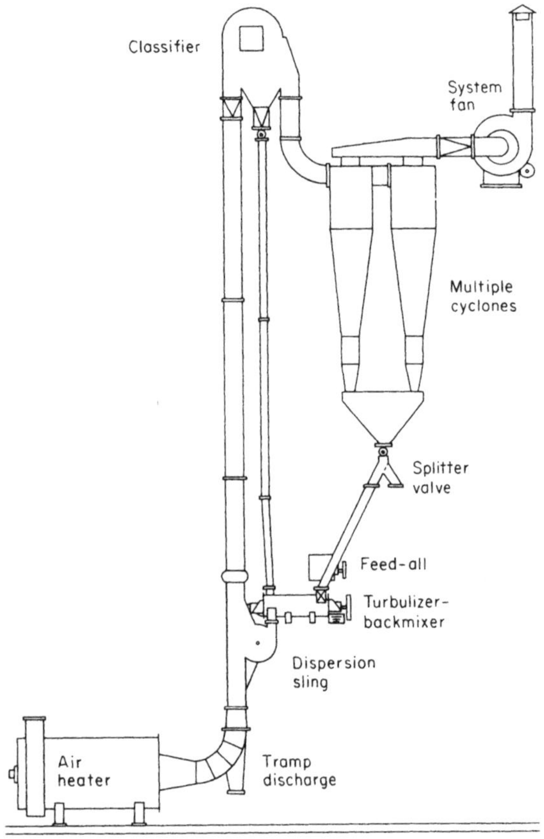

FIG. 12-102 Flow diagram of Strong Scott flash dryer with integral coarse-fraction classifier. (*Bepex Corp.*)

dryer. Many materials that have traditionally been regarded as difficult to dry can be processed to the required moisture content in a ring dryer. The recycle requirements of products in different applications can vary substantially depending upon the scale of operation, ease of drying, and finished-product specification. The location of reintroduction of undried material back into the drying medium has a significant impact upon the dryer performance and final-product characteristics.

Three configurations of the ring dryer have been developed to offer flexibility in design and optimal performance:

1. *Single-stage manifold-vertical configuration* The feed ring dryer (see Fig. 12-104) is similar to a flash dryer but incorporates a single-stage classifier, which diverts 40 to 60 percent of the product back to the feed point. The feed ring dryer is ideally suited for materials which neither are heat-sensitive nor require a high degree of classification. An advantage of this configuration is that it can be manufactured to very large sizes to achieve high evaporative capacities.

2. *Full manifold-horizontal configuration* The full ring dryer (see Fig. 12-105) incorporates a multistage classifier which allows much higher recycle rates than the single-stage manifold. This configuration usually incorporates a disintegrator which provides adjustable amounts of product grinding depending upon the speed and manifold setting. For sensitive or fine materials, the disintegrator can be omitted. Alternative feed locations are available to suit the material sensitivity and the final-product requirements. The full ring configuration gives a very high degree of control of both residence time and particle size, and is used for a wide variety of applications from small production rates of pharmaceutical and fine chemicals to large production rates of food products, bulk chemicals, and minerals. This is the most versatile configuration of the ring dryer.

3. *P-type manifold-vertical configuration* The P-type ring dryer (see Fig. 12-106) incorporates a single-stage classifier and was developed specifically for use with heat-sensitive materials. The undried material is reintroduced into a cool part of the dryer in which it recirculates until it is dry enough to leave the circuit.

An important element in optimizing the performance of a flash or ring dryer is the degree of dispersion at the feed point. Maximizing the product surface area in this region of highest evaporative

TABLE 12-44 Typical Products Dried in Pneumatic Conveyor Dryers (Barr-Rosin)

Material	Initial moisture, wet basis, %	Final moisture, wet basis, %	Plant configuration
Expandable polystyrene beads	3	0.1	Single-stage flash
Coal fines	23	1.0	Single-stage flash
Polycarbonate resin	25	10	Single-stage flash
Potato starch	42	20	Single-stage flash
Aspirin	22	0.1	Single-stage flash
Melamine	20	0.05	Single-stage flash
Corn gluten meal	60	10	Feed-type ring dryer
Maize fiber	60	18	Feed-type ring dryer
Distillers dried grains (DDGs)	65	10	Feed type ring dryer
Vital wheat gluten	70	7	Full-ring dryer
Casein	50	10	Full-ring dryer
Tricalcium phosphate	30	0.5	Full-ring dryer
Zeolite	45	20	Full-ring dryer
Orange peels	82	10	Full-ring dryer
Modified corn starch	40	10	P-type ring dryer
Methylcellulose	45	25	P-type ring dryer

driving force is a key objective in the design of this type of dryer. Ring dryers are fed using similar equipment to conventional flash dryers. Ring dryers with vertical configuration are normally fed by a flooded screw and a disperser which propels the wet feed into a high-velocity venturi, in which the bulk of the evaporation takes place. The full ring dryer normally employs an air-swept disperser or mill within the drying circuit to provide screenless grinding when required. Together with the manifold classifier this ensures a product with a uniform particle size. For liquid, slurry, or pasty feed materials, backmixing of the feed with a portion of the dry product will be carried out to produce a conditioned friable material. This further increases the versatility of the ring dryer, allowing it to handle sludge and slurry feeds with ease.

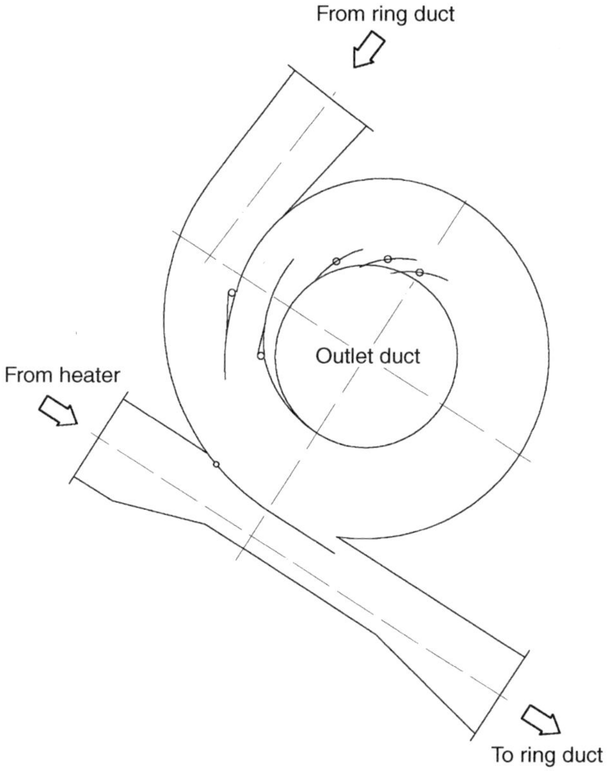

FIG. 12-103 Full manifold classifier for ring dryer. (*Barr-Rosin.*)

Dried product is collected in either cyclones or bag filters depending upon the product-particle properties. When primary collection is carried out in cyclones, secondary collection in a bag filter or scrubber is usually necessary to comply with environmental regulations. A rotary valve is used to provide an air lock at the discharge point. Screws are utilized to combine product from multiple cyclones or large bag filters. If required, a portion of the dried product is separated from the main stream and returned to the feed system for use as backmix.

Design methods for ring dryers Depending on the temperature sensitivity of the material to be processed, air inlet temperatures as high as 750°C can be utilized. Even with heat-sensitive solids, high feed moisture content may permit the use of high air inlet temperature since evaporation of surface moisture takes place at the wet-bulb air temperature. Until the surface moisture has been removed, it will prevent the solids temperature from exceeding the air wet-bulb temperature, by which time the air will generally have cooled significantly. Ring dryers have been used to process materials with feed moisture contents between 2 and 95 percent, weight fraction. The product moisture content has been controlled to values from 20 percent down to less than 1 percent.

The air velocity required and air/solids ratio are determined by the evaporative load, the air inlet temperature, and the exhaust air humidity. Too high an exhaust air humidity would prevent complete drying, so then a lower air inlet temperature and air/solids ratio would be required. The air velocity within the dryer must be sufficient to convey the largest particle, or agglomerate. The air/solids ratio must be high enough to convey both the product and backmix, together with internal recycle from the manifold. For estimating purposes a velocity of 25 m/s, calculated at dryer exhaust conditions, is appropriate both for pneumatic conveyor and ring dryers.

Agitated Flash Dryers Agitated flash dryers produce fine powders from feeds with high solids contents, in the form of filter cakes, pastes, or thick, viscous liquids. Many continuous dryers are unable to dry highly viscous feeds. Spray dryers require a pumpable feed. Conventional flash dryers often require backmixing of dry product to the feed in order to fluidize. Other drying methods for viscous pastes and filter cakes are well known, such as contact, drum, band, and tray dryers. They all require long processing time, large floor space, high maintenance, and aftertreatment such as milling.

The agitated flash dryer offers a number of process advantages, such as ability to dry pastes, sludges, and filter cakes to a homogeneous, fine powder in a single-unit operation; continuous operation; compact layout; effective heat- and mass-transfer short drying times; negligible heat loss and high thermal efficiency; and easy access and cleanability.

The agitated flash dryer (Fig. 12-107) consists of four major components: feed system, drying chamber, heater, and exhaust air system. Wet feed enters the feed tank, which has a slow-rotating impeller to break up large particles. The level in the feed tank is maintained by a

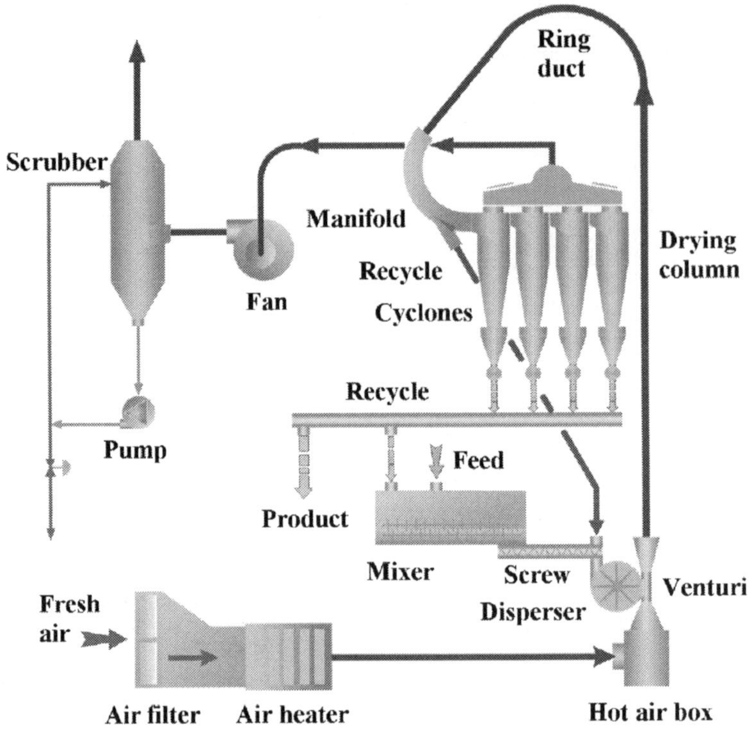

FIG. 12-104 Flow diagram of feed-type ring dryer. (*Barr-Rosin.*)

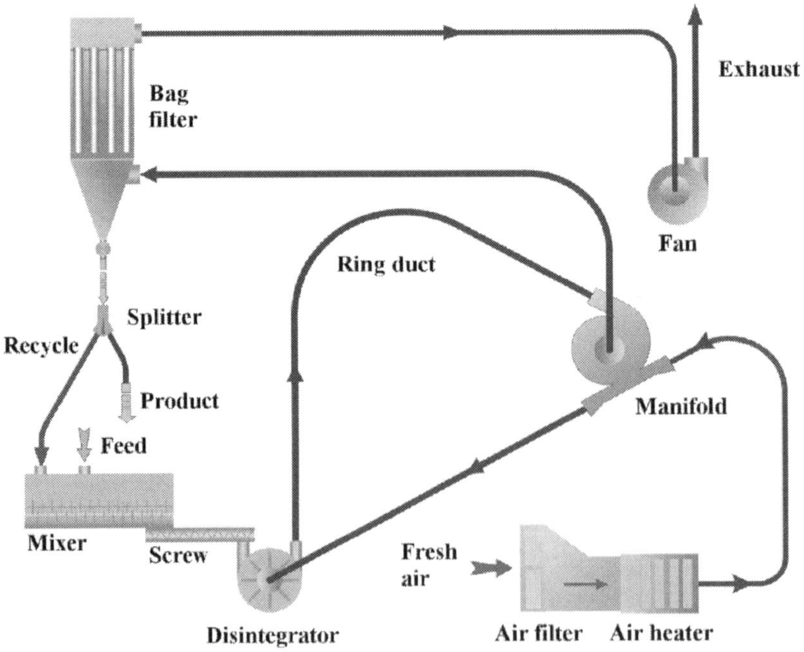

FIG. 12-105 Flow diagram of full manifold-type ring dryer. (*Barr-Rosin.*)

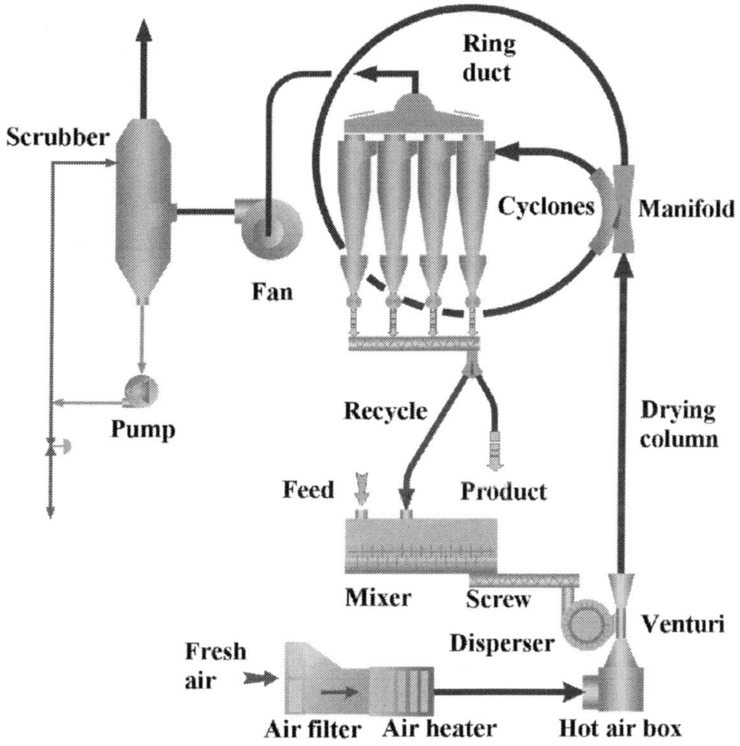

FIG. 12-106 Flow diagram of P-type ring dryer. (*Barr-Rosin.*)

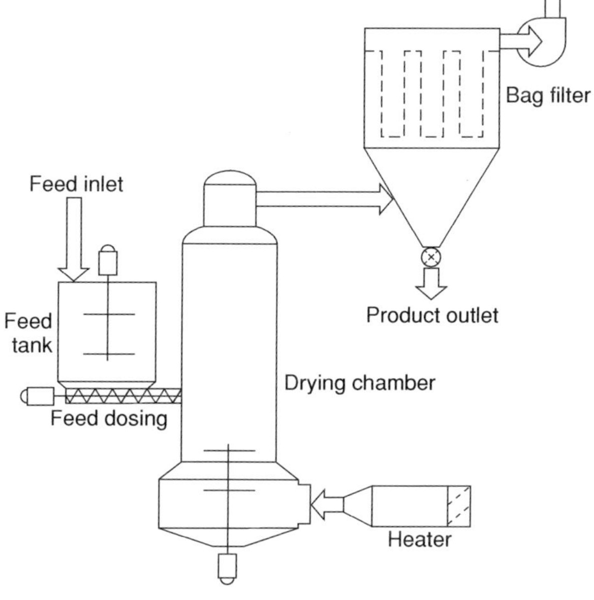

FIG. 12-107 Agitated flash dryer with open cycle. (*Niro, Inc.*)

level controller. The feed is metered at a constant rate into the drying chamber via a screw conveyor mounted under the feed tank. If the feed is shear thinning and can be pumped, the screw feeder can be replaced by a positive displacement pump.

The drying chamber is the heart of the system consisting of three important components: air disperser, rotating disintegrator, and drying section. Hot, drying air enters the air disperser tangentially and is introduced into the drying chamber as a swirling airflow. The swirling airflow is established by a guide-vane arrangement. The rotating disintegrator is mounted at the base of the drying chamber. The feed, exposed to the hot, swirling airflow and the agitation of the rotating disintegrator, is broken up and dried. The fine dry particles exit with the exhaust air and are collected in the bag filter. The speed of the rotating disintegrator controls the particle size. The outlet air temperature controls the product moisture content.

The drying air is heated either directly or indirectly, depending upon the feed material, powder properties, and available fuel source. The heat sensitivity of the product determines the drying air temperature. The highest possible value is used to optimize thermal efficiency. A bag filter is usually recommended for collecting the fine particles produced. The exhaust fan maintains a slight vacuum in the dryer, to prevent powder leakage into the surroundings. The appropriate process system is selected according to the feed and powder characteristics, available heating source, energy utilization, and operational health and safety requirements.

Open systems use atmospheric air for drying. In cases where products pose a potential for dust explosion, plants are provided with pressure relief or suppression systems. For recycle systems, the drying system medium is recycled, and the evaporated solvent recovered as condensate. There are two alternative designs. In the self-inertizing mode, oxygen content is held below 5 percent by combustion control at the heater. This is recommended for products with serious dust explosion hazards. In the inert mode, nitrogen is the drying gas. This is used when an organic solvent is evaporated or product oxidation during drying must be prevented.

Design methods The size of the agitated flash dryer is based on the evaporation rate required. The operating temperatures are product-specific. Once established, they determine the airflow requirements. The drying chamber is designed based on air velocity (approximately 3 to 4 m/s) and residence time (product-specific).

Other Dryer Types

Freeze Dryer Industrial freeze drying is carried out in two steps:

1. Freezing of the food or beverage product
2. Freeze drying, i.e., sublimation drying of the ice content and desorption drying of the bound or crystal water content

Freeze drying differs from conventional drying in that when ice is sublimated, only water vapor is transported within the product, causing no displacement of soluble substances such as sugars, salts, and acids. In all conventional drying systems in which water is dried, the water containing the soluble substances is transported to the product surface by capillary action. The water will evaporate from the surface, leaving the soluble substances displaced on the product surface. The major advantages of freeze drying are therefore

- Preservation of original flavor, aroma, color, shape, and texture
- Very little shrinkage, resulting in excellent and instant rehydration characteristics
- Negligible product loss
- Minimal risk of cross-contamination

The freeze drying process is today used widely for a number of products including vegetables, fruits, meat, fish, and beverage products, such as

- Instant coffee for which excellent flavor and aroma retention are of special importance
- Strawberries for which excellent color preservation is of special importance
- Chives for which shape preservation is of special importance

Freezing The freezing methods applied for solid products are all conventional freezing methods such as blast freezing, individual quick freezing (IQF), or similar.

The products maintain their natural cell structure, and the aim is to freeze the free water to pure ice crystals, leaving the soluble substances as high concentrates or even crystallized. To ensure good stability of the product during storage, a product temperature of −20 to −30°C should be achieved to ensure that more than 95 percent of the free water is frozen.

Liquid products have no cell structure, thus the structure of the freeze dried products is formed by the freezing process. The intercrystalline matrix of the concentrated product giving the structure of the freeze dried product is formed around the ice crystals. The size of the ice crystals is a function of the freezing time. Quick freezing results in small ice crystals, slow freezing in large ice crystals. The structure of the matrix determines the freeze drying performance as well as the appearance, mechanical strength, and solubility rate. Small ice crystals lead to light color (high surface reflection of light), diffusion restrictions for vapor transport inside the product, and a good mechanical strength of the freeze dried product. Large ice crystals lead to the opposite results.

Thus the freezing method must be carefully adapted to the quality criteria of the finished product. The preferred methods are

- Drum freezing, by which a thin slab of 1.5 to 3 mm is frozen within 1.5 to 3 min
- Belt freezing, by which a slab of 6 to10 mm passing through different freezing zones is frozen during 10 to 20 min
- Foaming, used to influence the structure and mainly to control the density of the freeze dried product

Freeze drying Freeze drying of foods takes place in a freeze dryer at vacuum levels of 0.4 to 1.3 mbar absolute, corresponding to sublimation temperatures from −30 to −17°C depending on the product requirements. The main components of the freeze dryer are

- The vacuum chamber, heating plates, and vapor traps, all built into the freeze dryer

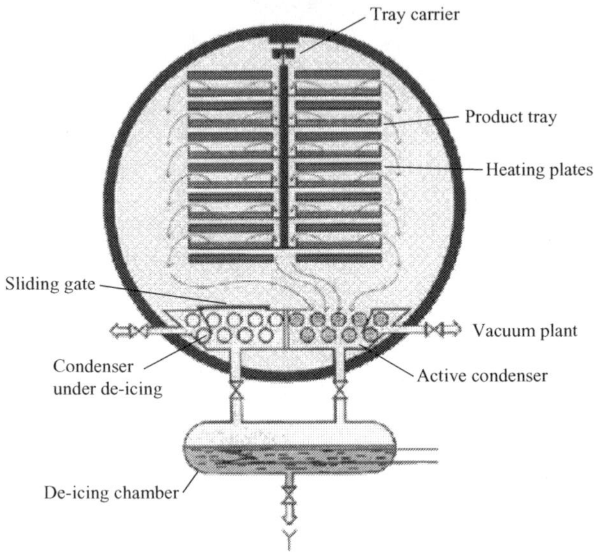

FIG. 12-108 Cross-section of RAY™ batch freeze dryer. (*Niro A/S.*)

- The external systems, such as the transport system for the product trays, the deicing system, and the support systems for supply of heat, vacuum, and refrigeration

Batch freeze drying The frozen product is carried in trays, and the trays are carried in tray trollies suspended in an overhead rail system for easy transport and quick loading and unloading. The freeze dryer as illustrated in Fig. 12-108 is charged with 3 to 6 trolley loads depending on the size of the freeze dryer. The trollies place the trays between the heating plates for radiation heat transfer. Radiation is preferred to ensure an even heat transfer over the large heating surface, typically 2× (70 to 140 m^2). The distribution of the heating medium (water or thermal oil) to the heating plates and the flow rate inside the plates are very important factors. To avoid uneven drying, the surface temperature difference of the heating plates should not exceed 2 to 3°C at maximum load.

When the loading is completed, the freeze dryer is closed and vacuum applied.

The operation vacuum should be reached quickly (within 10 min) to avoid the risk of product melting. For the same reason, the heating plates are cooled to approximately 25°C. When the operation vacuum is achieved, the heating plate temperature is raised quickly to the maximum drying temperature restricted by the capacity of the vapor traps, to perform the sublimation drying as quickly as possible for capacity reasons. During this period, the product is kept cool by the sublimation, and approximately 75 to 80 percent of the free water is sublimated.

The capability of the freeze drying plant to perform during this period is vital for efficient operation. To maintain the required sublimation temperature, the surface temperature of the ice layer on the vapor trap condenser must compensate for the pressure loss of the vapor flow from the sublimation front to the condenser.

The evaporation temperature of the refrigerant must further compensate for the temperature difference through the ice layer to the evaporating refrigerant.

With the flow rate at 1 mbar of approximately 1 $m^3/(s \cdot m^2$ of tray area), the thermodynamic design of the vapor trap is the main issue for a well-designed freeze dryer.

A built-in vapor trap allowing a large opening for the vapor flow to the condenser and a *c*ontinuous *de*icing (CDI) system, reducing the ice layer on the condenser to a maximum of 6 to 8 mm, are important features of a modern freeze drying plant. Approximately 75 percent of the energy costs relate to the refrigeration plant, and if the requirement

TABLE 12-45 Freeze Dryer, Performance Data, Niro RAY™ and CONRAD™ Types

	Tray area, m^2	Typical sublimation capacity		Electricity consumption, kWh/kg, sublimated	Steam consumption, kg/kg sublimated
		Flat tray, kg/h	Ribbed tray, kg/h		
RAY Batch Plant—1 mbar					
RAY 75	68	68	100	1.1	2.2
RAY 100	91	91	136	1.1	2.2
RAY 125*	114	114	170	1.1	2.2
CONRAD Continuous Plant—1 mbar					
CONRAD 300	240	240	360	1.0	2.0
CONRAD 400	320	320	480	1.0	2.0
CONRAD 500*	400	400	600	1.0	2.0

*Other sizes available.

of the evaporation temperature is 10°C lower than optimum, the energy consumption of the refrigeration plant will increase by approximately 50 percent.

At the end of the sublimation drying, the product surface temperature reaches the maximum allowable product temperature, requiring that the temperature of the heating plates be lowered gradually, and the drying will change to desorption drying. The temperature will finally be kept constant at the level of the maximum allowable product temperature until the residual moisture has been reduced to 2 to 3 percent, which is a typical level for a freeze dried product.

Continuous freeze drying From the description of batch freeze drying, it can be seen that the utility requirements vary considerably. During sublimation drying the requirements are 2 to 2.5 times the average requirement. To overcome this peak load and to meet the market request for high unit capacities, continuous freeze dryer designs have been developed. The special features are twofold:

- The tray transport system is a closed-loop system in which the trays pass one by one under the tray filler, where frozen product is automatically filled into the trays at a preset weight. The full tray is charged to the vacuum lock which is then evacuated to the drier vacuum level. Then the tray is pushed into the dryer and grabbed by an elevator which is filled stepwise with a stack of trays. Next a full stack of trays is pushed into the drying area whereby each of the stacks inside the drying area will move one step forward. Thus the last stack containing the finished, freeze dried product will be pushed out of the drying area to an outlet elevator which will be emptied stepwise by discharge of the trays through the outlet vacuum lock. From the outlet vacuum lock the trays are pushed to the emptying station for emptying and then returned to the tray filler.
- As the tray stacks are pushed forward through the freeze dryer, they pass through various temperature zones. The temperature zones form the heating profile, high temperatures during the sublimation drying, medium temperatures during the transition period toward desorption drying, and low temperatures during the final desorption drying. The temperature profile is selected so that overheating of the dry surface is avoided.

Design methods The size of the freeze drying plant is based on the average sublimation capacity required as well as on the product type and form. The external systems for batch plants must be designed for a peak load of 2 to 2.5 times the average capacity in the case of a single plant. Further, a batch plant is not available for drying all the time. A modern batch freeze dryer with the CDI system loses approximately 30 min per batch. Typically, 2 to 3 batches will be freeze dried per day. The evaporation temperature of the refrigeration plant depends on the required vacuum. At 1 mbar it will be −35 to −40°C depending on the vapor trap performance. Sample data are shown in Table 12-45.

Field Effects Drying—Drying with Infrared, Radio-Frequency, and Microwave Methods

Dielectric Methods (Radio-Frequency and Microwave) Schiffmann (1995) defines dielectric (radio-frequency) frequencies as covering the range of 1 to 100 MHz, while microwave frequencies range from 300 MHz to 300 GHz. The devices used for generating microwaves are called magnetrons and klystrons.

Water molecules are dipolar (i.e., they have an asymmetric charge center), and they are normally randomly oriented. The rapidly changing polarity of a microwave or radio-frequency field attempts to pull these dipoles into alignment with the field. As the field changes polarity, the dipoles return to a random orientation before being pulled the other way. This buildup and decay of the field, and the resulting stress on the molecules, causes a conversion of electric field energy to stored potential energy, then to random kinetic or thermal energy. Hence dipolar molecules such as water absorb energy in these frequency ranges. The power developed per unit volume P_v by this mechanism is

$$P_v = kE^2f\varepsilon' \tan \delta = kE^2f\varepsilon'' \qquad (12\text{-}117)$$

where k is a dielectric constant, depending on the units of measurement, E is the electric field strength (V/m^3), f is the frequency, ε' is the relative dielectric constant or relative permeability, $\tan \delta$ is the loss tangent or dissipation factor, and ε'' is the loss factor.

The field strength and the frequency are dependent on the equipment, while the dielectric constant, dissipation factor, and loss factor are material-dependent. The electric field strength is also dependent on the location of the material within the microwave/radio-frequency cavity (Turner and Ferguson, 1995), which is one reason why domestic microwave ovens have rotating turntables (so that the food is exposed to a range of microwave intensities). This mechanism is the major one for the generation of heat within materials by these electromagnetic fields.

There is also a heating effect due to ionic conduction, since the ions (sodium, chloride, and hydroxyl) in the water inside materials are accelerated and decelerated by the changing electric field. The collisions which occur as a result of the rapid accelerations and decelerations lead to an increase in the random kinetic (thermal) energy of the material. This type of heating is not significantly dependent on either temperature or frequency, and the power developed per unit volume P_v from this mechanism is

$$P_v = E^2qn\mu \qquad (12\text{-}118)$$

where q is the amount of electric charge on each of the ions, n is the charge density (ions/m^3), and μ is the level of mobility of the ions.

Schiffmann (1995) indicates that the dielectric constant of water is over an order of magnitude higher than that of most underlying materials, and the overall dielectric constant of most materials is usually nearly proportional to moisture content up to a critical moisture content, often around 20 to 30 percent. Hence microwave and radio-frequency methods preferentially heat and dry wetter areas in most materials, a process which tends to give more uniform final moisture contents. The dielectric constant of air is very low compared with that of water, so lower density usually means lower heating rates. For water and other small molecules, the effect of increasing temperature is to decrease the heating rate slightly, hence leading to a self-limiting effect.

Other effects (frequency, conductivity, specific heat capacity, etc.) are discussed by Schiffmann (1995), but are less relevant because the range of available frequencies (which do not interfere with radio transmissions) is small (2.45 GHz, 910 MHz). Lower frequencies lead to greater penetration depths into material than higher frequencies, with 2.45-GHz frequencies sometimes having penetration depths as low as 1 in. For in-depth heating ("volumetric heating"), radio frequencies, with lower frequencies and longer wavelengths, are often used.

Infrared Methods Infrared radiation is commonly used in the dehydration of coated films and to even out the moisture content profiles in the drying of paper and boards. The mode of heating is essentially on the material surface, and IR sources are relatively inexpensive compared with dielectric sources.

The heat flux obtainable from an IR source is given by

$$q = F\alpha\varepsilon\,(T^4_{\text{source}} - T^4_{\text{drying material}}) \qquad (12\text{-}119)$$

where q = heat flux, W/m^2; α = Stefan-Boltzmann constant = 5.67×10^{-8} W/(m$^2 \cdot$ K^4); ε = emissivity; F = view factor; and T = absolute temperature of the source or drying material.

The emissivity is a property of the material. The limiting value is 1 (blackbody); shiny surfaces have a low value of emissivity. The view factor is a fractional value that depends on the geometric orientation of the source with respect to the heating object.

It is very important to recognize the T^4 dependence on the heat flux. IR sources need to be very hot to give appreciable heat fluxes. Therefore, IR sources should not be used with flammable materials. Improperly designed IR systems can also overheat materials and equipment.

OPERATION AND TROUBLESHOOTING

Troubleshooting Dryer troubleshooting is not extensively covered in the literature, but a systematic approach has been proposed by Kemp and Gardiner (2001). The main steps of the algorithm are as follows:

- *Problem definition*—definition of the dryer problem to be solved.
- *Data gathering*—collection of relevant information, e.g., plant operating data
- *Data analysis*—e.g., heat and mass balance—and *identification of the cause of the problem*
- *Conclusions and actions*—selection and implementation of a solution in terms of changes to process conditions, equipment, or operating procedures
- *Performance auditing*—monitoring to ensure that the problem was permanently solved

There is often a danger in practice that the pressure to get the plant back into production as soon as possible may lead to some of these stages being omitted. Even if a short-term fix has been found, it is highly desirable to make sure what the problem really was, to see whether there are better ways of solving it in the long term, and to check that the problem really has been solved (sometimes it reappears later, e.g., when a temporarily cleaned heat exchanger becomes fouled again, or climatic conditions return to previous values).

The algorithm might also be considered as a "plant doctor." The doctor collects data, or symptoms, and makes a diagnosis of the cause or causes of the problem. Then alternative solutions, or treatments, are considered and a suitable choice is made. The results of the treatment are reviewed (i.e., the process is monitored) to ensure that the "patient" has returned to full health. See Fig. 12-109.

The algorithm is an excellent example of the "divergent-convergent" (brainstorming) method of problem solving. It is important to list all possible causes and solutions, no matter how ridiculous they may initially seem; there may actually be some truth in them, or they may lead to a new and better idea.

Problem Categorization In the problem definition stage, it is extremely useful to categorize the problem, as the different broad groups require different types of solution. Five main categories of dryer problems can be identified:

1. Drying performance (outlet moisture content too high, throughput too low)
2. Materials handling (dried material too sticky to get out of dryer, causing blockage)
3. Product quality (too many fines in product or bulk density too low)

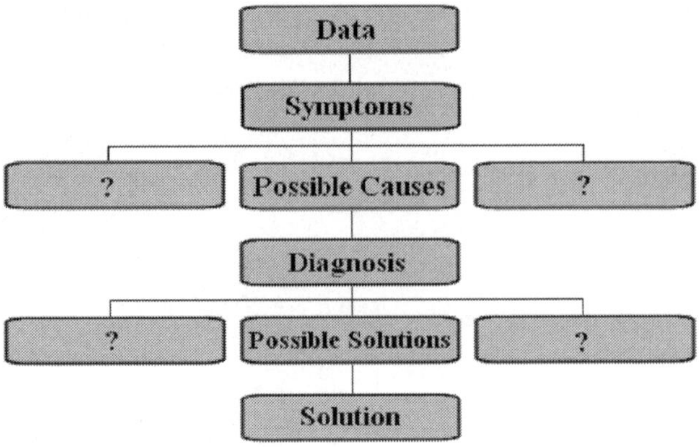

FIG. 12-109 Schematic diagram of algorithm for dryer troubleshooting.

4. Mechanical breakdown (catastrophic sudden failure)
5. Safety, health, and environmental (SHE) issues

Experience suggests that the majority of problems are of the first three types, and these are about equally split over a range of industries and dryer types. Ideally, unforeseen SHE problems will be rare, as these will have been identified in the safety case before the dryer is installed or during commissioning. Likewise, major breakdowns should be largely avoided by a planned maintenance program.

Drying Performance Problems Performance problems can be further categorized as

1. Heat and mass balance deficiencies (not enough heat input to do the evaporation)
2. Drying kinetics (drying too slowly, or solids residence time in dryer too short)
3. Equilibrium moisture limitations (reaching a limiting value, or regaining moisture in storage)

For the heat and mass balance, the main factors are

- Solids throughput
- Inlet and outlet moisture content
- Temperatures and heat supply rate
- Leaks and heat losses

As well as problem-solving, these techniques can be used for performance improvement and debottlenecking.

Drying kinetics, which are affected by temperature, particle size, and structure, are limited by external heat and mass transfer to and from the particle surface in the early stages, but internal moisture transport is the main parameter at lower moisture.

Equilibrium moisture content increases with higher relative humidity, or with lower temperature. Problems that depend on the season of the year, or vary between day and night (both suggesting a dependence on ambient temperature and humidity), are often related to equilibrium moisture content.

Materials Handling Problems The vast majority of handling problems in a dryer concern sticky feedstocks. Blockages can be worse than performance problems as they can close down a plant completely, without warning. Most stickiness, adhesion, caking, and agglomeration problems are due to *mobile liquid bridges* (surface moisture holding particles together). These are extensively described in particle technology textbooks. Unfortunately, these forces tend to be at a maximum when the solid forms the continuous phases and surface moisture is present, which is the situation for most filter and centrifuge cakes at discharge. By comparison, slurries (where the liquid forms the continuous phase) and dry solids (where all surface moisture has been eliminated) are relatively free-flowing and give fewer problems.

Other sources of problems include electrostatics (most marked with fine and dry powders) and *immobile liquid bridges*, the so-called sticky-point phenomenon. This latter is sharply temperature-dependent, with only a weak dependence on moisture content, in contrast to mobile liquid bridges. It occurs for only a small proportion of materials, but is particularly noticeable in amorphous powders and foods and is often linked to the glass transition temperature.

Product Quality Problems (These do not include moisture level of the main solvent.) Many dryer problems either concern product quality or cannot be solved without considering the effect of any changes on product quality. Thus it is a primary consideration in most troubleshooting, although product quality measurements are specific to the particular product, and it is difficult to generalize. However, typical properties may include color, taste (not easily quantifiable), bulk density, viscosity of a paste or dispersion, dispersibility, or rate of solution. Others are more concerned with particle size, size distribution (e.g., coarse or fine fraction), or powder handling properties such as rate of flow through a standard orifice. These property measurements are nearly always made off-line, either by the operator or by the laboratory, and many are very difficult to characterize in a rigorous quantitative manner. (See also "Fundamentals" Section.)

Storage problems, very common in industry, result if the product from a dryer is free-flowing when packaged, but has caked and formed solid lumps when received by the customer. Sometimes, the entire internal contents of a bag or drum have welded together into a huge lump, making it impossible to discharge.

Depending on the situation, there are at least three different possible causes:

1. *Equilibrium moisture content*—hygroscopic material is absorbing moisture from the air on cooling.
2. *Incomplete drying*—product is continuing to lose moisture in storage.
3. *Psychrometry*—humid air is cooling and reaching its dew point.

The three types of problem have some similarities and common features, but the solution to each one is different. Therefore, it is essential to understand which mechanism is actually occurring.

Option 1: The material is hygroscopic and is absorbing moisture back from the air in storage, where the cool air has a higher relative humidity than the hot dryer exhaust. *Solution:* Pack and seal the solids immediately on discharge in tough impermeable bags (usually double- or triple-lined to reduce the possibility of tear and pinholes), and minimize the ullage (airspace above the solids in the bags) so that the amount of moisture that can be absorbed is too low to cause any significant problem. Dehumidifying the air to the storage area is also possible, but often very expensive.

Option 2: The particles are emerging with some residual moisture, and continue to dry after being stored or bagged. As the air and solids cool down, the moisture in the air comes out as dew and condenses on the surface of the solids, causing caking by mobile liquid bridges. *Solution:* If the material is meeting its moisture content specification, cool the product more effectively before storage, to stop the drying process. If the outlet material is wetter than specification, alter dryer operating conditions or install a postdryer.

Option 3: Warm, wet air is getting into the storage area or the bags, either because the atmosphere is warm with a high relative humidity (especially in the tropics) or because dryer exhaust air has been allowed to enter. As in option 2, when the temperature falls, the air goes below its dew point and condensation occurs on the walls of the storage area or inside the bags, or on the surface of the solids, leading to caking. *Solution:* Avoid high-humidity air in the storage area. Ensure the dryer exhaust is discharged a long way away. If the ambient air humidity is high, consider cooling the air supply to storage to bring it below its dew point and reduce its absolute humidity.

See Kemp and Gardiner, "An Outline Method for Troubleshooting and Problem-Solving in Dryers," *Drying Technol.* **19**(8):1875–1890 (2001).

Dryer Operation

Start-up Considerations It is important to start up the heating system before introducing product into the dryer. This will minimize condensation and subsequent product buildup on dryer walls. It is also important to minimize off-quality production by not overdrying or underdrying during the start-up period. Proper control system design can aid in this regard. The dryer turndown ratio is also an important consideration during start-up. Normally the dryer is started up at the lowest end of the turndown ratio, and it is necessary to match heat input with capacity load.

Shutdown Considerations The sequence for dryer shutdown is also very important and depends on the type of dryer. The sequence must be thoroughly thought through to prevent significant off-quality product or a safety hazard. The outlet temperature during shutdown is a key operating variable to follow.

Energy Considerations The first consideration is to minimize moisture content of the dryer feed, e.g., with dewatering equipment, and to establish as high an outlet product moisture target as possible. Other energy considerations vary widely by dryer type. In general, heating with gas, fuel oil, and steam is significantly more economical than heating with electricity. Hence RF, microwave, and infrared drying is energy-intensive. Direct heating is more efficient than indirect in most situations. Sometimes air recycle (direct or indirect) can be effective to reduce energy consumption. And generally operating at high inlet temperatures is more economical.

Recycle In almost all situations, the process system must be able to accommodate product recycle. The question is, How to handle it most effectively, considering product quality, equipment size, and energy?

Improvement Considerations The first consideration is to evaluate mass and energy balances to identify problem areas. This will identify air leaks and excessive equipment heat losses and will enable determination of overall energy efficiency.

A simplified heat balance will show what might need to be done to debottleneck a convective (hot gas) dryer, i.e., increase its production rate F.

$$F(X_I - X_O)\lambda_{ev} \approx GC_{PG}(T_{GI} - T_{GO}) - Q_{wl}$$

Before proceeding along this line, however, it is necessary to establish that the dryer is genuinely heat and mass balance limited. If the system is controlled by kinetics or equilibria, changing the parameters may have undesirable side effects, e.g., increasing the product moisture content.

The major alternatives are then as follows (assuming gas specific heat capacity C_{PG} and latent heat of evaporation λ_{ev} are fixed):

1. Increase gas flow rate G—usually increases pressure drop, so new fans and gas cleaning equipment may be required.
2. Increase inlet gas temperature T_{GI}—usually limited by risk of thermal damage to product.
3. Decrease outlet gas temperature T_{GO}—but note that this increases NTUs, outlet humidity, and relative humidity, and reduces both temperature and humidity driving forces. Hence it may require a longer drying time and a larger dryer, and may also increase equilibrium and outlet moistures X_E and X_O.
4. Reduce inlet moisture content X_I, say, by dewatering by gas blowing, centrifuging, vacuum or pressure filtration, or a predryer.
5. Reduce heat losses Q_{WI} by insulation, removing leaks, etc.

Dryer Safety This section discusses some of the key considerations in dryer safety. General safety considerations are discussed in Sec. 23, "Safety and Handling of Hazardous Materials," and should be referred to for additional guidance.

Fires, explosions, and, to a lesser extent, runaway decompositions are the primary hazards associated with drying operations. The outbreak of fire is a result of ignition which may or may not be followed by an explosion. A hazardous situation is possible if

1. The product is combustible
2. The product is wetted by a flammable solvent
3. The dryer is direct-fired

An explosion can be caused by dust or flammable vapors, both of which are fires that rapidly propagate, causing a pressure rise in a confined space.

Dust Explosions Dispersion dryers can be more hazardous than layer-type dryers if we are drying a solid combustible material which is then dispersed in air, particularly if the product is a fine particle size. If this finely dispersed product is then exposed to an ignition source, an explosion can result. The following conditions (van't Land, *Industrial Drying Equipment,* Marcel Dekker, New York, 1991) will be conducive to fire and explosion hazard:

1. Small particle sizes, generally less than 75 μm, which are capable of propagating a flame
2. Dust concentrations within explosive limits, generally 10 to 60 g/m^3
3. Ignition source energy of 10 to 1000 mJ or as low as 5 mJ for highly explosive dust sources
4. Atmosphere supporting combustion

Since most product and hence dust compositions vary widely, it is generally necessary to do quantitative testing in approved test equipment.

Flammable Vapor Explosions This can be a problem for products wetted by flammable solvents if the solvent concentration exceeds 0.2% v/v in the vapor phase. The ignition energy of vapor-air mixtures is lower (< 1 mJ) than that of dust-air suspensions. Many of these values are available in the literature, but testing may sometimes be required.

Ignition Sources There are many possible sources of an ignition, and they need to be identified and addressed by both designers and operators. A few of the most common ignition sources are

1. Spontaneous combustion
2. Electrostatic discharge
3. Electric or frictional sparks
4. Incandescent solid particles from heating system

Safety hazards must be addressed with proper dryer design specifications. The following are a few key considerations in dryer design.

Inert system design The dryer atmosphere is commonly inerted with nitrogen, but superheated steam or self-inertized systems are also possible. Self-inertized systems are not feasible for flammable solvent systems. These systems must be operated with a small overpressure to ensure no oxygen ingress. And continuous on-line oxygen concentration monitoring is required to ensure that oxygen levels remain well below the explosion hazard limit.

Relief venting Relief vents that are properly sized relieve and direct dryer explosions to protect the dryer and personnel if an explosion does occur. Normally they are simple pop-out panels with a minimum length of ducting to direct the explosion away from personnel or other equipment.

Suppression systems Suppression systems typically use an inert gas such as carbon dioxide to minimize the explosive peak pressure rise and fire damage. Dryer operating pressure must be properly monitored to detect the initial pressure rise followed by shutdown of the dryer operating systems and activation of the suppression system.

Clean design Care should be taken in the design of both the dryer and dryer ancillary (cyclones, filters, etc.) equipment to eliminate ledges, crevices, and other obstructions which can lead to dust and product buildup. Smooth drying equipment walls will minimize deposits. This can go a long way in prevention. No system is perfect, of course, and a routine cleaning schedule is also recommended.

Start-up and shutdown Start-up and shutdown situations must be carefully considered when designing a dryer system. These situations can create higher than normal dust and solvent concentrations. This coupled with elevated temperatures can create a hazard well beyond normal continuous operation.

Environmental Considerations Environmental considerations are continuing to be an increasingly important aspect of dryer design and operation as environmental regulations are tightened. The primary environmental problems associated with drying are particulate and volatile organic compound (VOC) emissions. Noise can be an issue with certain dryer types.

Environmental Regulations These vary by country, and it is necessary to know the specific regulations in the country in which the dryer will be installed. It is also useful to have some knowledge of the direction of regulations so that the environmental control system is not obsolete by the time it becomes operational.

Particulate emission problems can span a wide range of hazards. Generally there are limits on both toxic and nontoxic particles in terms of annual and peak emissions limits. Particles can present toxic, bacterial, viral, and other hazards to human, animal, and plant life.

Likewise, VOC emissions can span a wide range of hazards and issues from toxic gases to smelly gases.

Environmental Control Systems We should consider environmental hazards before the drying operation is even considered. The focus should be on minimizing the hazards created in the upstream processing operations. After potential emissions are minimized, these hazards must be dealt with during dryer system design and then subsequently with proper operational and maintenance procedures.

Particle Emission Control Equipment The four most common methods of particulate emissions control are as follows:

1. *Cyclone separators* The advantage of cyclones is they have relatively low capital and operating costs. The primary disadvantage is that they become increasingly ineffective as the particle size decreases. As a general rule of thumb, we can say that they are 100 percent efficient with particles larger than 20 μm and 0 percent efficient with particles smaller than 1 μm. Cyclones can also be effective precleaning devices to reduce the load on downstream bag filters.
2. *Scrubbers* The more general classification is wet dedusters, the most common of which is the wet scrubber. The advantage of wet scrubbers is that they can remove fine particles that the cyclone does not collect. The disadvantages are they are more costly than cyclones and they can turn air contamination into water contamination, which may then require additional cleanup before the cleaning water is put to the sewer.
3. *Bag filters* The advantages of filters are that they can remove very fine particles and bag technologies continue to improve and enable ever-smaller particles to be removed without excessive pressure drops or buildup. The primary disadvantages are higher cost relative to cyclones and greater maintenance costs, especially if frequent bag replacement is necessary.
4. *Electrostatic precipitators* The capital cost of these systems is relatively high, and maintenance is critical to effective operation.

VOC Control Equipment The four most prevalent equipment controls are

1. *Scrubbers* Similar considerations as above apply.
2. *Absorbers* These systems use a high-surface-area absorbent, such as activated carbon, to remove the VOC absorbate.
3. *Condensers* These systems are generally only feasible for recovering solvents from nonaqueous wetted products.
4. *Thermal and catalytic incinerators* These can be quite effective and are generally a low capital and operating cost solution, except in countries with high energy costs.

Noise Noise analysis and abatement is a very specialized area. Generally, the issue with dryers is associated with the fans, particularly for systems requiring fans that develop very high pressures. Noise is a very big issue that needs to be addressed with pulse combustion dryers and can be an issue with very large dryers such as rotary dryers and kilns.

Additional considerations regarding environmental control and waste management can be found in Secs. 22, "Waste Management," and 23, "Process Safety."

Control and Instrumentation The purpose of the control and instrumentation system is to provide a system that enables the process to produce the product at the desired moisture target and that meets other quality control targets discussed earlier (density, particle size, color, solubility, etc.). This segment discusses key considerations for dryer control and instrumentation. Additional more detailed information can be found in Sec. 8, "Process Control."

Proper control of product quality starts with the dryer selection and design. Sometimes two-stage or multistage systems are required to meet product quality targets. Multistage systems enable us to better control temperature and moisture profiles during drying. Assuming the proper dryer design has been selected, we must then design the control and instrumentation system to ensure we meet all product quality targets.

Manual versus Automatic Control Dryers can be controlled either manually or automatically. Generally lab-, pilot-, and small-scale production units are controlled manually. These operations are usually batch systems, and manual operation provides lower cost and greater flexibility. The preferred mode for large-scale, continuous dryers is automatic.

Key Control Variables Product moisture and product temperature are key control variables. Ideally both moisture and temperature measurement are done on-line, but frequently moisture measurement is done off-line and temperature (or exhaust air temperature) becomes the primary control variable. And generally, inlet temperature

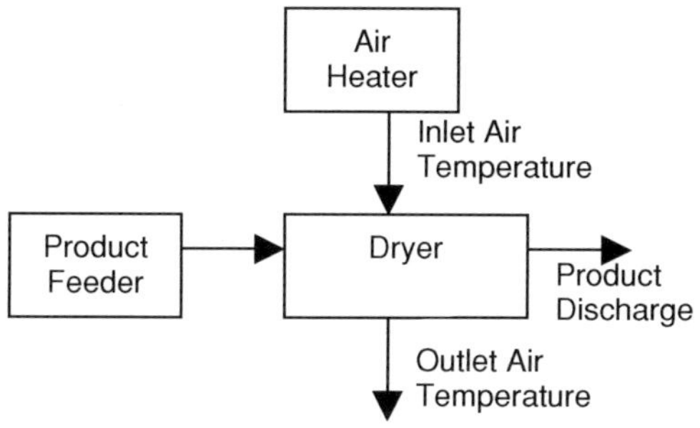

FIG. 12-110 Typical dryer system.

will control the rate of production and outlet temperature will control the product moisture and other product quality targets.

Common Control Schemes Two relatively simple, but common control schemes in many dryer systems (Fig. 12-110) are as follows:

1. Outlet air temperature is controlled by feed rate regulation with inlet temperature controlled by gas heater regulation.
2. Outlet air temperature is controlled by heater regulation with feed rate held constant.

Alternatively, product temperatures can replace air temperatures with the advantage of better control and the disadvantage of greater maintenance of the product temperature sensors.

Other Instrumentation and Control

Pressure Pressure and equipment pressure drops are important to proper dryer operation. Most dryers are operated under vacuum. This prevents dusting to the environment, but excess leakage in decreases dryer efficiency. Pressure drops are especially important for stable fluid-bed operation.

Air (gas) flow rate Obviously gas flows are another important parameter for proper dryer operation. Pitot tubes are useful when a system has no permanent gas flow sensors. Averaging pitot tubes work well in permanent installations. The devices work best in straight sections of ductwork which are sometimes difficult to find and make accurate measurement a challenge.

Product feed rate It's important to know that product feed rates and feed rate changes are sometimes used to control finished product moistures. Weigh belts are common for powdered products, and there is a wide variety of equipment available for liquid feeds. Momentum devices are inexpensive but less accurate.

Humidity The simplest method is sometimes the best. Wet- and dry-bulb temperature measurement to get air humidity is simple and works well for the occasional gas humidity measurement. The problem with permanent humidity measurement equipment is the difficulty of getting sensors robust enough to cope with a hot, humid, and sometimes dusty environment.

Interlocks Interlocks are another important feature of a well-designed control and instrumentation system. Interlocks are intended to prevent damage to the dryer system or to personnel, especially during the critical periods of start-up and shutdown. The following are a few key interlocks to consider in a typical dryer system.

Drying chamber damage This type of damage can occur when the chamber is subjected to significant vacuum when the exhaust fans are started up before the supply fans.

Personnel injury This interlock is to prevent injury due to entering the dryer during operation, but more typically to prevent dryer start-up with personnel in the main chamber or inlet or exhaust air ductwork on large dryers. This typically involves microswitches on access doors coupled with proper door lock devices and tags.

Assurance of proper startup and shutdown These interlocks ensure, e.g., that the hot air system is started up before the product feed system and that the feed system is shut down before the hot air system.

Heater system There are a host of important heater system interlocks to prevent major damage to the entire drying system. Additional details can be found in Sec. 23, "Process Safety."

Drying Software Several software programs for psychrometric charts and calculations are available and are described in the "Psychrometry" section.

Dryers are included as modules in standard process simulators such as Aspen Plus and HYSYS (Aspen Technology), Pro/II (Simsci/Invensys) and Unisim (Honeywell), and the prototype Solidsim solids process simulator. These are confined (as of 2006) to heat and mass balances or, at most, simple scoping design.

Many higher-level dryer models have been produced by researchers and universities, but they are not commercially available. Windows-based drying programs are available in the Process Tools (Aspen Technology), including a psychrometric chart, dryer selection expert system, dryer scoping design, and fluid-bed dryer simulation. Some CFD programs (e.g., Fluent, CFX) include a module for spray dryers.

In addition to textbooks, a detailed online knowledge base, the Process Manual, is available from Aspen Technology (see www.processmanual.com). This covers equipment, scientific background, design, and operation for drying and 10 other technical areas in solids and separation processes. Company and university licenses are available.

Detailed reviews of drying software packages given by Menshutina and Kudra, *Drying Technol.* **19**(8):1825–1850 (2001); and by Kemp, Chap. 7 in *Modern Drying Technology*, vol. 1, Wiley–VCH (2007), and *Drying Technol.* **25** (2007).

Made in United States
Orlando, FL
04 November 2026

90200187R00061